MILDENHALL TO MELBOURNE

*Dedicated to Melissa,
thanking her for her help.*

With love

STUART McKAY MBE

MILDENHALL TO MELBOURNE
The World's Greatest Air Race

THE TIGER HOUSE PRESS

MILDENHALL TO MELBOURNE
© Stuart W. McKay MBE, 2009
ISBN 978-0-9563981-0-9
First published in 2009 by
The Tiger House Press
Staggers, 23 Hall Park Hill, Berkhamsted,
Hertfordshire HP4 2NH, England
Tel: 01442 862077. Fax: 01442 862077
Printed in Great Britain by
Stephens & George Ltd.
Endpaper illustration by Andrew Hay,
Flyingart, in association with Nick
Spencer, Mildenhall Museum.
Goat Mill Road, Dowlais, Merthyr Tydfil,
Mid Glamorgan CF48 3TD
All rights reserved. No part of this publication may be reproduced, stored in a retrieval system, transmitted in any form or by any means, electrical, mechanical or photocopied, recorded or otherwise, without written permission of the copyright owners.

"The very fact that the International Air Race is creating such a world-wide interest, is certainly going to make thousands upon thousands of people study their atlas and learn what they do not at present know, namely, where Melbourne is."

SIR MACPHERSON ROBERTSON

Acknowledgements

THE CORE OF THIS book was completed 25 years ago to celebrate the 50th anniversary of the MacRobertson International Air Races but due to a number of technical difficulties was never published. In the intervening years rich new seams of information have been discovered and explored with the result that this volume, published to celebrate the 75th anniversary of 'the world's greatest air race' is now the most comprehensive study of the event and the people who took part.

It is impossible to complete a work of history without the willing support of an army of helpers scattered, in this case, all over the world. Sadly, due to the time frames, it is inevitable that many have passed on to join the principal players in this unique drama, all of whom are now directing operations or taking centre stage on a flight deck elsewhere.

To everybody who has provided any snippet which has helped bind all the individual elements of this story into the whole I offer my grateful appreciation. I am particularly indebted to the staff in a number of institutions for the supply of illustrations and all the necessary permissions: Guildhall Library, London, Suffolk Record Office, Royal Aero Club archive, Royal Air Force Museum, National Library of Australia, Canberra, Northern Territory Archive Service, Northern Territory Library, State Library of Victoria, State Library of New South Wales, Museum Victoria, Queensland Museum and the State Library of Queensland.

In no particular order and for a hatful of the best reasons I offer my special thanks to the following:

Dr Colin Dring, Miss Annie Edwards, Peter Gould, Mike Hooks, Michael F. Jerram, Peter Murton, Nick Redman, Michael Redmond, Andrew Dawrant, Nick Stroud, Nick Spencer, John W. Underwood, Peter Elliott, Paul and Liz Gliddon, Adele Stephenson, Richard T. Riding, Desmond Penrose, Michael Vaisey, Neil Lewis and George Mackie.

Contents

CHAPTER ONE
Celebrating the Centenary p.9

TWO
The Plan grows Wings p.12

THREE
The Airway to Australia p.18

FOUR
Who is Racing to Melbourne? p.24

FIVE
The Six and Four p.30

SIX
Entrants with Authority p.65

SEVEN
Reality and Disappointment p.76

EIGHT
Final Adjustments p.94

NINE
Take-off at Dawn p.105

TEN
Ever Eastward p.112

ELEVEN
In Search of a Prize p.119

TWELVE
A Dream comes True p.130

THIRTEEN
Incident at Albury p.139

FOURTEEN
Not Disgraced p.144

FIFTEEN
A Victory for Melbourne p.158

SIXTEEN
And After p.163

SEVENTEEN
Final Straight p.172

EIGHTEEN
Postscript p.196

INDEX
p.199

Foreword

THE MILDENHALL to Melbourne Air Races of October 1934 spawned some of the boldest of all air adventures. The idea was conceived when the world needed positive relief from the misery of the worst trade depression ever to hit the industrialised countries and the entire organisation was skillfully steered by dedicated men through the necessary complications of political and diplomatic processes which stretched not only from Australia to Great Britain but across the Atlantic to the United States of America.

At a time when long distance communications were still primitive, negotiations, successfully completed over a range of 12,000 miles resulted in a workable set of Regulations; the nomination of numerous official landing places which were surveyed, prepared and staffed to handle men and machines in a hurry, and the opening of channels which worked quietly to arrange free passage of racing aeroplanes through the airspace of 16 different countries. The provisioning of supplies to remote areas, especially fuel and oil for what was expected to be a great fleet of thirsty machines, was an exercise in clear forward planning assessed against pure guesswork.

Although the winning aeroplane completed the course in less than 72 hours, the back marker arrived 100 days after the cut-off. The Races (Speed and Handicap run concurrently) were certainly some of the safest and the carefully considered framework of the governing Regulations was largely responsible for that. The guide lines had been proposed by the sponsor of the Races, Sir MacPherson Robertson, whose whole hearted acquiescence had been the steady platform upon which the organising committees were able to build.

Those who railed against the Races on the grounds that they would be a complete waste of time and effort and that nothing would be learned from them were forced to hold their peace as the weight of public interest swung solidly behind the event. The world demanded all possible information about the crews, the aeroplanes and their technical merits. New terminologies were explained: retractile undercarriages; variable pitch propellers; flaps; direction finding wireless apparatus; stressed skin construction.

With most projects taking years to reach maturity, how was it possible for a completely new and original design be conceived and built and flown with such success in just ten months? Why was there no British or European airliner even vaguely similar to the Boeing 247 or Douglas DC-2 in which passengers travelled in comfortable, heated cabins? When would there be a regular passenger through air service between Europe and Australia? Why should international air travel formalities become so tedious and tiresome again following the end of the declared race period?

The MacRobertson International Air Races, Mildenhall to Melbourne, stimulated thought and confirmed hope, enterprise, initiative, courage and tenacity, all qualities enjoyed by the founding fathers of the Australian State of Victoria and the City of Melbourne whose Centenary celebrations they were designed to crown. Not least, the event reflected the greatest personal credit displayed by the benevolent confectioner who had travelled the same route, found fame and fortune, and whose name will forever be linked with what was arguably the greatest air race in the history of aviation.

Stuart McKay
Berkhamsted, Hertfordshire.
October 2009

1

Celebrating the Centenary

Sir Harold Gengoult-Smith, Lord Mayor of Melbourne, is believed to have made the first suggestion of an Air Race during discussions about the Centenary Celebrations.

AS DAWN BROKE over rural Suffolk on Saturday, 20 October 1934 and the sun rose to paint the high clouds with vivid pink and orange and to herald another cool autumnal morning, the normally quiet country village of Mildenhall was alive with pedestrians; the narrow streets choked with traffic slowly edging its way towards the new aerodrome where the whole procession ground to a halt, merged into the greatest traffic jam the area had ever known.

The aerodrome, built adjacent to the hamlet of Beck Row, was later identified as RAF Mildenhall by the pundits at the Air Ministry, probably in an effort to associate with the nearest administrative centre for civic affairs, and had not yet been occupied by heavy bomber squadrons of the Royal Air Force whose home it was to be. For much of the previous week the two completed hangars, the barrack blocks and unfinished administrative buildings, contractors' huts, marquees and caravans, had all vibrated in tune with full power engine runs as civilian pilots and engineers applied themselves to the delicate adjustments required by a fleet of racing aeroplanes.

Now, in the half-light, ladies in ball gowns with escorts dressed incongruously in white tie and tails mingled easily with the more conventional trench coats and country tweeds as an estimated 60,000 spectators, conveyed to the area by 8,000 vehicles, jostled towards the airfield's perimeter fence eager to be in advantageous positions by 6.30am.

The most fortunate watched in awe, waving and cheering and clapping as, heading into a stiff westerly breeze, 20 aeroplanes of varied design took off in as many minutes, turned south and raced away into the early morning. Less than three days later, the first of them arrived in Australia; others would take weeks or even months, and some were destined never to reach their goal.

The multitude which had gathered in the heart of East Anglia on that cool morning was a witness to history; each spectator was privy to the first act in a drama whose plot had been fashioned for almost two years by a dedicated team stationed along a 12,000 mile line stretching through the Empire from England to Australia. The fact that any aeroplanes had left England at all was in some eyes a remarkable achievement and each of the 20 departures was an individual victory for pilots and engineers, designers, builders, administrators and backers. There might have been many reasons why the village of Mildenhall should not have woken to a perfectly ordinary Saturday; why for the following week the world should not have gone about its business without an ear cocked to the wireless or an eye glued to the newspapers, but the organisers of the MacRobertson International Air Races from London to Melbourne had steered round the objections and quarrels, the politics and jibes, the criticisms and misinformed speculation and had achieved their singular objective.

The application to their deputed tasks by the many committees had been total but the conception of the event had been inspired, the backing immediate and secure, thanks largely to a patriotic Australian confectioner whose name was to be irrevocably linked with what many still regard as the greatest air race in the history of aviation.

Sir MacPherson Robertson, who enjoyed affectionate recognition as 'MacRobertson' was the son of an immigrant Scottish family, born at Ballarat, Victoria, in 1860. At the age of eight, Robertson and his mother returned to Scotland whilst father sought his fortune in Fiji. The boy immediately busied himself in an effort to raise income with early morning newspaper deliveries followed by an hour assisting a local barber. From 9.00am until 4.00pm he attended school at the insistence of his mother, and rounded off his day at 10.00pm after a further four hours of mixing soap and lathering chins.

In 1874 he travelled 'home' to Australia where his enterprises ranged from shovelling coal to cutting corks until a year later he secured a five year apprenticeship with a confectionery manufacturer. He was paid two shillings and sixpence a week for a 13 hour day but the experience was worth much more. When he was only 20 Robertson set up his own confectionery manufacturing business with production based on a few simple ingredients and using utensils located in the bathroom of his parents' house. The business prospered and set the foundation of the empire which eventually included the manufacture of his own wrappers and boxes, and ownership of extensive estates directed towards cultivation of raw ingredients. In the early 'thirties Robertson rejected an English bid of more than £3,000,000 for the business which then employed 3,200 workers, mingling amongst whom a grey-haired gentleman in a white dustcoat was frequently to be seen, maintaining personal contact and concern.

Robertson displayed his philanthropic nature in 1928 by backing a motor expedition which circumnavigated the Australian continent and he was responsible for covering expenses of Sir Douglas Mawson's joint British/Australian/New Zealand Antarctic Expedition the following year. For support of the motor circumnavigation he was knighted and a year later, elected a Fellow of the Royal Geographical Society. Between 1927 and 1934 Sir MacPherson Robertson donated more than £360,000 towards exploration, education and charity, in addition to gifts in kind and numerous smaller bequests from what he called 'pocket money expenses'. His gift of £100,000 to the Melbourne Centenary Fund was sufficient for the provision of a girl's high school, a bridge over the City's Yarra River and foundation of a Temple of Youth. His involvement with other major charitable works continued under a blanket of secrecy and only years later did the benefactor's identity leak out.

Robertson's business was an early user

Mindful of the approaching Centenary of the City of Melbourne and conscious of the possible wider implications, influential and confident citizens, convinced that the Depression was soon to lose its grip, established a Festival Committee in 1932. Ideas were floated, dreams were dreamed, optimistic budgets were drafted and rejected. A varied programme of sport and culture was proposed; good, honest, basic family entertainment. But was there any single idea amongst the recommendations which would shake the whole country; that might grip the imagination of the world and focus all attention on a City and a State which were hungry for recognition? What sort of event could be organised which would totally embrace the dynamic spirit of the celebrations?

Although there have been other claimants, it is generally recognised that Sir Harold Gengoult-Smith, Lord Mayor of Melbourne, was first to consider the prospect of an international air race. He discussed the matter with his friend Wing Commander Adrian Cole, lately Commander of Melbourne's RAAF Base at Laverton. What a prospect! A great race to Melbourne, the southern-most capital city in Australia, but from where? It had to be London, capital of the mother country and hub of the Empire. An air race would attract the world's greatest aviators in the best aircraft, flying halfway round the world. Melbourne's name, location and enterprise would become household knowledge as the publicity gathered momentum, reaching a climax as the winning aircraft flashed across the line.

Gengoult-Smith and Adrian 'King' Cole pursued the matter at some length; the City Father with the interests of his State at heart, and the aviator who brought practical and professional judgement to bear. That such a race could be run was not in doubt, but flying was an expensive busi-

of aircraft, purchasing a DH.61 Giant Moth in 1928 to transport staff between the widely dispersed growing areas and processing units, and in the same year he financed a commercial aviation business. Sir MacPherson Robertson the philanthropist was no stranger to aeroplanes.

At the height of the Depression in the 'thirties, with the concentration of Australians massed in the south eastern corner of the continent, the population began to show concern that the vast areas to the north were exposed, almost totally undefended and in practical terms, probably undefendable. Any major southern migration by the expansionist regimes of South East Asia would be almost impossible to repel using nationally based forces, but were not reinforcements stationed throughout the Empire, along a line stretching right back to Britain? Some comfort might be wrung from the assurance that assistance could be quickly provided if a permanent air route be established, along which rapidly and efficiently, all urgent supplies could be moved by aeroplane!

E V E Neill represented Sir MacPherson Robertson's interests and was elected as Secretary of the Centenary Celebrations Race Committee.
(National Library of Australia, Canberra.)

The prototype DH.61 Giant Moth G-EBTL was sold to Australia in 1928 and operated by the MacRobertson Miller Aviation Company on services between Adelaide and Broken Hill.

ness and a race spanning half the globe would take much time, effort and money to organise. To attract the cream of the world's aviators, some special incentive was essential. A race in celebration of Victoria and Melbourne demanded original ideas and prizes from within the state. Gengoult-Smith arranged a meeting with one of the City's great entrepreneurs, the 72-year-old confectionery millionaire, Sir MacPherson Robertson.

Early in the New Year of 1933, Sir MacPherson had already given the plan his wholehearted support and had agreed to put up generous cash prizes. He was particularly concerned that the organisation should be as efficient as possible; entries would be accepted on an international basis, and safety was to be a primary concern.

The Lord Mayor conveyed his thoughts and those of his financial backer to other essentially interested parties: the State Government, Department of Aviation, and the Centenary Committee. There were no objections, and a Race Committee was established on 1 April 1933 under the chairmanship of Sir Harold, enrolling the talents of 'King' Cole as vice-chairman, E V E Neill, representing Sir MacPherson, as secretary, together with, in its original constitution: Wing Commander A H Cobby, Controller, Department of Civil Aviation (later superseded by Captain E C Johnston); the British Trade Commissioner; Dean of the Consular Corps; Secretary of Associated Aero Clubs of Australia with Guy Moore representing the Victorian section; Squadron Leader F F Knight, lawyer; W E Bassett, lecturer in Aviation Engineering at Melbourne University; and Alderman Woolten, a financial expert.

Sub-committees were formed to deal with the route, regulations, information briefs for competitors, finance and the control of receptions. The Lord Mayor took chairmanship of the latter. Air Commodore Richard Williams, Chief of the Australian Air Staff, was conveniently placed in England attending the RAF Staff College and investigating possible new equipment for the RAAF, and he agreed to act as liaison officer in London. Agreement on basics was unanimous and on 3 April Major W T Conder, Chairman of the Centenary Committee, announced provisional details of the route. Shortly after, he confirmed that there would be no limit on size, power or type of machines; no restrictions on crew complement except that pilot changes en-route would not be permitted; what mix of emergency gear and rations should be carried and where the finishing line was to be established.

Furthermore, at a time when the existing England to Australia record stood at almost nine days, Conder's statement added, "It is expected that the winner will complete the flight in four days." He also advised the world that aircraft would leave England on Saturday, 20 October, 1934. The Air Race was to be the stimulus for additional celebrations which were planned to continue well into the following year.

But as time was to tell, matters were to be far from simple.

2
The Plan grows Wings

Lindsay Everard, an enterprising private owner with his own aerodrome in Leicestershire, and who was nominated as Chairman of the Race sub-committee, in the company of HM King George V, who disliked aeroplanes.

ALTHOUGH ACTIVE since 1926 and highly motivated, by 1933 the Australian Aero Clubs had not developed sufficiently for their authority to be recognised by the world governing body of sporting aviation, the Paris based Fédédération Aéronautique Internationale (FAI). Therefore, it was necessary to vest basic control of the prospective Air Race with the Royal Aero Club, operating from its Piccadilly headquarters in London.

As the Race was planned to start from England, much of the practical organisation would need to be originated from there anyway, so it was something of an embarrassment to the Melbourne Committee to discover that following their public announcement of the Race, including basic route details, prizes and dates, nobody had even consulted the Royal Aero Club. Urgent discussions immediately took place between the Centenary Committee and the Club's representative in Australia with the result that a letter outlining the proposals was despatched with some haste to London.

On 26 April 1933, Commander Harold 'The Hearty' Perrin, Secretary of the Royal Aero Club, read the correspondence to Committee members assembled for their regular monthly meeting. Their immediate decision was to defer consideration of the matter until further details could be provided from Melbourne.

Air Commodore Williams, taking leave from his official duties at the RAF Staff College, contacted Perrin prior to the Committee meeting in May for urgent discussions on the whole project. Later he introduced details of a preliminary handicap formula which had been devised in Melbourne by Walter Bassett.

The Royal Aero Club decided that its Technical and Racing Committees should consult jointly, but before this was possible, Centenary Race Secretary E V E Neill released further proposals to the Australian Press. Amongst these was the first public indication that two Races were planned to be flown simultaneously: a Speed Race and a Handicap Race based on Bassett's formula.

Neill's Press statement had been published at the end of May but it was not until 14 June 1933 that the Racing and Technical Committees of the Royal Aero Club under the chairmanship of Lindsay Everard, met Air Commodore Williams and Flight Lieutenant T A Swinbourne RAAF, acting as liaison officers. For the first time, all the details which had been released were discussed formally and thoroughly between representatives of the two organising authorities and, apart from some objections to Bassett's proposed handicap formula, there were no major disagreements. Robert Mayo was asked to establish a sub-committee to study the handicap proposals which Melbourne insisted were critical if any substantial entry of commercial or light sporting aircraft was to be encouraged in addition to the speedsters.

Williams was persuaded to provide positive assurance that the prize money really had been offered and, having completed his assignment in England, returned home from where he was able to report that the Australian Committee was already in possession of Sir MacPherson's £10,000. On 29 March 1933, prior to any public announcements, he had presented his cheque to the Lord Mayor of Melbourne. Written evidence of the security of the prize fund and guaranteed cover of the Royal Aero Club's anticipated expenses of £2,000 were sent from Melbourne to London by mail. Only then did the Royal Aero Club make any official announcement to the British press but there were few banner headlines.

Why did Sir MacPherson Robertson consider sponsoring the event at all? In a letter to the Advertising Association of Australia he set out some of his thoughts:

"With regard to the Aerial Races that I am sponsoring in connection with the Centenary Celebrations, my object in promoting these contests is due to several reasons.

Australia is greatly handicapped by her isolation, and I am definitely convinced that if some means can be devised to overcome the long lapse of time in the exchange of ideas, the completing of contracts, financial settlements, *et cetera*, tremendous benefit will certainly accrue to the primary producer whose market is abroad, and to manufacturers and other members of the commercial community with business ramifications outside of Australia. In these Aerial Races I see chances of bringing about these benefits.

Commander Harold Perrin, Secretary of the Royal Aero Club since 1903, in the company of HM Queen Mary. The bulk of the organisation of the Races fell squarely on the shoulders of Perrin and his small staff.

In one or other of the Races will be demonstrated how speedily contact between the United Kingdom and Australia can be made. Efficient aerial mail service and passenger carrying should be demonstrated as entirely practicable, and so the way is open for the obtaining of the three benefits that I have referred to.

What is more, it seems to me that one can logically conclude that an efficient goods service should follow a good passenger service, and so, if only in a small way to start with, a lot of Australia's perishable products should be able to get to their marketable destinations in very much less time than they do now, with consequent minimised deterioration, and an enhanced chance of reaching a favourable market.

Again, the very fact that the International Air Race is creating such a world-wide interest, is certainly going to make thousands upon thousands of people study their atlas and learn what they do not at present know, namely, where Melbourne is. Summarised, I think this Aerial Race will do a great deal to put Melbourne on the map.

Again, the arrival of the last word in aeroplanes in such numbers as we can reason-

Sir MacPherson Robertson's fortune came from his confectionery empire whose products were marketed under the 'MacRobertson' brand. The wrappers for 'Air Race' milk chocolate bars carried the Melbourne Centenary dates.

ably expect must help greatly to wake up the people in Australia to the fact that we are completely outside the picture in regard to modern aviation, which is painfully manifested in the fact that for neither of these Races have we a machine in Australia that can enter with any chance of success, and this is a country that has produced some of the finest airmen the world has ever known.

I am an Australian, tremendously proud of my country and keenly desirous for its welfare, and whenever I see an opportunity to do what I think will further its welfare, as long as I have money to expend, I will use it in backing my judgment in this regard."

In deference to the sponsor, and using his preferred method of address, this pinnacle part of Melbourne's Centenary Celebration was to be called *The MacRobertson International Air Races*, with emphasis on the plural. Once detailed planning had begun Sir MacPherson increased his offer of prize money from £10,000 to £15,000 (Australian) allowing £10,000 (Australian), about £7,500 sterling, to be awarded to the winner of the Speed Race who would also take a solid 18 carat gold cup commissioned from Melbourne jewellers, Hardy Bros., valued at not less than £500. (The cup, when delivered to the Melbourne Committee, was declared to contain 18 carat gold to the value of £650.) Second and third places were allocated prizes of £1,500 and £500 respectively. Two prizes were offered for the Handicap Race: £2,000 and £1,000, and all crew members completing the course within 16 calendar days would each receive a solid 18 carat gold medallion.

As the year turned and the spring and summer of 1934 progressed, correspondents to the aviation papers were suggesting that big countries who might have provided nationally sponsored entrants were not doing so because they feared loss of prestige if they did not win whilst some of the smaller nations simply had not thought of taking part. In the event there were no

The obverse and reverse faces of the solid gold medallions which were to be presented to all competitors completing the course within the prescribed time and the solid gold cup to be claimed by the winner of the Speed Race.

military entrants of any sort nor were there any ostensibly private entries enjoying covert government, institutional or corporate funding and support.

Other lines of discussion suggested there might have been more entries had there been additional prizes and that 'the whole event was Sir MacPherson Robertson's and that other people must not be allowed to advertise themselves or their products by taking part of his thunder'.

The Draft Rules stated that nobody could win more than one prize. If a Speed machine flew sufficiently well to qualify for a Handicap prize too, then the entrant would be offered the award of the higher value and relinquish all title to the other, which would pass to the next placed machine. Any aircraft, powered by any engine, built or operated by any country, was eligible for entry. C G Grey, Editor of *The Aeroplane*, made clear his feelings about certain rumours which had been reaching him. A suggestion that the Conditions be altered, leaving the Race open only to British aeroplanes, he described as a 'fatal error'.

"We in this country would be guilty of a grave discourtesy if we took upon ourselves to cavil at any conditions which the donor of such a munificent prize might wish," he wrote, and continued, "our Australian relatives who are notable for their sturdy independent spirit, would be entitled to resent interference of any kind from this country."

Perhaps the greatest fear from adopting any such amendment was the predictable taunt that the British aircraft industry was afraid of facing foreign competition, with the damage that such a reputation might generate. Similar sentiments were expressed over a submission that a large cash prize should be awarded to the first British machine to arrive. Grey suggested that if there was a wealthy sportsman willing to put up the cash, it might be more wisely invested in the production of a special British machine. Contrary to Grey's opinion, the French Government publicised an offer of 150,000 Fr. to the French pilot making the fastest time, and 5,000 Fr. to the highest placed French pilot in the Handicap Race.

Although basic agreement on the Rules had been reached in June 1933, the Royal Aero Club decided that several clauses required clarification and possible amendment. The Club was still unconvinced about the mechanics of running the two Races concurrently, neither were they confident that with the expectancy of a large entry proper physical control could be exercised. Certainly the original intention was to have run a single Race, but Melbourne had insisted that more entries and greater interest would be generated by a Handicap section, a view readily endorsed by the sponsor who had immediately increased the prize fund by a further £5,000 as an incentive.

The Australian Committee members could not be dissuaded from their convictions and 'King' Cole reiterated that they were the governing body for the Races and not the Royal Aero Club, whose responsibilities were, in the letter of the law, restricted to originating and enforcing the Rules for the start and control and conduct of the Race as far as Koepang. Useful observations would always be considered, however, similar to those received from the Aero Club of India and Burma who advised that during October Calcutta was prone to flooding and should be deleted. Allahabad was more suitable and Karachi would be preferable to Gwandar, they suggested.

The estimate of the Royal Aero Club's expenses, £2,000 in January, had doubled by the middle of June and by the end of the month were assessed at £5,050, a dramatic increase which required Melbourne's agreement. The Australians were spending a further £2,000, which also covered a jamboree after the finishers had crossed the line to be established at the famous Flemington Racecourse. Australian newspapers were critical when they balanced expenditure against total prize money, but Cole was optimistic that given proper business management, the public in England and Australia could be easily parted from its shillings in order to cover the burdens of administration.

The Handicap formula again was closely examined by Mayo's sub-committee but they could suggest no improvement apart from a better definition of the horsepower factor. The Club's major objection had been that aircraft with cruising speeds of less than 150mph were likely to be penalised, a criticism levelled by a Committee of sportsmen whose personal flying achievements lay mostly within that very speed bracket. Presented as a mathematical equation, the Handicap formula was probably meaningless to most who saw it, including some of the crews who would be competing against the algebraic jumble in search of a valuable prize.

Solution of the equation provided a value for 'V': airspeed in miles per hour. The airspeed divided into the route mileage produced an estimated flying time. The aircraft arriving overhead Flemington in the least time in excess of its declared handicap or by the greatest margin less than its allocated time, would win. The Race Committee defined 'Handicap Time' as actual flying time less the handicap allowance. In the context of the Handicap Race, flying time was all time spent between commencement of the Race in England and the finishing line in Melbourne, less all of the time spent on the ground at the authorised Checking Points. It was later confirmed that time spent on the ground anywhere else (aircraft putting down en-route with mechanical or navigational problems) would be considered as flying time, and added to the totals. This important point was, perhaps, not emphasised sufficiently and nobody seems to have requested clarification. The lack of appreciation cost one finisher a well deserved prize.

THE HANDICAP FORMULA

Walter Bassett's proposal finally accepted by the Mayo sub-committee and published in the Rules was as follows:

$$V = \frac{140\,(1 - 0.2L)\,(P)^{1/3}}{(W-L)\,(A)}$$

W: The all-up weight of the aeroplane weighed in pounds, complete with petrol, oil (full tanks assumed), personnel and freight.

A: Area of wings in sq. ft. including ailerons but exclusive of all fairings on axles, struts and the like, and wing surfaces forming part of the top and bottom surfaces of the fuselage.

P: For engines rated at sea level, the maximum hp of the prototype engine at normal rpm at sea level. For engines rated at height above sea level, the maximum hp of the prototype engine at normal rpm at the rated height divided by the corresponding height/power factor. 'P' was the only part of the formula which required clarification as to what constituted 'normal' rpm. A note was added:

"*Normal* rpm is defined as the maximum rpm authorised for continuous running as established during the approval tests of the engine."

L: Payload. The Rules were most specific about 'L' and although the weight limit of 200lb per person was maintained, those responsible for computing the individual handicap figures thought it should have been reduced by at least 20lb. When published, the definition of the payload remained unchanged:

"The following only will be recognised as payload: 200lb for each person carried, together with his unsealed baggage, and all sealed packages of any description, which may include excess baggage, spare parts, ballast, printed matter, etc., but not petrol, or oil. The payload at the finish of the Race must agree with the certified log-book entry and conform to the declared payload throughout the Race.

If at the finish of the Race the seals of the payload are broken or the payload does not agree with the log-book entry, handicaps will be re-adjusted in respect of the deficiency, and such deficiency will be deemed to have existed from the start.

Any person certified as payload may be disembarked at a Control or Checking Point and a substitute embarked, provided that the fact is recorded in the pilot's log-book by the responsible authority at that point.

Any person disembarked and not replaced means a payload deficiency of 200lb, and the consequent handicap re-adjustment. Dead weight may not be carried in place of disembarked personnel."

In the case of the Speed Race it was very much a question of first across the finishing line. A pre-Race compendium of facts published in *Flight* explained how the differences in starting times would be handled:

"In the interests of safety it has been decided to send the machines off at short intervals, so that there may be no risk of collision. In the case of the Speed Race, this would impose unfairness on late starters, and in order to remedy this the necessary time allowance will be corrected on the arrival of competitors in Singapore. For example, if the last machine in the Speed Race is started 30 minutes after the first one, it will not, upon reaching Singapore, be held up at all, but the first machine to leave England will be held up for 30 minutes at Singapore. During that period no work on the machine will be permitted. That is to say as far as working on it is concerned, the machine will be treated as if it were flying."

With regard to cancellation and the traumatic impact any such decision would have on entrants, the Melbourne Centenary Council confirmed that its right of cancellation would only be exercised in the event of war or national emergency and that the starting date would be 20 October 1934, very late autumn in England but early spring in Victoria.

This date was significant as the anniversary of the first attempt to fly from England to Australia in 1919, when Captain G C Matthews and Sergeant T Kay had left Hounslow Aerodrome in a Sopwith Wallaby. However, the Wallaby flight had actually started on 21 October which in 1934 fell on a Sunday. Deferring to Observers of the Lord's Day, the start of the MacRobertson Races was advanced by 24 hours.

Matthews and Kay had reached Bali where the Wallaby crashed on 14 April 1920, having covered 10,000 miles in almost six months. On 12 November 1919, Captain Ross Smith and his crew aboard a Vickers Vimy left Hounslow Heath and arrived in Port Darwin on 10 December. Perhaps this would have been a more auspicious date upon which to start, a celebration of the first *successful* flight between the old and the new, but October was already considered late by observers of European weather and the very worst of the monsoon was likely to be encountered between Karachi and Singapore. Designers were already contemplating, amongst other problems, the effects of torrential rain on high speed, highly stressed racing aircraft.

Frailty of long distance communication between start and finish and protracted discussion over the facilities to be offered in between caused growing frustration amongst aviation writers, potential competitors and designers. If the intention of the Races was to provide not only a great public spectacle but an opportunity for the aviation industry to improve the product, then available time had to be used most efficiently and time was already short.

Adrian Cole's Australian Committee had suffered a blow early in November 1933 when two members had tendered their resignations over the thorny matter of airworthiness. Although regulations and certification had been of major concern for almost a year, only basic Rules and Supplementaries had been published to date. Captain E C Johnston, representing the Department of Civil Aviation, raised the matter of the clause relating to certification conforming *substantially* or being *equivalent to* the International Commission on Air Navigation (ICAN) standard. In Australian airspace, he argued, such concessions were not acceptable, and in any case some countries, the USA for example, had not even signed the Convention. Competitors might produce documents which, therefore, conformed to nothing at all.

Johnston was supported by Guy Moore representing the Associated Aero Clubs of Australia, but opposed by the remainder of the Committee who were astonished and angry, aware of the many potential entrants who had based their racing assessments on previously published information.

Cole refused to agree to any further alterations to the Rules and initially the Lord Mayor would not accept the resignations, but Johnston's unbending attitude prevailed and both he and Moore departed. In view of his administrative position, Johnston was offered, and accepted, a vague non-executive post as advisor.

On 24 November the Royal Aero Club in London convened the first meeting of the MacRobertson International Air Race sub-committee, a body which inherited total responsibility from the Racing, Technical and Competition Committees, all of which had contributed specialist assistance, and was to report through the Club to the Centenary Celebrations Council (Victoria). With some urgency, the new sub-committee met frequently and at least twice a week during the late summer months. Under the chairmanship of Lindsay Everard, the members were all experienced aviators: Major John Buchanan, Lieutenant Colonel Maurice Darby, Flight Lieutenant Christopher Clarkson, Major Alan

Captain E C Johnston, representing the Australian Department of Civil Aviation on the Centenary Air Race Committee, objected to the clause in the Rules relating to certification and chose to resign his position when advised there would be no change.
(National Library of Australia, Canberra.)

Goodfellow, Major Robert Mayo, and the Australian Representative of the Melbourne Committee, Flight Lieutenant T A Swinbourne RAAF. Inevitably, the utterly dependable Commander Harold Perrin was appointed Secretary.

By early January 1934 the London and Melbourne Committees between them still had not reached complete agreement on the intermediate compulsory Controls and Checking Points. Only ten months separated the debate and officially approved aeroplanes from appearing at the 'commencement point' and even that location had not been established.

Having advised all interested parties that full details of the event were available from its offices in London's fashionable Piccadilly, the Royal Aero Club was still unable to confirm some of its own provisional requirements, a fact that was seized upon by an Australian magazine to exercise its opinion that it was "unreasonable and regrettable that Britain should have interfered in the first instance in the drawing up of the conditions." Although Rules and Regulations had been published by the beginning of October 1933, it was generally acknowledged that the final consolidation of the requirements was slow to emerge from the corridors of power.

Did C G Grey have prior knowledge or was his Editorial broadside, fired from the front pages of *The Aeroplane* on 24 January 1934 condemning the apparent lack of progress, a slice of journalistic luck? From the very first announcement of the Races, Grey had kept his readers fully informed of his personal views, which were customarily swift to condemn, to applaud or to lampoon. On 25 January 1934 at a price of one shilling per copy, the definitive Rules finally were published, but by mutual agreement, were only available in English.

CONDITIONS AND REGULATIONS

The Regulations of the Fédération Aéronautique International, world governing body in competitive flying, and the Competition Rules of the Royal Aero Club, senior flying organisation of the British Empire, have been drawn on to frame the Race conditions, which specify that:

Both Races are open to any individual, organisation, or nation. There is no limit to the type or power of the aircraft.

In the event of an aircraft winning and/or being placed in both Races, the nominator will elect which amount of prize money he will receive, and the alternative amount of prize money which he forfeits thereby shall be payable, subject to this rule, to the nominator whose aircraft is next placed in the Race to which such forfeited money relates.

The Centenary Council reserves the right to refuse any entry and/or prohibit the flight of any aircraft.

The nomination fee is £50 for the Speed Race and £10 for the Handicap Race. Where an aircraft is entered in both Races, only £50 is payable.

The nomination fee is returnable to the nominator if the aircraft commences its flight in accordance with the conditions, and reaches Baghdad within 16 calendar days.

A nominator's aircraft, pilots, crews, representatives, and servants shall at all times during the Races be at the sole risk in all respects of the nominator.

In all matters relating to the Races, the decision of the Centenary Council (which will act on the advice of its experts) shall be absolutely final and binding on all persons concerned.

The Centenary Council reserves to itself the right to add to, amend or repeal these Conditions should it think fit, or to cancel the Races, or either of them, at any time.

The right of cancellation will only be exercised in the event of war or national emergency.

There is no limit to the numbers of the crew. The term 'crew' includes passengers.

Each aircraft shall carry sufficient food and water to maintain life for the pilot and each member of the crew for three days, and one life-belt for the pilot and each member of the crew, and in addition not less than six smoke signals (land and water).

Each aircraft shall bear the certificate from its country of registration that it conforms substantially to the minimum airworthiness requirements of the ICAN (International Commission on Air Navigation) normal category, and shall conform to such requirements throughout the Race.

The pilot must produce, at least seven clear days before the commencement of the Race, a pilot's certificate and documentary evidence that he has had sufficient practice in the use of the required instruments; also that he has flown not less than 100 hours solo. (*Author's note*: there was no requirement for a minimum number of hours to have been flown in the type of aircraft entered for the Races.)

Each aircraft must carry the following instruments: an adequate compass, an adequate turn and bank indicator, an adequate drift indicator, an adequate pitch indicator and an adequate altimeter.

Each pilot must carry adequate maps, charts and aerodrome plans and locations. A nominated official at the commencement point is to be the sole judge as to their adequacy.

The United Kingdom representative may, in his discretion, allow the pilot of any aircraft to be changed, provided that he is notified not less than 24 hours before the start of the Race.

The pilot (and where more than one pilot is carried, the pilot designated [1] on the entry form) shall be deemed to be the 'pilot in charge' of the aircraft, and must be carried in the aircraft throughout the Race.

The pilot and each and every member of the crew shall obtain a passport, franked by the proper representative of each and every country to be visited and/or flown over, and shall ascertain and observe the laws of such countries with respect to flight over their territories.

The pilot shall report, with his aircraft completely erected, and bearing the certificate of airworthiness valid for the period of the Race, seven days before the Race, to the commencement point. The aircraft must not leave the commencement point until the start of the Race.

The authorised official at the commencement point shall affix a seal to each engine and a seal to the airframe (fuselage) of each aircraft at least 24 hours before the start of the Race, and such seals shall remain unbroken during the Race. The act of third parties will not excuse an infringement of this rule. No spare engine or engines may be carried or substituted during the Race.

Replacements are allowed provided the provisions of this rule (those relating to the sealing of the engine and the airframe) are not infringed.

Refuelling in the air is allowed.

Night flying is allowed provided that night flying equipment is carried.

The finishing line is a marked line drawn approximately east and west through the centre of the racecourse at Flemington, Melbourne. The extremities of this line will be marked by two pylons.

Unless specially instructed, aircraft shall not land on the racecourse, but shall proceed immediately to, and report at, such aerodrome as is notified to the pilot in charge.

'Handicap time' means flying time, less the handicap allowance time. 'Flying time' means the time spent between the commencement point in England and the finishing line in Melbourne, less time spent at the authorised Checking and/or Control Points.

Each pilot shall carry a log-book or other approved document showing his time of landing and taking off, and certified by the authorised official of the Centenary Council at each Checking and/or Control Point visited.

Competitors and all occupants of competing aircraft must have a medical certificate proving that they have been vaccinated against smallpox within the past three years.

All competitors must make their own arrangements for fuel and oil, replacements, food and accommodation.

Protests may be lodged up to eight days after the decision of the Race judges has been declared.

Completed entry forms were to be in duplicate; one set for the Royal Aero Club in London and the other addressed to the Secretary, Centenary Council, the Town Hall, Melbourne, a request which was sure to confuse matters due to the prospect of double counting, and partially succeeded. Furnished details were to include the name, nationality and flying experience of the pilots and crew together with details of their flying and FAI Competition licences; the manufacturer, type and configuration of the aircraft (biplane or monoplane), registration and type and number of engines. The entry forms were to be accompanied by drawings and where possible photographs of airframe and engine illustrating the aircraft type full size or in model form or failing that, general arrangement drawings were acceptable. In addition entrants were asked to sign a declaration that all Rules, Regulations and Conditions would be complied with.

Early in their deliberations the Race Committees had discussed and confirmed a clause in the Regulations which later resulted in resignations in Australia and concerned the certification standard of the competing machines. Aircraft not issued with a full Certificate of Airworthiness prior to the start would need, according to the clause, *'to conform substantially to the minimum airworthiness requirements of the ICAN (International Commission on Air Navigation) and shall conform to such requirements throughout the Race.'* On 19 April 1934 the following statement was issued in London:

"It is intended that all competing aircraft shall comply strictly with all major requirements affecting airworthiness and safety, such as load factors and take-off requirements, the certifying authorities being left a discretionary latitude as to exact compliance with minor requirements.

For the guidance of manufacturers and competitors it is pointed out that the responsibility for issuing the necessary certificates of compliance under the Rules, rests with the respective certifying authorities."

The aim was to offer hope to those entrants who would, perhaps, not have sufficient time to qualify for a full certificate. But the word *substantially* was technically vague and open to mischievous misinterpretation and, more than any other, this clause was to continue to cause confusion and ill feeling.

The inclusion of a clause permitting aerial refuelling was almost certainly due to the influential nature of prospective entrant, Sir Alan Cobham. Never far from the innovative face of aviation throughout his entire life, Cobham had established a company based at Ford in Sussex from where early in the spring of 1933, trials were flown over the south coast involving a DH.9 acting as tanker aircraft and a DH.60G Moth chartered from the Rollason Aviation Company, as receiver. The theory was to allow an aircraft to take off lightly loaded for a long journey when otherwise it would almost certainly have been performance limited, and to receive fuel once it had achieved initial cruising altitude. The prize, Cobham stated, was the prospect of a non-stop flight from England to Australia.

Having published the Regulations a host of additional paperwork was to follow including three sets of Supplementaries, the last dated as late as September 1934. With extra and revised information being published almost until the start, no competitor could possibly complain that they had not been kept fully advised.

3

The Airway to Australia

INTERNATIONAL co-operation was of critical importance if there were to be few problems for aeroplanes racing along a route halfway around the world. The normally accepted political and bureaucratic barriers had to be breached until all competitors had passed safely by. Although some countries were slow in responding to over-flight requests, most signified immediate interest, realising perhaps that the world at large was becoming steadily more aware of the Races following each public pronouncement.

At a time when international communications were still fragile, dogmatic diplomatic channels conspired further to slow approvals and clearances at national level. Some observers considered it remarkable that working under such a system any progress was made at all. Cole's Committee's requests had first to be passed to the Centenary Council who sent them to the State Premier from where they were forwarded to the Prime Minister in Canberra. From the Australian capital cables were despatched to the High Commissioner in London, then, via Dominion and Foreign Offices, messages were relayed to respective ambassadors, envoys and *chargés d'affaires* for the attention of appropriate government departments. Any replies or queries were transmitted along a reciprocal path.

But generally the rate of progress was encouraging. Potentially time-consuming Port Health and Customs formalities were to be set aside at most places for competitors in transit apart from a mandatory smallpox vaccination check at Darwin where authorities stated that their examination would be "very rapid, lasting not longer than ten minutes, and in all probability, less than three." Most landing fees were to be waived and the Australian authorities announced abolition of import duties for all entrants on the understanding that their machines would leave the country within 12 weeks of arrival. After some delay, it was also confirmed that all prize monies would be exempt from income tax.

Even after the finite Rules and Regulations had been declared, it was still not possible to confirm the route, to the evident displeasure of the press and many potential competitors. From Melbourne, Cole emphasised that 19 countries were required to grant over-flight or landing permissions for aeroplanes and crews of unspecified nationality. Most had agreed provisionally but some approvals were still required including ratification from Turkey, a country which appeared to resent any violation of her airspace by foreign aircraft. Towards the end of February, despite reminders, Turkey had still to provide answers, and it seemed likely that the proposed route would need further modification. But at the eleventh hour the Turkish authorities granted clearance for Race aircraft and even offered landing facilities for British registered machines without the usual prior permission. Other nationalities were required to make advanced application through their own government agencies.

Cole astonished everyone when he revealed that apart from landing permission

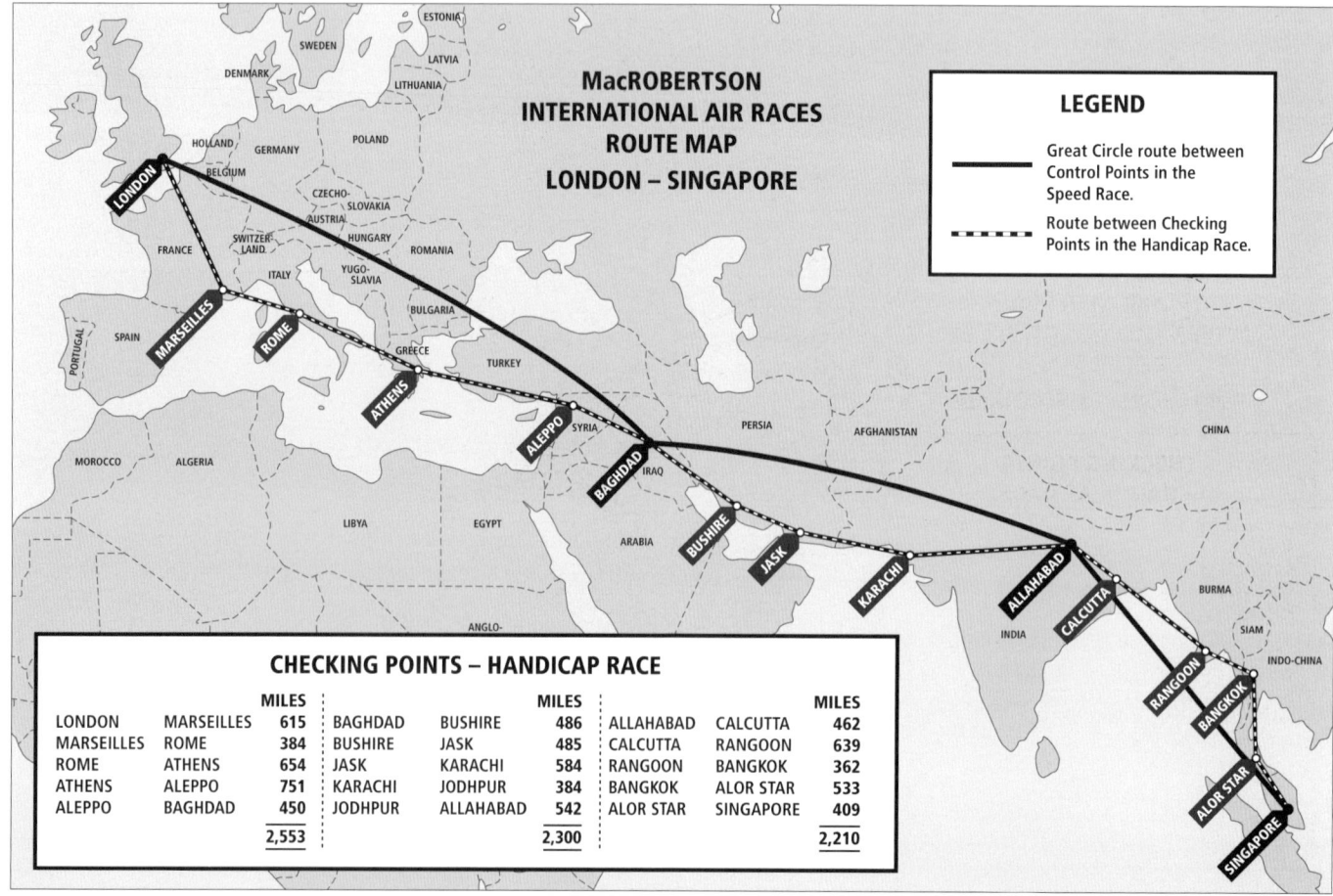

Turkey was now the *only* country to confirm over-flight clearance. There was no doubt that all the others would agree eventually but, until confirmation was received, there was no possibility that comprehensive route details could be published.

Competitors in the Speed Race provisionally were scheduled to make compulsory stops at Baghdad, Allahabad, Singapore, Darwin and Charleville. These stops were known as Controls. Aircraft flying against Handicap were also to land at the Controls but, in addition, were offered a number of optional stops called Checking Points. These were to be situated at Marseilles, Rome, Athens, Aleppo, Bushire, Jask, Karachi, Jodhpur, Calcutta, Rangoon, Bangkok, Alor Star, Batavia, Rambang, Koepang, Newcastle Waters, Cloncurry and Narromine. An Australian suggestion that Sourabaya should be substituted for Rambang was rejected by London.

The framing of the Rules and the introduction of compulsory stops had successfully denied any entrant the opportunity of taking off in a grossly overloaded condition in a quest for extreme range. Distances between Controls were still substantial with no sector less than 2,000 statute miles prior to arrival at Darwin.

Answering criticism that the Rules of the Races made it virtually imperative 'to take risks in night landings' and that this could be avoided if there were no set Control Points, the aviation correspondent of the *Daily Telegraph* wrote:

"Had the competitors been free to choose their own stations, many would elect to land and take-off sometimes at night. The present development of the aeroplane and aerodrome make it reasonable to permit a risk to experienced pilots. Moreover, this is essentially a contest in which skill and sound judgement, and well-found craft should prove more effective than a mere display of desperate courage."

The shortest distance between any two points on the earth's surface is measured in relation to what is known in the technicalities of navigation as a 'Great Circle'. Two definitions of a Great Circle are the intersection of Earth's surface and plane passing through Earth's centre or, arc on a spherical surface whose plane passes through the centre of the sphere. In a speed race it is obviously beneficial to travel the Great Circle track (or route) but, when applied to some sectors in the MacRobertson Races, this passed over areas of almost total desolation, no facilities, minimal population, and the prospect of enormous difficulties for any crew that was unlucky enough to be forced down. The pioneering and most regularly used routes from Europe to Australia followed convenient chains of physical features, services and civilisation. In almost every case these were not the shortest routes, and still presented hazards to travellers. Although they were tried and tested they were up to 1,000 miles longer than the sum of the Great Circle sectors and for the aeroplanes in the Races necessarily added between five and ten hours of flight time.

Crews would be required to exercise fine judgement in their decision making. Given that an aircraft was capable of flying nonstop directly between the compulsory Controls, sufficient extra fuel would be required for possible diversions and adjustments enroute. Cutting corners off the established tracks to fly what amounted to the closest equivalent of a Great Circle route appeared to be the best method of trimming distances and time, whilst maintaining tenuous contact with the most direct line to Australia. The results would be decided by pitting an efficient aeroplane against scant and prob-

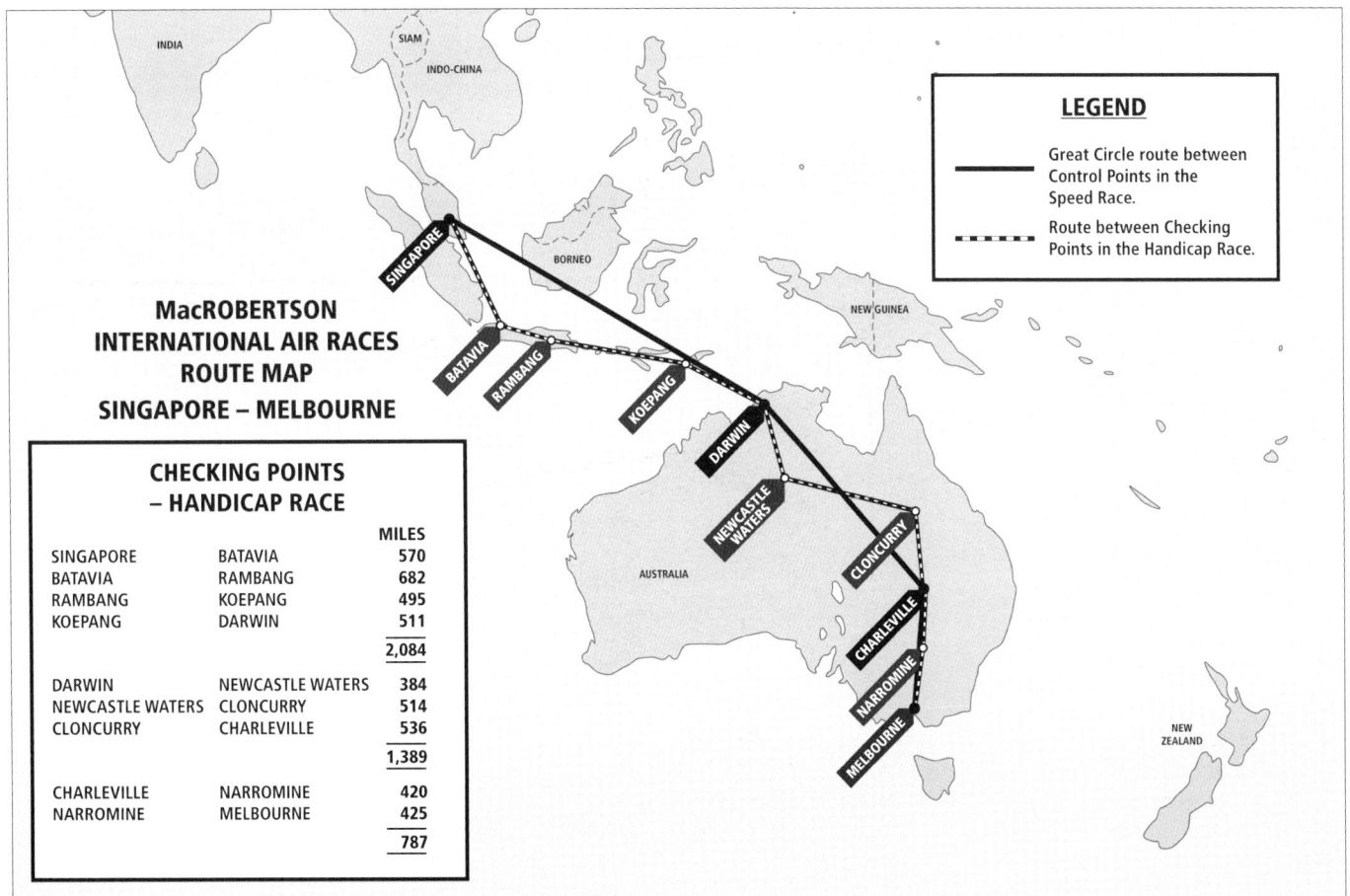

The Australian survey vessel HMAS Moresby *was scheduled to take station in the Timor Sea mid-way between Koepang and Darwin, showing all lights as an aid to competitors tracking between the two airfields.*
(National Library of Australia, Canberra.)

ably inaccurate weather reporting, some luck, and the experience and determination of the crew.

Although the direct route occasionally passed over prohibited areas and was, therefore, in need of 'adjustment' in practice, most authorities voiced no objections, safe in the knowledge that sensitive territory would be circumnavigated with normal respect as any reported infringement would result in immediate disqualification of the offenders.

The sub-committee invited a number of pilots familiar with the routes to their London meeting at the end of February where their views were canvassed. Joining Mr Aler from KLM, Lord Apsley and Major Brackley were three entrants: Charles Scott, Owen Cathcart-Jones and Jim Mollison. In addition to other business, all agreed that the starting time would be 06.30 GMT on Saturday 20 October.

In spite of the preparation of a pamphlet by the Aero Club of India and Burma containing details of the local arrangements for technical and other services and a practical examination of the weather conditions to be expected in October, plus a confirmation that the Government of India had exempted the competitors from all Customs formalities, four weeks before the start no anticipated concession had been granted for the planned route to pass through restricted airspace in the Indian North West Frontier Province, and a more southerly passage was substituted, skirting the Persian coast on the Gulf. The distance between Baghdad and Allahabad was increased by 200 miles which, in the case of the Speed Race, was a severe penalty but applied equally to all contestants. Karachi or Jodhpur were expected to be attractive alternatives, both offering good day/night landing facilities and experienced manpower.

Noting that whilst a British 'commencement point' still had not been identified, the Great Circle distances between Control Points nominated for the Speed Race (calculated in statute miles) were published, probably assuming Marble Arch as the anchor point for 'London' as follows:

London	Baghdad	2,553
Baghdad	Allahabad	2,300
Allahabad	Singapore	2,210
Singapore	Darwin	2,084
Darwin	Charleville	1,389
Charleville	Melbourne	787
	Total	11,323

The distance between Control Points deliberately was lessened towards the end of the route as a concession to tired crews and overworked engines.

Competitors in the Handicap Race found that their route was divided into comparatively easy stages almost parallel with the Speed course and with Checking Points situated at convenient intervals mostly less than 600 miles apart. Nobody was obliged to follow this route but they were compelled to land at all Control Points and they could follow the Speed route if their range permitted:

London	Marseilles	615
Marseilles	Rome	384
Rome	Athens	654
Athens	Aleppo	751
Aleppo	Baghdad	450
Baghdad	Bushire	486
Bushire	Jask	485
Jask	Karachi	584
Karachi	Jodhpur	384
Jodhpur	Allahabad	542
Allahabad	Calcutta	462
Calcutta	Rangoon	639
Rangoon	Bangkok	362
Bangkok	Alor Star	533
Alor Star	Singapore	409
Singapore	Batavia	570
Batavia	Rambang	682
Rambang	Koepang	495
Koepang	Darwin	511
Darwin	Newcastle Waters	384
Newcastle	Cloncurry	514
Cloncurry	Charleville	536
Charleville	Narromine	420
Narromine	Melbourne	425
	Total	12,277

Harold Perrin acknowledged that running two races within one operation had more than doubled the work of the Royal Aero Club sub-committee established to oversee preparations along the route as far as Koepang, from where matters became the responsibility of the Centenary Committee in Melbourne.

In January it was suggested to the same Committee that a subscription list to a 'prize fund' should be opened to provide cash prizes for a race to be organised along the reverse route to assist the expenses of those crews intending to return by air to Europe. Perrin must have been extremely relieved to see the idea voted down.

Officials of the National Aero Club or the Civil Aviation Directorate were requested to offer their services at each aerodrome within their jurisdiction and generous RAF/RAAF support was already promised. Apart from Turkey, which had no National Aero Club, the Royal Aero Club was quickly able to confirm with the governing bodies in France, Italy and Greece that their members would oversee appropriate Checking Points. Additional facilities and manpower arranged by the British Air Ministry and at national level were confirmed elsewhere in Europe and Asia.

Jask Aerodrome on the Gulf of Oman on the southern coast of Persia had little to offer apart from the Cable and Wireless Company office managed by Mr G B Gelly and the KLM bungalow run by Mr Ottens. The Cable and Wireless head office in London volunteered their man as Control Officer after Perrin's organisation made approaches and soon a stream of copies of Race Rules, Regulations and all the Supplements and general updates were on their way to the Gulf. Gelly was asked to provide a dimensioned sketch map of the airfield but he knew that any attempt accurately to record distances would probably put him in jail, so he walked the site in a casual manner noting what was required by measured tread and the data he posted to London was what appeared in the Race briefings.

At Karachi an automatic, electrically illuminated landing 'Tee' was installed in the south western corner of the aerodrome, already welcomed by the crews of an increasing number of daily mail flights and at Jodhpur, the aviation minded Maharaja, His Highness Unmaid Singh Bahadur, offered to pay all local expenses incurred in connection with the organisation and to accommodate at no cost any crew members night-stopping there. At Allahabad, the lately confirmed Indian Control Point, the government had made preparations via Mr M G Pradhan, the aerodrome manager, for fuel to be available from special tanks strapped to lorries in the interests of flexibility during turnround. A generator was hired to provide power for floodlights and everyone was assured that, in spite of five inches of rainfall during the corresponding week the previous year, the surface of the landing ground should bear up well.

The Royal Singapore Aero Club set up facilities at their Control Point with the co-operation of Group Captain Sydney Smith, Officer Commanding Far East Command, Royal Air Force, who arranged for a small detachment from the locally based No. 100 Squadron to be positioned at Alor Star, an aerodrome regularly used by Armstrong Whitworth Atalantas of Imperial Airways. Their job was to try to ensure that at least part of the landing area was kept free from the interminable flooding, and to maintain an operational wireless link.

The RAF Station at Singapore requested technical details of all participating aircraft to ensure they had adequate repair facilities to hand but were advised by London that such matters were the sole responsibility of the competitors.

Dr W Groenefeld-Meyer, Director of Civil Aviation in the Dutch East Indies, was in London in June during which time he confirmed that any competitor, most probably those in the Handicap section, using any aerodrome within his jurisdiction, would be relieved of all charges for landing, weather reports and transport and there would be no delay for medical inspection.

Once airborne from the start, individual flag-off times were to be cabled directly to Singapore where, as agreed, time adjustments were to be incorporated based on an assumed 06.30 GMT start for every aircraft. In one of its many handouts the London sub-committee had reminded crews that they were responsible for the origination

Two shots of Darwin Aerodrome, the top picture showing the site in relation to the settlement. The lower picture reveals a permanent hangar and some facilities in the south west corner. The promised expansion of the landing area was not fulfilled before the arrival of the first competitors. (Northern Territory Archive and National Library of Australia, Canberra.)

and despatch of 'arrival' messages whilst transit times at each aerodrome were to be telegraphed ahead and back to London by Control officials, from where they would be distributed worldwide following an agreement with Reuters. Darwin requested that a wireless message from aerodromes despatching competitors directly to Australia should advise anticipated arrival times and ensure that the message would

In September 1934 the RAAF despatched a mobile W/T outfit to Charleville in Queensland where it took up position on 10 October. (National Library of Australia, Canberra.)

be received some hours before the aircraft.

The nomination of a compulsory Control Point at Charleville in Queensland, slightly less than 800 miles from the finish and 1,400 miles from Darwin, was considered by some to be unnecessary. The argument was reinforced by the length of the previous four sectors to Darwin, each in excess of 2,000 miles, but the Australian Committee was adamant. The imposition of Charleville, if imposition it was, was a deliberate attempt to route competitors around the Central Desert which lay beneath the direct track from Darwin to Melbourne. Only five years previously, Charles Kingsford Smith and his crew had force landed the *Southern Cross* on the north west coast of Australia and were stranded there for 13 days. During the search one of Kingsford Smith's former business partners was killed. Melbourne remembered, and acquiesced with Sir MacPherson's dictate that safety was paramount.

Delayed due to late confirmation of the route, the Centenary Council was at last able to publish its *'Guiding Brochure and Route Handbook'*, a comprehensive looseleaf publication which was the inspiration of the Melbourne Committee. Not only did it include location and detail maps of every Control and Checking Point, together with lucid notation, but conversion and navigation tables, and details of all wireless facilities available along the way. Pilots were warned of the dangers of carrying arms and ammunition across international frontiers, how some countries forbade photographic apparatus in aircraft, where permits were required and special authorisations necessary. There was even a section which carried useful translations of some heartfelt phrases including 'how much?' in seven languages.

By the middle of September 1934 when the last of the Supplementary Regulations were published, each competitor had been furnished with a veritable library of common sized booklets prepared by the sub-committee, each with a different colour cover for easy identification and all printed by Harrison and Sons ('Printers in ordinary to His Majesty') more famously known for the production of postage stamps:

Instructions and Control Points.
Supplementary Regulations I, II and III.
Conditions (original and re-issue).
Pilots' Brochure and Handbook of the Indian Section.
Amendments to the Handbook of the Indian Section.
Rules and Regulations. Netherlands East Indies.
Important Notices to Competitors I and II.

Taking the most direct route, the Speed contestants were likely to encounter few geographical problems within Europe although the October weather was expected to be unpredictable. Mountains north west of Bucharest rose sharply to almost 5,000ft, falling away to the Black Sea beyond the city, but rising again to touch 6,000ft on the Turkish coast. Beyond Aleppo in Syria it was a straightforward run into Baghdad and the first compulsory Control Point. Peaks of 10,000ft. were to be expected on the second sector shortly after passing near Isfahan in Persia, and there was rugged country in prospect until abeam Delhi from where the remainder of the sector to Allahabad was across the Indian plains. Stage three left the Indian coast south of Calcutta to cross the Bay of Bengal on a track which would intercept the Malay Peninsula near Phuket island, after which it was a simple matter to progress south east on a direct route to Singapore. Two thousand miles of water dotted, initially at least, with jungle covered islands, greeted the Speed racers between Singapore and Darwin, from where the stages to Charleville and Melbourne were but a short dash by comparison, although in sum total further than the distance from Singapore to the Australian north coast.

All Control aerodromes along the route were considered satisfactory for the machines in the Speed Race with the possible exception of Darwin which offered a longest run of only 500 yards. Local officials were aware of the shortcomings and had already promised prospective entrant Bernard Rubin that some ground outside the immediate boundary would be levelled and incorporated to provide a run of 1,000 yards. In the event an area of only 600 yards by 650 yards was available for October which was barely adequate.

The Royal Aero Club arranged for listening watches to be maintained by wireless units stationed between India and Singapore at a cost of £40 per unit for the duration and, in addition, RAF short-wave radio stations in Iraq, India and Singapore agreed to pass on all messages.

From Sunday, 21 October, the coal burning survey vessel HMAS *Moresby*, previously the Royal Navy destroyer HMS *Silvio*, was tasked to patrol on a direct line between Koepang and Darwin, showing lights to assist competitors with their track checks. The Australians, who assumed absolute responsibility for the organisation between Koepang and the finish, were determined to equal or better the planning of the London Club's admittedly much longer section and had air-surveyed their part of the route. In association with the Shell Oil Company, special maps showing all useful facilities were prepared with a promise of delivery to Singapore or Koepang should they miss the last possible pre-start airmail service to London.

Light beacons and stocks of landing flares were positioned along the track from Darwin, where special radio direction finding equipment was to be installed. Additional communications were set up at Rockbank and Charleville by the RAAF and RAAF Wireless Reserve, with further direction finding apparatus at Charleville and Melbourne operating on the 36 to 38 metre short wave band and installed by Amalgamated Wireless. An intricate primary web of communications linking Singapore, Batavia and Koepang with all the designated Control and Checking Points on Australian soil utilising landline, wireless telegraphy and cable was reinforced by 'alternative routes' to be activated in case of breakdown. Free of charge, the RAAF donated 600 Very cartridges to the Checking Points although additional supplies were thought necessary, and were ordered by the Race Committee directly from the Munitions Supply Board. Later, some competitors were to reveal that they had received demands for payment to cover the cost of flares which had been ignited at some airfields to assist with night landings.

The differences of attitude between energetic Adrian Cole and his intransigent Civil Aviation committee member, Captain E C Johnston, might be gauged from a comment that the Department of Civil Aviation was notable for its masterly policy of consistent inactivity. Reports from Australia indicated that the general state of aerodromes chosen as Checking and Control Points left something to be desired, that nothing very startling was likely to be achieved by the Department before the start of the Races and that it might be more productive if the local authorities were encouraged to get out their garden rollers.

4

Who is Racing to Melbourne?

C C Walker, Technical Director of the de Havilland Aircraft Company Ltd who was deeply involved in the origination of the specification for the DH.88 Comet.
(BAE Systems.)

EVER SINCE THE official confirmation that there would be a Race (or Races), and long before the final Rules were confirmed, rumour and speculation were rife concerning prospective entrants and the machines which might be built, modified or taken off the shelf and flown from England to Australia.

And, of course, it was a Race *from* England *to* Australia. Any entrant who operated in the southern hemisphere, or the USA from where the majority of entries was expected to be received, were faced with a considerable journey and expense to reach the starting line. Unlike the rules for many other types of race for which competitors were expected to journey from home base to the start unaided, no such stipulation was made for the MacRobertson, other than that machines were to fly into the starting aerodrome. Large packing crates and strange deck cargoes were, therefore, expected to be received at a number of British ports during the autumn of 1934.

In November 1933 it was reported that Captain J P Saul, who flew the Atlantic from west to east in 1930 with Sir Charles Kingsford Smith, had stated his intention to compete. A wealthy Irish syndicate was believed to be financing the entry and Saul had implied that he would be flying a Fokker F.XX, a machine capable of 180mph and the ability to carry petrol for hops of 3,000 miles. Captain van Dyke of KLM was expected to fly as co-pilot.

In January 1934, a Reuters' correspondent confidently cabled London from Melbourne with names of assured competitors. These included the Mollisons, Edgar Percival, Neville Stack, a Flight Lieutenant Jones and Messrs Cathcart, Marcendale and Nelson, all representing Great Britain. From Australia there would be Sir Charles Kingsford Smith, from India Mr Aspi Engineer and the United States' entries would include Wiley Post, Roscoe Turner, Jimmy Wedell, Jimmy Doolittle and James Hazlip.

The Royal Aero Club commented that the list was 'premature' and some of the names 'erroneous'. Certainly the Mollisons had declared an intention to race as a crew, and Neville Stack was talking with Airspeed about a special machine. Aspi Engineer should have read Mr Aspi and Mr Engineer, two pilots who were potential de Havilland customers.

The Hon. Mrs Victor Bruce was listed as a 'probable' entry flying a Miles Hawk. Edgar Percival retorted that he was not even considering the prospect of entering. Nobody could identify Marcendale; Jones and Cathcart were probably the same man, looking for a machine, and Nelson was Director General Designate of Civil Aviation in Hong Kong, to which appointment he declared he would be travelling by sea.

The Press regarded every movement of any well known aviator as worthy of their further investigation. Machines flown by them were analysed for speed and endurance and theoretical improvements after a change of engine or following significant modification were examined and explained.

Donald Bennett was offered the loan of a Rolls-Royce engine through old friends associated with the Schneider Trophy contest. Unfortunately, he was unable to secure the loan of a suitable airframe in which to install it. With active participation still very much in mind, Bennett studied for the civilian First Class Navigator's Licence, working through an extremely difficult and complicated syllabus which resulted in a highly respected professional qualification for which few applicants were successful.

C C Walker, Technical Director of the de Havilland Aircraft Company, was bold enough to expound his theories on the best type of aeroplane suited to a race which was run under a positive set of well-defined rules. He said in part that the best choice would seem to lie between combining all-out speed with long range, and utilisation of a pure racing machine which would need frequent fuelling stops. But night landings on tropical aerodromes would seem to be asking for trouble, he added. He was satisfied with the Rules too, and the influence they surely had on aircraft selection. Were it not for the take-off limitation, a really fast aeroplane could be entered and overloaded with enough petrol to cause a very flat take-off then be flown as fast as possible between the compulsory stops. Unless every competitor was equally treated in the matter of this permissible total weight to secure the required take-off performance, the contest would be a failure.

Z D Granville, creator of the Gee Bee racing aircraft expressed his opinion in the USA early in 1934 that the best aircraft would be either very fast with medium range or a long range aircraft with a high cruising speed and crewed by two or more. Hedging bets the Granville Miller and de Lackner Company laid down the International Supersportster R5, a single seater cruising at 260mph for 1,850 miles, and the International Courier carrying a crew of three and capable of cruising at 200mph for 16 hours. Granville was killed in a crash in South Carolina in February 1934 and neither aircraft was completed.

Charles Walker made it clear to everyone that the MacRobertson was a totally different proposition from other famous flights, where machines had been loaded far beyond any normally certifiable limits. And he regarded the casual speculation generated by the press which surrounded certain prospective entries as unfair to the public mind, while being dismissed by those with any practical knowledge or employed in aviation.

The time allowance within which all competing aircraft were to arrive at Melbourne was 16 calendar days. Outside of that limit would mean disqualification and no presentation medallion. After the first Handicap aircraft had completed the course, would it then be necessary to wait for the last machine to come in before the winner could be announced? Bearing in mind that only the actual flight time would be considered, there was nothing to stop any entrant from spending a leisurely day or so at Checking Points along the route. But all Handicap times were to be announced prior to the start, and the Rules were clear enough about allowances. By the time Race machines were a day or so out from England, there would already be a good indication of their relationship with published information, and nobody was expected to be panicking at Flemington ten days after the arrival of the first aircraft.

Some concern was expressed over the fact that no time allowances would be credited where machines arriving simultaneously might be in competition for the attention of officials and refuelling parties. Early starters would have an obvious advantage, it was alleged. Apart from costly duplication of manpower and facilities, and the hope that the length of the sectors could be relied upon to stretch out the field, there was little more that could be done.

Anthony Fokker chose his moment to declare that neither the aircraft industry nor commercial aviation could have any use for Races such as the London-Melbourne. It was a remarkably inept view. As early as March 1934 KLM had, subject to the resolution of financial matters, considered entering two or three different Fokker aircraft types in the Handicap Race but these were later withdrawn due, mostly, to late delivery from the Fokker factory. Instead, under the direction of the airline's founder and General Manager, Dr Albert Plesman, the company put all resources into securing the best chances in the Speed Race for its new Douglas DC-2, a machine which astonished all who first saw it, and for which Fokker had already sought a European manufacturing licence.

Aviation journalists, digging deep into their files, discovered that the first international air race was in 1909 at Rheims when American aviator Glen Curtiss had carried off the Gordon Bennett Cup. Since then speeds and weights and distances and altitudes had been relentlessly increased; aeroplanes were more practical and efficient and national aviation industries had grown into prestigious and essential assets employing thousands of skilled workers.

The Rules for the MacRobertson Races ensured that lessons would be learned from the winning designs, even though the large and the small, the frail and the all powerful, would be competing alongside one another. There had certainly been criticisms of Sir MacPherson and questions about his motives, but it was quickly established that he was acting as a sporting patriot and nothing else. This was to be the longest race; had attracted more entries than for any other air race and was more international and had captured more interest than all previous speed events combined. C G Grey, in generous mood, put it plainly enough: "Sir MacPherson Robertson deserves the hearty thanks of everyone who is concerned with aviation for having made possible the world's greatest Air Race."

Maps of the Race routes had appeared in almost every aviation publication in the world at some time during the months of discussion, and were republished frequently and more widely when some new feature caught the eye of a correspondent. One Editor's appraisal was that the Great Circle track drawn between each Control was close to the ethnological frontier, behind which the yellow races were massed, although not all should be considered a danger.

The Royal Aero Club became the administrative hub through which prospective entrants were advised that passports could be renewed, visas obtained, licences reviewed, overflight and landing clearances arranged and essential Customs formalities and carnets organised. In addition to the plethora of booklets published by the Club, purchase or even hire of route maps was facilitated through the Automobile Association's Aviation Department at Heston.

Harold Perrin also was expected to answer all the questions of a general nature that came flooding in often from those who might have been expected to be more self-reliant on matters of accommodation, food and drink, mosquito nets, heights to fly, log books, Very lights and smoke signals!

The Race Secretary received the first official entry papers early in February 1934 when the Aircraft Exchange and Mart Ltd. handed in their forms. A L T Naish was to fly a Cheetah-powered Airspeed Courier and he told a reporter that in 130 hours flying only nine pence had been spent on maintaining his company's demonstrator. No doubt he meant well, but the statement was open to mischievous interpretation.

Francis St. Barbe, Business Manager of the de Havilland Aircraft Company, ever anxious to wring publicity from any opportunity, was present at the Royal Aero Club in March when Mr Arthur Edwards handed in the forms as first entrant in the Speed event. Edwards, Chairman and Managing Director of Grosvenor House in Park Lane, London, had contracted to make a private purchase of one of the three de Havilland DH.88 Comets designed specially for the Races and under construction at the Stag Lane factory. His nominated crew of Tom Campbell Black and Charles Scott were in attendance also, to be photographed as the paperwork and £50 fee were registered.

Another confirmed entry had been made by Colonel 'Jesse' James Fitzmaurice, in London during April seeking a suitable mount for sponsorship by the Irish Hospitals Sweepstakes. He left soon afterwards for the USA and was then intending to fly by the KLM route to Batavia to study the conditions likely to be encountered during the monsoons.

Squadron Leader Brady of the Airspeed distributors, Aircraft Exchange and Mart, had introduced Neville Stack to Airspeed. He was hoping to raise sponsorship, possibly from Lady Houston, but his ambitious plans to purchase an AS.6, a design project intended to be powered by two Wolseley engines, were dented when he was quoted a price of between £12,000 and £15,000. Airspeed offered an alternative aircraft powered by two Cheetah engines and Alan Cobham, supported by Airspeed director Lord Grimthorpe, expressed interest in racing a similar machine himself. In spite of the Airspeed company's financial difficulties, not resolved until the following August, the board decided to go ahead with the AS.6 which was launched in November 1933 as the Envoy.

Stack proposed ideas for a number of improvements to the AS.6 to transform it into a high-specification racing aeroplane. The design effort was lead by Hessell Tiltman and introduced so many variations from that of the basic Envoy that the type was re-designated AS.8, powered by a pair of Cheetah VI engines due to a slippage in the supply of Wolseley motors, and eventually named Viceroy. Nominating Sydney Turner as his co-pilot, Stack's was the fourth confirmed entry.

Several sporting pilots had approached Airspeed after the announcement of the Race including Amy Mollison, a shareholder in the firm, whose specification for a single engined machine of high speed, long range and 30,000ft service ceiling was at that time beyond the capacity of the design team.

Bernard Rubin, an Australian born, Brit-

In March 1934 prospective entrant Bernard Rubin flew a route survey with Ken Waller in DH.85 Leopard Moth G-ACLX. Following their return to England in May, Bernard Rubin confirmed his entry of a DH.88 Comet. (National Library of Australia, Canberra.)

ish based millionaire ex-racing driver who owned property near Melbourne, left England in March on a route survey flight with Ken Waller, a flying instructor at the Cinque Ports Flying Club based at Lympne. Flying DH.85 Leopard Moth G-ACLX they arrived in Melbourne, completed their business and returned to England on 1 May where Rubin confirmed his entry shortly afterwards when he purchased one of the three de Havilland DH.88 Comets.

Perhaps the growing interest in the Race and the prospective entrants, denied Rubin and Waller the acclamation due for their survey flight which was in itself a significant achievement. Photographs of their social involvement with the Aero Club at Essendon arrived in England by sea nine weeks later, a fact much laboured when eventually they were published. The Leopard Moth had borne the two pilots back to Lympne in an unobserved time some ten hours less than Jim Mollison's official record set two years previously and the crew was most apologetic as their sole interest had been to collect route information.

Whilst Rubin and Waller were established on their reconnaissance it was suggested that the Dutch Pander Postjager, a tri-engined mail-carrying monoplane with a top speed of over 220mph, might compete. From Argentina, the American airwoman Laura Ingalls announced her participation, whilst in London, an unidentified RAF officer was advertising in an attempt to persuade an owner to hire him a 'fast machine' with the possible temptation of sharing expenses.

In Australia, Sir Charles Kingsford Smith was reported to be looking at the prospects of utilising a special Lockheed Altair with a 700hp motor, having approached de Havilland too late to purchase one of the three Comets the firm had laid down.

From an early stage, KLM, Royal Dutch Airlines, had seen the enormous publicity potential of an entry, particularly as much of the route to be followed approximated that of their own regular service from Amsterdam to Batavia. The company was anxious for political clearance to push on to Australia, although access to that continent had so far been denied. Operating an empty passenger aircraft to Australia would be expensive, and prove little. However, the Rules permitted the carriage of passengers, and as a commercial operation KLM was entitled to make a charge for any load. The company's new DC-2, a fast, all-metal American production aircraft powered by a pair of Wright Cyclone engines, was entered for both Speed and Handicap Races, although the machine would be flown at nothing more than continuous cruising power. The passengers would be treated like normal air travellers accompanied by the usual crew complement of four. It was to be a self supporting publicity exercise, flown in the glare of what was hoped would be massive and sustained international press coverage. Having completed the flight, the passengers were to be invited to spend a week in Australia and a further few days in the Dutch East Indies before returning to Europe. The estimated fare of 5,000 Dutch guilders (£420 at 1934 rates) included first class overnight accommodation and organised visits, and appeared to be set for enthusiastic patronage.

Jim Mollison put in his official entry with the third de Havilland Comet to be crewed by himself and his wife Amy. Jim estimated that in order to win a maximum of £7,500 sterling, and possibly nothing at all, involvement in the MacRobertson was expected to cost exactly twice as much. Short of money, his decision to buy a Comet was supported with some reluctance by Amy who was persuaded to provide half the purchase price. Jim's unlikely appeal for financial assistance from Blackpool Town Council was probably at the suggestion of William Courtenay, Mollison's contracted press co-ordinator. Blackpool's Deputy Town Clerk, Trevor Jones, was married to Amy's sister, Molly. The Council's refusal came as little surprise to those who were even aware of the approach.

The fact that a French pilot had returned from the USA having purchased a Lockheed Orion, was taken as indication of a near certain entry, and another five competitors were said to be waiting for French Ministerial consent. At the 1932 Paris Aero Show The Blériot Company had exhibited a low wing monoplane with single engine and retractable undercarriage; a machine surely destined for the Speed Race? Multi entries were expected from Italy and Germany where Heinkel was reported to be preparing something fast. The Dutch military authorities were interested in an entry for the Speed Race and had requested assistance from the Dutch Automobile Club which, for political reasons, refused to co-operate!

From the USA, a lengthy circular dated 6 April 1934, was received by the British press. The American National Aeronautic Association (NAA) was saying in effect that under the current Rules, US manufactured aircraft without a full Certificate of Airworthiness were to be barred from entering the Races. In some quarters this statement was taken as an attempt to portray the Royal Aero Club as protectionist and unsporting, yet no copy was sent to the Club directly. A letter including the substance of the circular but dated three days after its publication was eventually received by the President of the Royal Aero Club, the Duke of Atholl, a discourteous sequence which received blunt comment in the columns of the British aviation press which gave no support to the American claims.

In spite of the criticisms, matters were stirring in America. Before confirming his entry, Roscoe Turner, a familiar figure at most American aviation meetings and on film sets where aeroplanes were involved, made contact with the Royal Aero Club in an attempt to clarify the situation regarding the certification requirements. He reminded officials that American aircraft flew on either commercial or racing certificates, and in view of the regulations, which of these would be acceptable?

The system, administered by the Department of Commerce, was unknown in England and, following consultation between the Australian liaison officers, Air Ministry and Royal Aero Club, the British Air Attaché in Washington was asked to investigate the matter before further approaches were made to the Race sub-committee or higher authority. Unfortunately, news of the situation was leaked and views were aired in the press which suggested that all American aircraft would have difficulty in complying with the Regulations.

Nations which were signatories to the ICAN were permitted to establish their own standards and these were perfectly acceptable, but the USA had not signed the Convention, and American aircraft in the Races operating outside the ICAN requirements would be crossing national boundaries within which official paperwork was a cherished commodity.

The following opinion was expressed by *Observer* in the Christmas 1933 issue of *Flying* magazine:

"Unless the necessity for a C of A for each competing machine be waived there is no obvious means of preventing some stealing a march on others, and the cleverest interpreter and/or evader of rules will improve his own chances; by no means an uncommon thing in any form of enterprise.

It is a great pity that ICAN regulations would not permit of waiving all Cs of A for the occasion. It would be an excellent trial and we should learn a lot. There is every indication that America at least looks upon the event as considerably more important than the value of the first prize.

To have the Race spoiled, and international feeling embittered, by reducing a sporting contest to a technical debate would be a tragedy of the first magnitude, irrespective of wasting the chance of a decade for Aviation."

The position was considered so serious that James Rae, an official from the National Aeronautic Association, the US equivalent of the Royal Aero Club, travelled to London in April and together with officials of the sub-committee and Air Ministry, studied ICAN documentation. A cable was sent to Washington requesting that the Department of Commerce should confirm that each Race entry met the ICAN requirements as published, but the Department refused, and instead demanded a definition of the term 'substantially' called for within the Regulations.

ICAN requirements to which each aircraft was to be tied demanded compliance with an established set of performance figures relevant to take-off and landing. In the take-off case, the aircraft was required to clear an obstacle 20 metres in height above a reference point 600 metres from the aircraft's standing start. The whole purpose was to eliminate any possibility of grossly overloaded aircraft which were unable to maintain a minimum gradient of climb having left the ground. Machines which were capable of complying whilst carrying petrol for extreme range were considered both aerodynamically and fuel efficient, and structurally strong. Similar requirements were demanded with respect to landing distances after approaching over a theoretical screen.

The Americans were keen to adopt rules identical to those employed during the Bendix Races, where participating aircraft were approved outside of a 'Normal Category' certificate, providing that no commercial passengers were carried. Some sections of the Australian aviation community appeared to agree with this view on the grounds that the MacRobertson and Bendix Coast-to-Coast Races were similar, but the London sub-committee could not be swayed. A meeting in Melbourne on 23 April followed by another in London three days later endorsed the official stand: unless American entries conformed 'substantially' to the ICAN requirements, they would be ineligible.

The President of the NAA, Senator Hiram Bingham, blamed the Australians and the British for the situation and said that all American entries would be barred if the sub-committee insisted on maintaining its position. He alleged that the initial Draft Rules as circulated, and under whose requirements several aircraft were currently being built, had subsequently been changed. The 'Normal Category' requirement, which had been applied to Handicap entries only, now included Speed entrants also. Perrin was forced to admit that the Air Ministry had indeed objected to the Draft Rules, and yes, they had been changed. Within three weeks of the entry deadline of 1 June, it seemed unlikely that any further American competitors would be involved, a situation which was disastrous for the unique spirit of this much publicised international event. On 10 May the Royal Aero Club offered a concession in that although the make, type and model of aircraft had to be declared by 1 June, the supply of more detailed information called for on entry papers could now be delayed for a period of one month. By the middle of May there had been no change in the position, and neither the Club nor US Commerce Department were prepared to make public statements.

At the eleventh hour a brilliantly simple solution was offered by the British Air Ministry. The ICAN requirements had been drafted in 1932, since when a great deal of technical progress had taken place. The US Department of Commerce should be permitted the necessary latitude to decide whether American aircraft operating under their own 'Race' certification conformed 'substantially' to ICAN requirements, where 'substantial' was taken to cover the major items affecting airworthiness, particularly take-off and landing performance. Minor differences of opinion were open to discussion between the owners and the Department of Commerce. The Royal Aero Club added a further concession in that certificates need not be presented for scrutiny until the machines actually arrived in England and the British Air Attaché in Washington cabled London on 29 May confirming acceptance of the plan by the United States' authorities. The problem appeared to have been solved, and almost immediately, Perrin received entries from ten American crews, but it quickly became apparent that several of the entrants were suffering difficulties in raising financial backing and were later withdrawn or simply never followed up.

Harold Perrin's cabled requests for confirmation of intentions had almost without exception gone unanswered and he pursued the matter through the NAA in the light of published newspaper reports that even having achieved concessions, most American entries probably would be withdrawn. Perrin alleged that the situation was being used as a ploy in the hope of raising additional sponsorship, although everybody denied the accusation.

The Aero Clubs in New Zealand, situated at the furthest possible point from the starting line were almost unanimous in agreeing that the vast financial outlay required by any national entrant could be better employed at home. Disturbed at his country's attitude, Mr H Oram of the Manawatu Aero Club called an emergency meeting. He persuaded the Club that it should be represented, and within two hours of the deadline, an entry was telegraphed to Perrin in London. Major G A Cowper, a former Club instructor and the current Club Captain, Henry 'Johnnie' Walker, were nominated as crew. Phillips and Powis at Woodley Aerodrome, Reading, England, subsequently received a cable with a provisional order for a new Miles Hawk monoplane.

The Manawatu Aero Club Committee was not unanimous over the choice of Cowper as a pilot and dissatisfaction festered for six weeks before they approached Squadron Leader Malcolm McGregor, chief pilot instructor of the Club, to act as aircraft commander. McGregor accepted their offer after some consideration and an announcement was made on 1 August following which Oram and some opposing Committee members became embroiled in a bitter public dispute which eventually reached the Supreme Court in Wellington. Chief Justice Sir Michael Myers ruled that contract had not been proved and dismissed Oram's claim for an injunction against McGregor's nomination.

Having settled the politics, it was necessary to raise finance. A number of guarantors were secured following swiftly upon the formation of the Manawatu Centenary Air Race Committee and another cable was despatched to Woodley Aerodrome, this time to confirm the order on the Hawk. Just before he left home for England, McGregor posted a new Club flying regulation: no stunting below 1,000ft. He wanted to ensure that the Club would still be in business when he returned from England.

On 1 June at 12.00 noon Harold Perrin was able to announce that a provisional total of 72 entries had been received from

After considerable thought Freda Thompson travelled from Australia to England early in 1934, determined to be a Race entrant. She purchased a new DH.60G Moth Major from the de Havilland Aircraft Company but due to her late decision she missed the cut-off date for the presentation of entry forms.
(Via Alan Archer.)

13 countries: Australia, Denmark, France, Germany, Great Britain, Holland, India, Italy, New Guinea, New Zealand, Portugal, Sweden and the USA. The list was subject to amendment as eight entries had been received in Melbourne which needed checking for duplication, but nobody really expected all the places to be taken up. Most of the big names were entered as anticipated and some exciting new aeroplanes listed in addition to some types whose completion of the course would be a victory in itself.

Freda Thompson, the 28 year old daughter of a Melbourne banker, had qualified for her pilot's 'A' licence in 1930 and her commercial 'B' licence two years later. When the MacRobertson Races were announced she showed interest but not until she realised there were so few women entrants did she consider taking part herself.

Miss Thompson took passage to England early in 1934 where she bought a new DH.60G Moth Major from the de Havilland Company for £1,000. The aircraft was fitted with fuel tanks to increase the capacity from 19 to 66 gallons and the endurance to almost seven hours. The Moth was registered G-ACUC in June 1934 and named *Christopher Robin* although the Certificate of Airworthiness was not issued until the end of August. Sadly, after all the effort, it was Freda Thompson's late decision to take part which prevented her from making a formal Race entrant's application before the closing date of 1 June. Undeterred, she decided to fly her new Moth home to Melbourne anyway, making an attempt on Jean Batten's record solo time of 14 days.

On 28 September she left Lympne Aerodrome and safely arrived in Marseilles that same evening. The following day she flew via Rome and Brindisi but miscalculated the end of daylight time in Greece and damaged the aircraft during a precautionary landing in near darkness 30 miles short of Athens. The Moth Major was repaired locally and continued along the route of the MacRobertson Races, arriving in Darwin on 10 November, her pilot becoming the first Australian woman to fly solo from England to her home country.

As the entry lists closed an approach was made to the US Navy Department asking officialdom to explore the possibilities of free use of an aircraft carrier in and upon which American aeroplanes and crews might be transported from the USA, across the Atlantic to England. At the end of August the proposal was said to be 'still under consideration' but the plan was never sanctioned and the organisation and expense of directing competing aircraft to the starting line fell squarely on the shoulders of the entrants.

5

The Sixty and Four

No. 1. Messerschmitt Bf-108V1 Taifun D-ILIT was withdrawn before the start. Note the three-bladed propeller and Rally number.
(John W Underwood Collection.)

No. 2. This photograph of the Miles Hawk prototype was submitted to the Royal Aero Club by Miles Aircraft on behalf of the Manawatu Aero Club as part of the requirements of the Race entry form.
(Royal Aero Club Archive.)

AS HE CHECKED through his accumulation of red 'Forms A' (Speed Race) and blue 'Forms B' (Handicap Race) and reconciled the payments received and due, Harold Perrin could identify 71 entries. Only ten were solely for the Speed Race, 31 were Handicap machines and the remaining 30 were entrants in both Races. The Centenary Committee considered that these figures fully justified their determination to concur with Sir MacPherson's aims of establishing a Race of truly international character which would be conducted under conditions designed to minimise all foreseeable risks.

The entries were reviewed by Robert Mayo's sub-committee in London on 4 July as a result of which those for R W H Everett, H R A Kidston, G Huffmann, Pond and Sabelli, R Ammel and *Australian Women's Weekly* were rejected for non-compliance. H F Broadbent had retired at his own request and the Speed Model Fox Moth he was expected to fly was being prepared by de Havilland for sale to India.

The allocation of Racing numbers 1-64 was by draw although the order of take-off was not decided until 12 September when it was confirmed that Speed contestants would leave first followed by those in the Handicap Race. All positions were to be determined by notified speeds and aircraft would departing at 45 second intervals.

The Rules declared that allocated Race numbers were to be painted in black on either side of the aircraft's fin and rudder and underneath a wing or the fuselage. Their allocation, published in a list on 23 July, confirmed the prospect of a spectacular line-up at the start, the likes of which had never before been witnessed on any aerodrome, and in all probability would never be equalled. By 4.30pm on Sunday 14 October 1934, it was anticipated that each racing aeroplane would have been presented to scrutineers at the starting point.

1. Messerschmitt Bf-108V1. D-ILIT. (Germany)

Nominated by Wolfram Hirth, to be flown by himself and Herman Illg in the Handicap Race.

Wolfe Hirth was brother of the aero engine designer Helmuth Hirth and, paradoxically, a champion glider pilot in spite of having lost a leg during the First World War. Illg, also an accomplished glider pilot, was an aircraft engineer.

The aircraft was first registered in August 1934 but by late September following participation in the 1934 Rundflug, the entrants had been unable to provide details of their aircraft, pending approval by the German Air Ministry. The aircraft did not arrive for the start of the Race and her absence was never explained, although the reason was thought to have been a combination of technical problems and a shortfall in performance with the HM8A engine.

2. Miles Hawk Major. ZK-ADJ. *Manawatu*. (New Zealand)

Entered by the Manawatu Aero Club in the Handicap Race. Following local controversy over crew composition, which resulted in a court case, the pilots were named as Squadron Leader Malcolm McGregor and Henry 'Johnnie' Walker.

Malcolm McGregor was born in New Zealand in March 1896 but learned to fly

No. 2. Henry Walker and Malcolm McGregor at Woodley Aerodrome with their Miles Hawk on 11 October 1934.
(The Aeroplane.)

in England in 1916 and served on the Western Front flying Sopwith Pups and SE5As, returning to New Zealand in 1919. After operating joyriding, passenger and mail services, and surviving a crash which all but ended his career, he was appointed Chief Instructor at Manawatu in 1932.

At the age of eight in 1916, Henry Walker's family left Edinburgh to settle in New Zealand. He was taught to fly at the Manawatu Aero Club by Major G A C Cowper, the ousted contender for a crew seat in the Race, and qualified for his 'A' Licence in 1931. He was Club Captain at the time of the decision to enter the MacRobertson Races.

When the entry was first mooted it was based on the provisional order for a Miles M2C Hawk, a low-wing monoplane built of wood with two open cockpits in tandem, and powered by a 120hp Gipsy III engine. When Miles introduced the much improved M2F Hawk Major in the early summer of 1934 the New Zealand order was changed for one of the new machines with a Gipsy Major engine and aircraft No. 119 was reserved, later registered G-ACXU. The previously provisional order was confirmed by telegraph and a deposit of £100 wired to the manufacturer, Phillips and Powis, at Woodley Aerodrome near Reading, but the aircraft was far from ready when McGregor and Walker arrived in England on 21 September.

Due to a clerical error at Woodley the New Zealand order had not been treated as firm although the deposit had been banked but, crucially, the Royal Aero Club was not advised of the important change of intended powerplant and a young clerk in Fred Miles' office was found at fault after he attempted to falsify documents to cover his shortcomings. He was instantly dismissed from the company but Harold Perrin took an avuncular view and as a result of his personal intervention the young man was reinstated following a severe reprimand.

Phillips and Powis were so embarrassed that maximum effort was put into completion of the aeroplane, which was re-registered ZK-ADJ on 12 October but only handed over three days before the crew was expected to check-in at the starting point, allowing no time for proper endurance testing. Complete with the two sets of 'optional' blind flying instruments, the Hawk had cost the Manawatu Club £1,000.

3. Airspeed AS.6 Envoy. G-ACVI. *Miss Wolseley.* (GB)

Entered by Lord Nuffield in the Handicap Race and crewed by his personal pilot George Lowdell with Flight Lieutenant D F Anderson, the aircraft was registered on 9 October 1934 in the name of William Richard Morris (Lord Nuffield) at his Huntercombe address near Henley-on-Thames.

George Lowdell was an experienced ex-RAF pilot and had been an instructor with the Suffolk Aero Club at Ipswich and later with Brooklands Flying Club at Weybridge. He was currently retained as chief test pilot for Wolseley Motors, a company which had recently entered the aero engine market.

In order to save time, David Anderson, previously personal pilot to the Viceroy of India and now a Martlesham Heath test pilot, had been permitted to travel to Portsmouth to fly the machine at Airspeed's home airfield and where he was offered the chance to act as co-pilot in the Race. Not wishing to turn down such an opportunity, Anderson agreed and was immediately disqualified from testing the machine himself, which due to limitations on the weighing apparatus in use at the start was, rather ironically, instructed to travel to Martlesham Heath after all.

The Envoy was kept at Castle Bromwich for the personal use of Lord Nuffield and, powered by two Wolseley AR9 Aries III radial engines developing 185hp, cruised at 150mph and had a top speed of 170mph. Additional tankage in the fuselage extended the range to 1,500 miles.

4. Douglas Transport. (USA)

Entered by Australian Harold Gatty, a naturalised American living in Washington DC. The machine was to be flown by Jack Frye and H Hull in both the Speed and Handicap Races.

Gatty had accompanied Wiley Post in the Lockheed Vega *Winnie Mae* during their famous round-the-world flight in June 1931 when the standing record had been lowered by eleven days to 8 days 15 hrs and 15 mins. Gatty was later referred to as 'The Prince of Navigators'.

Jack Frye in his capacity as Vice-President in charge of operations for Transcontinental and Western Air (TWA), was recognised as the originator of the specification which led directly to the birth of the Douglas DC-1 and later DC-2, but his company stood to gain little from the operation of a commercial aeroplane in the MacRobertson Races.

There is some evidence to suppose that Frye and Hull's 'Douglas Transport' was in fact the prototype DC-1, powered by two Wright Cyclone engines, operated experimentally by TWA on some of the company's internal routes. The machine was fitted with long range tanks and flew extensive trials with new navigation equipment, an assignment much suited to the philosophy of the Race.

The DC-1 was also used by the US Department of Commerce and Army Air Corps in 1934 to test the new Sperry automatic pilot which was linked to a radio compass, but reports reached the Royal Aero Club in September indicating that TWA had lost interest in the Race and Gatty's proposed charter of the machine had been withdrawn.

An alternative view is that millionaire businessman Floyd Odlum had invited Gatty to enter the Races with a DC-2 for which he was prepared to cover all expenses. What Odlum failed to tell Gatty was that his close friend, Jackie Cochran, would be part of the crew. Gatty is believed to have disliked most women and especially women pilots and when he discovered the plan he immediately left the team. He was unable to secure a DC-2 for himself, an aircraft he still believed was his best choice for the Races, and nothing further was heard of the entry.

5. Boeing Transport Type B247D. NR257Y. *Warner Brothers' Comet.* (USA)

Entered by Colonel Roscoe Turner to be flown by Turner and Clyde Pangborn in the Speed Race with Reeder Nichols acting as operator of the wireless equipment which he had selected and installed himself.

The Boeing 247, an all-metal, low wing twin with retracting undercarriage, designed as an air-conditioned ten seat airliner, flew on 8 February 1933. NR257Y, a type 'D', was the first of 17 production aircraft of the new marque fitted with Pratt and Whitney geared and supercharged Wasp 'H' series engines, each rated at 550hp, enclosed in new NACA cowlings and driving three-bladed, Hamilton Standard controllable pitch propellers. Other modifications distinguishing the D model from the ten aircraft of the first batch helped to increase range, ceiling, climb, single engine performance and an increase in top speed to 200mph, an improvement of 20 mph. Turner's choice of aircraft was driven primarily by its good speed and long range, which was increased further by the addition of six extra tanks of about 1,100 gal-

No. 5. Colonel Roscoe Turner with co-pilot Clyde Pangborn and a model of the Boeing 247D which they had chartered from United Airlines to take part in the Races.
(Royal Aero Club Archive.)

Two views of the Boeing 247 NX13301 supplied by the manufacturer which accompanied the submission of Roscoe Turner's Race entry forms.
(Royal Aero Club Archive.)

No. 5. The fuselage of Boeing 247D NR257Y being loaded onto the boat deck of the SS Washington *in New York Harbour. The lifting derrick was specially modified.*
(National Library of Australia, Canberra.)

The fuselage of NR257Y was positioned fore and aft between the two funnels of the SS Washington. The photograph was taken during the ship's scheduled visit to Plymouth on 2 October before leaving for Le Havre.
(National Library of Australia. Canberra.)

Prior to unloading at Southampton Docks on 7 October where the aircraft arrived on board the SS Westernland, *inbound from Antwerp.*
(The Aeroplane.)

lons capacity installed in the fuselage, raising the still-air range from about 750 miles to 2,300 miles.

To prove her systems the aircraft flew non-stop Seattle-Los Angeles in five hours, and Los Angeles to New York via Wichita and Pittsburg prior to shipment to England on board the SS *Washington*. The ship had been selected specially as she was capable of accommodating the aircraft as deck cargo with the wing panels outboard of the engines removed. In New York Harbour the aircraft was positioned alongside the ship on a lighter but the lifting derricks needed to be increased in length by 20ft at a cost to Turner of $500.

What happened after that is confusing. Turner is quoted as saying that, unable to land at Southampton Docks due to storms in the English Channel, the ship was diverted to Le Havre, where on 3 October French Customs demanded import duty of the equivalent of £400 if the aircraft touched French soil. Ever resourceful, he chartered an American owned ocean-going lighter onto which the dismantled aircraft was lowered after the captain arranged for obstacles and obstructions to be removed by blow-torch. With the nose and tail of the fuselage overhanging the vessel's rails, the unlikely combination was headed towards England.

However, after calling at Plymouth the SS *Washington* was always scheduled to dock at Le Havre and from there to sail to Hamburg, not visiting Southampton at all. But on 7 October the Boeing was unloaded at Southampton from the SS *Westernland* which had arrived from Antwerp. Just where and when the transfer was made are facts lost to history. To complicate the story further contemporary press reports in the USA suggest the aircraft left New York on

No. 5. Having been assembled at Hamble, on 8 October the Boeing 247 arrived at Heston where she was prepared for the Races.
(Ted Vaisey.)

For the benefit of the press Roscoe Turner introduced a toy version of his pet lion Gilmore. The growing animal had been left in the USA.
(National Library of Australia. Canberra.)

the SS *President Roosevelt* which routed to Hamburg via both Plymouth and Le Havre but a week earlier. To catch that ship might have been the original intention.

Once safely arrived at Southampton, the aircraft was transferred to a lighter, towed to the Hamble River and, under the supervision of Trevor Westbrook, Manager of the Supermarine works at Woolston, unloaded and re-assembled in the old Avro hangar on the adjacent aerodrome, where a signwriter embellished the fuselage with the Stars and Stripes.

Possibly not realising that there were two airfields laid out at Hamble, created when a public road had been built, bisecting the original site, Turner taxied out on 8 October and lined up on the southern field, an uneven stretch of grass that had not been used for flying movements for some years. To the amazement of many who thought he would never get airborne, he bounded the aircraft off the ground between the Avro hangar and Southampton Water, and turned on course for Heston where she was to be prepared before moving on to the starting point.

Roscoe Turner who had already been credited with opening the discussion on conformity of certification requirements and who had estimated, like Jim Mollison, that his participation would cost twice as much as the value of the first prize, originally had entered an unspecified Douglas transport aircraft but negotiations with that company broke down when it became evident that the delivery date could not be achieved. Instead he arranged a pre-delivery charter from United Airlines, October to December, of one of their latest Boeing airliners, the Model 247D, to be partly sponsored by Warner Bros., but principally by his previous employer, Gilmore Oil.

Turner cabled details of his alternative aircraft to London where the information was received with dismay. This late change of type was contrary to the Regulations and liable to disqualification. The attitude of the sub-committee was greeted with equal disbelief by Turner whose innermost thoughts were widely publicised even protesting directly to the Centenary Council in Melbourne for which impetuosity he later apologised.

The sub-committee too was unhappy about its own inevitable decision and following consultation with Adrian Cole in Melbourne, they determined to do all possible to try to accept Turner's entry. The solution was quite simple. Would the change in aircraft type be unfair to other entrants? Their estimation was that it would not. Roscoe Turner and his Boeing would be welcome in England.

The Heinz food company failed to confirm their earlier interest although their '57' trademark was already painted on the fuselage, a fact which was to cause some confusion later when observed by race officials.

For publicity purposes Turner convinced the Gilmore Oil Company to provide him with a lion cub which he promptly named 'Gilmore'. The animal was treated like a domestic cat and often travelled in the aircraft with Turner, assuring him of press attention. He arrived in England with a soft-toy representation of *Gilmore*, leaving the real cub at home in the USA, much to the disappointment of many.

Born on 29 September 1895 at Corinth, Mississippi, Turner was one of the oldest and best known of American civil pilots having trained as a balloon observer with the Army Signal Corp's Aviation Section in October 1917. Following an honourable discharge in 1919 he joined a barnstorming act as a parachutist but left the show late in 1921 to open a garage business in his home town. There he rebuilt a Curtiss Jenny which had been deliberately crashed as part of the barnstorming routine and taught himself to fly on it. By 1923 he was carrying passengers and offering flying lessons. In 1932 having taken up air racing he won the Bendix Trophy and was later to be the only three-time winner of the Thompson Trophy Air Race. From 1929 Turner, an extrovert and a showman, sporting a waxed moustache, always introduced himself as 'Colonel' and spectators might have been forgiven any suspicion that he designed the badges and decorations for his Hollywood style uniform himself (which he had!) but he had been invested with his rank by the State Governor in the Nevada Division of the National Guard, and was not the complete celluloid impostor that some had impishly decided must surely be the case.

In describing 'the Turner uniform' - robin's-egg-blue tunic, fawn coloured whipcord breeches, Sam Browne belt, black riding boots and a gold and crimson flying helmet - aviation writer Cy Coldwell commented: "A pilot with nerve enough to wear that uniform and kick a half-grown lion in the pants is bound to come in first eventually!"

In September, Turner contacted Harold

No. 6. Built from wood, the Pander Postjager was designed as a fast mail carrier from Europe to the East Indies and had already created a record for the route.
(John W Underwood Collection.)

No. 7. The photograph of Desoutter II OY-DOD which was attached to the entry form submitted by Lieutenant Michael Hansen of the Danish Army Air Service, shows the aircraft prior to her sponsor's titling being applied.
(Royal Aero Club Archive.)

Perrin with a request that the sub-committee seek a free airline ticket to allow him to survey the route. Perrin replied that the airlines flew only as far as Singapore, the trip schedule was nine and a half days each way, the fare was about £380 and seats were at a premium!

Clyde Pangborn, born in Bridgeport, Washington on 28 October 1894, learned to fly with the Aviation Section of the US Army Signal Corps in 1917 but due to his abilities was retained as an instructor at Ellington Field, Houston, Texas until he was demobilized in 1919. He too was involved in various barnstorming and joyriding businesses across America until 28 July 1931 when he and Hugh Herndon Jr set out from Floyd Bennett Field, New York on an attempt to beat the round-the-world flight record only recently established by Wiley Post and Harold Gatty. Due to a combination of mishaps all attempts on the record were abandoned after they were delayed in Russia but eventually they did set a record by flying their overloaded Bellanca from Sabishiro Beach in Japan, non-stop 5,500 miles across the Pacific to Wenatchee, Washington in 41 hrs 13 mins.

Pangborn flew as a night mail and demonstration pilot until details of the MacRobertson Races were announced. Unable to raise enough finance in support of his entry of the Granville Monoplane, Race No. 46, he had subsequently teamed up with Turner and at short notice, the Granville entry was made over to Jacqueline Cochran.

6. Pander S-IV. PH-OST. *Panderjager*. (Holland)

Entered in the Speed Race by 'The Dutch Syndicate' (Hollands Syndicaat) based in The Hague, with Gerritt 'Geys' Geysendorffer as captain, described as a KLM pilot of the 'old school', together with Flight Lieutenant Dirk L. 'Dick' Asjes, a flying instructor granted special leave from the National Aviation School at Waalhaven as co-pilot. Simon van Straten, nominated as wireless operator, was later replaced by Mr P Pronk, more usually employed by Radio Holland.

Built speculatively by Dutch furniture manufacturer H Pander and Sons, who had constructed light aircraft since 1924, the aircraft was designed specifically as a high-speed mail carrier powered by three 420hp Wright Whirlwind engines capable of a maximum speed of 223mph and a normal cruise speed of 186mph. Flown by the same pilots nominated for the MacRobertson Races but with van Straten as 'sparks', she had covered the 9,000 miles from Holland to Batavia in The Netherlands' East Indies carrying Christmas mail at the end of 1933 in 46 hours. Under the command of her captain who was experienced on much of the route, and who had agreed to take part in the Race at very short notice, the entry was considered a potential winner. Named *'Postjager'* ('Mailhunter' or 'Rapid Post' in free translation) for her 1933 flight, the aircraft was renamed *'Panderjager'* for the MacRobertson.

7. Desoutter II. OY-DOD. (Denmark)

Entered by Danish Army Air Service Lieutenant Michael Hansen, to be flown by himself in the Handicap Race with Lieutenant Daniel Jensen acting as navigator and engineer.

The high winged cabin monoplane powered by a Gipsy III engine was used by Danish Air Lines from July 1931 for taxi and pleasure flying work, but it was not an economical proposition and in May 1934 with only 84 hours logged, she was bought by Michael Hansen with the specific intent of competing in the MacRobertson Races.

Hansen, who had arranged a months leave from the Army Air Service, was un-

No. 8. This photograph of a model of the Fokker FXXXVI, PH-AJA, was submitted by KLM with their entry papers, but due to late delivery and a shift in policy the aircraft was withdrawn.
(Royal Aero Club Archive.)

No. 9. The Keith Rider R-3 NR14215 at Clover Field Airport, Santa Monica, home of test pilot J E Granger and where the aircraft crashed on take-off on 2 October.
(The Aeroplane.)

able to fund the entry himself and arranged sponsorship from the influential Danish daily newspaper *Berlingske Tidende* whose name was dutifully painted on the fuselage sides.

8. Fokker XXXVI. PH-AJA. (Holland)

Entered by KLM in the Handicap Race to be flown by Captain I W Smirnoff, an ex-Russian Air Force Officer or Captain G M H Prijns.

Still under construction in March 1934, the big aircraft was a high-wing cantilever monoplane weighing over 16 tons, powered by four Wright Cyclone engines each developing 650hp, providing a cruising speed of 142mph over a range of 870 miles, and a maximum speed of 168mph. The original intent was to operate the Fokker in an airline configuration with two pilots, wireless operator, engineer and a second engineer acting as steward, carrying 16 passengers with sleeping berths 'or equivalent', and to allow another company entry, the Douglas DC-2 (Race No. 44), fitted with extended range tanks and blanked-out windows, to carry mail on long sectors at a high cruising speed. As the Fokker aircraft was delivered late, KLM management decided the crew would have insufficient time in which to familiarise themselves with the new machine before the start. The entry was withdrawn on 19 September and the passenger capacity transferred to the DC-2.

9. Keith Rider R-3. NR14215. (USA)

An entry in both Speed and Handicap Races, to be flown by James Granger.

Keith Rider was a well known designer of racing aircraft which had proved popular and successful on the American circuits and included the B-1, R-1 and R-2. The new aircraft designed specifically for the MacRobertson Races, the R-3, a single seat, low-wing monoplane of all metal construction with a fixed undercarriage, was built in a disused coffin factory in Santa Monica, financed by Mrs Edith Clark. The R-3 was registered X14215 as a Model A-1 to Keith Rider on 6 May 1934, only three weeks before the MacRobertson entry nomination list was closed.

James Granger was chosen as test and Race pilot and in June 1934 he travelled to Europe to study what was known of the proposed route. The aircraft was delivered to Clover Field, Santa Monica, in August and entered for the Bendix Races as a preliminary to shipment to England.

It was intended to fit the Pratt and Whitney Wasp Junior engine previously installed in Jimmy Doolittle's Laird Super Solution but having decided that a more powerful engine was required an experimental Series A Wasp Junior was converted into a supercharged Model SB developing 400hp and driving a two-bladed Hamilton Standard propeller. Fuel capacity was 170 US gallons and Rider anticipated that the Race could be completed in a flight time of about 44 hours.

The delay in confirming selection of the engine caused the aircraft to miss the Bendix and it was not until 2 October 1934 that Granger started his first take-off run at Clover Field. The tail was raised too high; Granger lost control and the aircraft turned over. Although the R-3 was relatively undamaged the pilot sustained a fractured skull, an injury from which he died.

10. Lockheed 9D Orion. F-AKHC. (France)

An entry in both Speed and Handicap Races by Michel Détroyat who planned to fly solo.

The Orion, a low-wing, wooden monoplane with retracting undercarriage and powered by a single Pratt and Whitney Wasp engine, was delivered by sea to France and first flown by its owner on 10 August 1934.

Détroyat, Chief Test Pilot of the Morane-Saulnier Company, was one of Europe's outstanding aerobatic pilots and had once flown a challenge against his equally famous compatriot, Dewoitine's Chief Test

No. 10. Michel Détroyat who was not satisfied with the performance of his Lockheed Orion with a Hispano-Suiza engine, withdrew his entry in September and sold the aeroplane to the French Government.

No. 11. Varney Speed Lines withdrew Lockheed Orion NC12225 from commercial service in Mexico and the aircraft was shipped to Europe in July but never appeared on the starting line.

Pilot Marcel Doret, in front of a crowd estimated at 150,000.

Following delivery of the Orion, it was decided to replace the Wasp with a locally built Hispano-Suiza HS9V radial engine developing 700hp, a licence-built Wright Cyclone with a variable pitch propeller. However, Michel Détroyat was not satisfied with the performance of his modified machine and a month before the start he withdrew his entry and offered the aircraft for sale whereupon it was purchased by the French Air Ministry.

11. Lockheed 9 Orion. NC12225. *West Wind.* (USA)

Entered in both Speed and Handicap Races by Walter T Varney to be flown by himself and Captain Franklin Rose.

The aircraft, operated by Varney Speed Lines, was withdrawn from commercial work on Mexico's Lineas Aereas Occidentales SA service for pre-race 'reconditioning' and overhaul of her Pratt and Whitney 410hp Wasp A engine. The Orion already held the San Francisco-Los Angeles record at a speed of 263mph and with some enthusiasm was shipped to Europe in July with the intention of flying along part of the Race route but nothing further was heard and the Orion did not appear at the start. It was later reported she had been sold to a customer in Romania.

12. René Couzinet 150. F-ANEX. (France)

Entered by Vicomte Jacques de Sibour in the Speed Race to be flown by himself and Captain M Rossi.

Jacques de Sibour, accompanied by his wife Violette, daughter of the American department store magnate Gordon Selfridge, had left the de Havilland Aircraft Company's aerodrome at Stag Lane in North London in a DH.60 Moth on 14 September 1928 to start a world tour 'in easy stages'. They returned on 19 July the following year having covered 33,000 miles. In February 1930 de Sibour accepted a new DH.60 Moth, delivered to him in Paris, and the following month flew the aeroplane to Addis Ababa where he sold her to the Regent of Abyssinia.

The René Couzinet 150 was a low wing cabin monoplane powered by a single Hispano Suiza engine developing 575hp.

This entry was cancelled within two weeks of application when Rossi telegraphed from the USA to confirm he would not be available but in September there was speculation that he had reconsidered. Rossi had set several records with his partner Cordes and there were rumours he had withdrawn rather than fly with de Sibour. Eventually it was left to a spokesman from the manufacturers to confirm that the aircraft would not be ready, and that the entry had been 'premature'.

The company was badly under-financed and any considered prospect that the government might sponsor the aeroplane to fly in the Races was short lived. Apart from wind tunnel models the full-sized aircraft was never completed.

13. Wibault 366. F-ANEN. (France)

An entry into Speed and Handicap Races by Captain Edouard Corniglion-Molinier, to be flown by himself and Captain Leon Challe.

Corniglion-Molinier, a First World War fighter pilot and regular air force officer on leave of absence, was asked about superstitions and the Race number 13. He replied that it was a good number as he had been born on 13 January.

Challe was another regular officer on leave, having joined the air force in 1921, since when he had been involved in several long-distance and world record flights including Paris-Saigon and return in 1927 and a crossing of the South Atlantic in 1930, in recognition of which a suite of music was composed in 1939.

Although the aeroplane had first flown at the beginning of September, it was realised that the 366, an all-metal, low-wing five seater, powered by a single 650hp Hispano-Suiza 12Y engine, would not be ready for the Races and the crew transferred their interests to Race entry No. 54, the Blériot III/6 of Charles Quatrematre currently flight testing at Villacoublay near Paris.

14. Airspeed AS.5A Courier. G-ACJL. (GB)

Entered in the Handicap Race by Aircraft Exchange and Mart Ltd. to be flown by Alan Naish, a director of the company and a former RAF officer with Middle East experience, but with no nominated co-pilot.

Following the confusion surrounding the nominee's withdrawal of Race No. 26, the first production AS.5A Lynx-powered Courier G-ACJL, registered to Aircraft Exchange and Mart Ltd. in May 1933, was instead offered to Squadron Leader David Edmund Stodart. On the original entry form the aircraft was described as an AS.5 powered by a Cheetah V engine, properly an AS.5B variant.

Australian by birth, David Stodart had

lived in England for many years and was where he learned to fly in 1913. He qualified for a medical degree in Edinburgh but in 1914 joined the Royal Flying Corps rather than the Royal Army Medical Corps as might have been expected. In 1918 as commanding officer of an RFC Reconnaissance Flight with the rank of Major, David Stodart maintained the Flight's operations single-handed for three weeks when all other officers were incapacitated through sickness or wounds, devotion which resulted in the award of a DSO to add to his DFC. He retired from the RAF in 1931 having spent some of his post-war Service career attached to the Royal Australian Air Force. He was to be accompanied in the Race by a distant cousin, Kenneth Gerald Stodart, who, at the age of 15 in 1925, had joined the RAF as a Halton Apprentice. He learned to fly at No. 5 FTS, Sealand, in 1931 and between March 1932 and April 1934 was posted as a Sergeant Pilot to No. 43(F) Squadron flying the Hawker Fury 1 based at Tangmere.

Arrangements were made for the AS.5A Courier airframe to be loaned to the Stodarts, who had no financial backers, by Airspeed agent Aircraft Exchange and Mart Ltd., and the 240hp Lynx IVC engine was similarly loaned free of charge by Air Marshal Sir John Higgins, Chairman of the board of Air Service Training, a subsidiary of engine makers Armstrong Siddeley Motors Ltd. Various systems on the aircraft also were loaned, notably by the Vickers company. Airspeed director Neville Shute Norway (the author Nevil Shute) later claimed that the entry had been largely funded by the personal generosity of his fellow directors while he was resolute in refusing any injection of company money. An otherwise standard production aircraft, G-ACJL had been fitted with a fuselage tank of 129 gallons capacity extending her still-air range to 1,200 miles, and she was credited with a cruising speed of 132mph.

15. Fairey IIIF. G-AABY. *Time and Chance.* **(GB)**

An entry made by Flying Officer C G Davies to be flown by himself and Lieutenant Commander Clifford Hill RN, and entered in both Speed and Handicap Races.

The aircraft was actually owned by the Shanghai merchants Arnhold and Company, and was registered to them at their London address on 21 September 1934. Originally, the entry had been sponsored by Vincent Kelly of Bridlington, Yorkshire. Later, Kelly was insistent that although he had paid the £50 entry fee he was anxious not to incur any liability whatsoever and unless Davies took on the title of nominee the entry would be cancelled.

Born in Pendleton, Manchester, in February 1899 and a pilot in the Royal Flying Corps in 1917, Cyril Davies was a 'B' licensed pilot with 1,400 hours. The two crew had met when Davies was on detachment

No. 14. The first production AS.5 Courier G-ACJL was fitted with a fixed undercarriage as standard and painted in the colours of Aircraft Exchange and Mart Ltd., sales and distribution agent for the Airspeed company. (Richard Riding.)

No. 15. These two views of Fairey IIIF G-AABY accompanied the entry forms submitted by Cyril Davies. The photographs were probably taken to special order at North Weald Aerodrome where the aircraft was prepared for the Races. (Royal Aero Club Archive.)

No.15. Lieutenant Commander Clifford Hill RN and Cyril Davies first met on board HMS Glorious when part of the Mediterranean Fleet. (The Aeroplane.)

No.16. James 'Jimmy' Melrose, 'The Young Australian', and his DH.80A Puss Moth VH-UQO in which he circumnavigated Australia before setting-off for England. (National Library of Australia. VN3723085.)

nouncements issued by de Havilland and was not in a position to contact the company until 1 March, one day after closure of the order book. He admitted that his reason for entering the Races was simply for fun. He wanted a relaxed flight to a region he had never been to, and a holiday when he got there.

The Fairey IIIF was a standard military General Purpose type fitted with a 530hp Napier Lion engine. A range of 600 miles and a top speed of 140mph were estimated for the aircraft in her Race configuration.

G-AABY was a demonstration aircraft fitted with a tropical radiator which had been on business to China where she is reported to have suffered several accidents and was finally returned to Great Britain for repair. She was Race-prepared at North Weald Aerodrome, an RAF station to the north east of London. Davies admitted that the crew's financial outlay for the IIIF had been substantial and that after paying off all their pre-Race bills they were left with £8. 8s 0d between them.

16. DH80A Puss Moth. VH-UQO. *My Hildergarde.* (Australia)

Entered in the Handicap Race by Charles James (Jimmy) Melrose of Glenelg, Adelaide, South Australia, and sponsored by his mother, Mrs Hildergarde Melrose.

Having flown his Puss Moth VH-UQO on an 8,000 mile circumnavigation of Australia in August 1934, Melrose set off from Parafield Aerodrome on 13 September, where he had learned to fly the previous June, bound for England. It was his 21st birthday and he arrived at Croydon at 7.00am, eight days and nine hours later. It was the fastest and least publicised of all Australia-England flights, but was not officially observed and therefore not recognised. The journey was almost entirely uneventful, and his arrival gave *'The Young Australian'* as he was known, plenty of time for rest and organisation for the return journey. Melrose was described by C G Grey as "the youngest pilot in the Race, but very wise, cautious and clever."

to a Fleet Air Arm unit on board HMS *Glorious* in the Mediterranean Fleet between 1930 and 1933. They had often flown sorties together on reconnaissance Fairey IIIFs and had maintained contact after Davies left the Service to become warden of the SoS Society's Camden Hostel, a charitable institution for the homeless in London and for which he earned the nickname *'The Missionary'*.

Clifford Hill, granted special leave from the Royal Navy to act as navigator, was referred to as *'Burglar Bill'* on account of the swarthy complexion he had cultivated from long hours on board ship in sunny climes. He had served in Australia in 1926 as an exchange officer on HMS *Delhi* and had joined the Fleet Air Arm in 1929.

Davies would have been interested in a DH.88 Comet but he missed the press an-

No. 19. At the time of submission of Bernard Rubin's entry forms, the de Havilland Aircraft Company was able only to furnish a postcard copy of a painting of a DH.88 Comet commissioned from the well-known aviation artist Stanley Orton Bradshaw.
(Royal Aero Club Archive.)

There were several suggestions for pilots to accompany Ken Waller (left) in the Races after Bernard Rubin was forced to withdraw on medical advice. Having been unsuccessful in attempting to raise an entry himself, Owen Cathcart-Jones made a direct approach to Bernard Rubin and was offered the position.
(Flight.)

Melrose's entry forms had been submitted by his mother who had described the aircraft type as a DH.60 Moth. He advised the Royal Aero Club by letter on 3 September of a change as it was now his intention to fly a DH.85 Leopard Moth, a high-wing cabin type similar in appearance to the Puss Moth, a new aircraft he was to deliver to Australia on behalf of the de Havilland Company. In the event the new machine was not ready in time so it was at the de Havilland factory at Hatfield, where the DH.88 Comets were being erected and prepared, that the manufacturer 'cleaned up' Puss Moth VH-UQO, removing the air brake fairings from the undercarriage legs and the wing folding mechanism, sealing gaps, fitting wheel spats, extra tankage in the cabin and a Fairey metal propeller. The 120hp Gipsy III engine which had powered the aeroplane safely to England was replaced by a 130hp Gipsy Major I for the return journey.

Following the Roscoe Turner precedent the sub-committee accepted these late changes and agreed that the entry should read: "DH Moth. Gipsy engine."

17. Caudron C.530 Rafale Monoplane. F-ANAM. (France)

Entered by pilot André Gueit in the Handicap Race.

The Rafale, a two-seat, low-wing monoplane powered by a Renault engine, was capable of a top speed of 189mph. M. Gueit who was known to be living in Algeria, declared notice of his withdrawal on 5 September.

18. Potez 39. (France)

Entered by Marcel Fréton in both Speed and Handicap Races to be flown by himself and co-pilot D'Estailleur Chanteraine.

Both pilots had considerable experience in long distance flying, particularly in Africa where in 1931 they had made a 22,000 mile coastal tour.

The Potez was a high wing military observation monoplane with a Lorraine Pétrel engine replacing the more usual 580hp Hispano Suiza 12H. The entrants subsequently encountered difficulty in obtaining a suitable variable pitch propeller, and without one realised there was little chance of gaining a good position in either Race category and withdrew.

19. DH.88 Comet. G-ACSR. (GB)

The entry was made from his suite at the Berkeley Hotel in Piccadilly by Bernard Rubin, to be crewed by himself and the 26 year old Ken Waller in Speed and Handicap Races. Waller was Assistant Chief Pilot Instructor at the Cinque Ports Flying Club, based at Lympne.

At the beginning of September, just as the Comet was getting into its test flight pro-

No. 22. The Northrop Delta SE-ADI was to have been flown by a crew of three from Swedish Air Lines but was withdrawn for unspecified reasons thought to have been technical problems or damage caused by a fire on the ground.

gramme, Bernard Rubin was taken seriously ill. The prescribed rest cure not only prevented him from flying to Australia, where he owned sheep and cattle stations, but postponed his wedding plans too.

The services of Flight Lieutenant E H 'Mouse' Fielden, an RAF officer seconded as personal pilot to HRH The Prince of Wales, were sought as replacement crew but a news release suggesting his positive engagement was described as 'premature'. The post was thought also to have been offered to Flight Lieutenant George Stainforth, and the KLM pilot I W Smirnoff, but eventually was taken up by Lieutenant Owen Cathcart-Jones.

Cathcart-Jones, a 34 year old ex-Royal Marines officer, had been seconded to the RAF in 1924, when he had taken the opportunity to learn to fly. He later served in carriers with the Fleet Air Arm in the Mediterranean and on the China Station, and had flown several long distance and record breaking trips.

Cathcart-Jones, 'Seajay' to his friends, had established a freelance air charter business in 1933 with Miss Marsinah Neilson, a former pupil and now a young Dutch commercial pilot. When details of the MacRobertson Races were announced they determined to enter as a team providing a suitable aircraft and backer could be found. Cathcart-Jones believed that the only aircraft with a winning chance had to be American and he settled on a Northrop Delta. At a total price of $48,000 plus delivery it was expensive and the team spent a whole year trying to raise finance with a total lack of success.

Reading a newspaper report about Bernard Rubin's illness and withdrawal from the Races caused him to press hard his case for a seat in the Comet. Rubin ignored his direct approaches but working through Waller whose friendship and trust he had cultivated at Lympne, Cathcart-Jones was granted an interview and when he confirmed that he had 'no conditions' when seeking to fly the Comet, Rubin offered him the vacant seat "rather than trying to find someone else."

20. Lockheed 5C Vega. NC105W. *Winnie Mae*. (USA)

'Entered by Wiley Post, the aircraft was to be flown solo by the entrant in the Speed Race.

From the experience gained during his record breaking round the world flight with Harold Gatty in June 1931 and on a solo trip in July 1933, Wiley Post had decided to fly the MacRobertson at a high altitude, between 30,000ft and 40,000ft, avoiding most of the weather encountered at low level. He added a pair of superchargers to the Wasp C engine and experimented with a rubberised suit worn with aluminium helmet and incorporating oxygen, pressure and temperature regulation. Cruising speed of the *Winnie Mae* was expected to be more than 275mph, with a top speed of about 350mph.

An innovation was a landing gear constructed of tubular steel which, to improve weight and drag, was released immediately after take-off, the aircraft landing on a fixed wooden skid. Although the scheme worked for high altitude research it was hardly likely to have been a practical asset during the Race.

Extensively modified, the all-wood, high-wing monoplane named after the daughter of her first owner, F C Hall, was to have made an attempt on the world's altitude record, relying on the Wasp's sea level power output of 420hp to be maintained at 30,000ft by the twin super-chargers mounted in series, but rumours reached England that Post had wrecked his engine during trials and before the arrival deadline in England the entry was withdrawn.

The truth was that Wiley Post was so focused on perfecting his high-altitude pressure suit, on which he collaborated with the B F Goodrich Rubber Company, that by the time he was satisfied with the results of a 40,000ft trial flight over Chicago, the MacRobertson had already been decided.

21. Percival P.1B Gull Mk. IIA. (USA)

An entry was made by Salvador Farre to fly in the Handicap Race.

The Gull was designed by Edgar Percival, a native of Sydney who had set up his own business in England building efficient touring and racing monoplanes. This entry which was planned to be fitted with a 150hp Napier Javelin engine was unique in that it was the only American entry to specify a foreign-built aircraft type. Nothing further was heard of or from Mr Farre, except that the Percival Company confirmed he had never placed an order for a Gull or any other type of aircraft.

22. Northrop Delta 1C. SE-ADI. *Halland*. (Sweden)

The entry made by Marshall Lindholm was to be flown by him and co-pilot Georg Lindow in both Speed and Handicap Races, with Alfred Serander completing the crew as wireless operator.

Both Lindholm and Lindow flew the mail service with Swedish Air Lines, and AB Aerotransport sportingly offered their recently acquired Delta, an all metal, low-wing monoplane, powered by a single Pratt and Whitney Hornet engine. The fixed undercarriage was heavily trousered, and the aircraft was reported to be one of the most expensive of its type in the world.

Although both pilots continued to prepare for the Race, Aerotransport were quoted in early October as saying that the aircraft would not compete, and it did not appear at the start in England. Technical problems which had afflicted other Deltas were thought to have been the most likely cause for withdrawal and, in addition, there were suggestions that *Halland* had been damaged by fire on the ground.

No. 23. This photograph of a model of the Fokker XXII PH-AJP was supplied by KLM but the type was late in delivery and eventually withdrawn in favour of the DC-2.
(Royal Aero Club Archive.)

No. 25. The Wedell-Williams 303 was designed specifically for the MacRobertson Races but had not been completed by June 1934 when Jimmy Wedell was killed in a take-off accident in a Gipsy Moth. The entry was withdrawn in September. This photograph is of an earlier Wedell-Williams project, the Type 44 built for another MacRobertson pilot, Roscoe Turner.
(National Library of Australia. Canberra.)

No. 26. Sir Alan Cobham's entry of an Airspeed Courier was withdrawn by the manufacturer due to financial restructuring within the company. This photograph of an aircraft supplied to the Air Ministry was included with the original entry forms.
(Royal Aero Club Archive.)

23. Fokker XXII. PH-AJP. *Papegaai*. (Holland)

An entry by KLM in the Handicap Race with either Smirnoff or Prijns (see Race No. 8) nominated as captain.

The aircraft, powered by four Pratt and Whitney Wasp engines, was intended as a back-up to Race No. 8, carrying ten passengers. On 19 September the entry was withdrawn due to late delivery of the aircraft from Fokker and all effort was re-directed towards the company's Douglas DC-2, Race No. 44.

24. DH.60 Moth. Unspecified. (GB)

This entry was made by the well-known aviation journalist William Courtenay nominating himself as pilot for the Handicap Race, although later it was expected that Wally Hope would fly the aircraft. Hope, a veteran air taxi and racing pilot, had won the King's Cup Air Race on an unprecedented three occasions and was an accomplished navigator.

An alternative theory is that the aircraft was DH.60G Moth Major G-ACNS (No. 5068) which was offered for sale by Brian Lewis and Company during the spring of 1934 and extensively flown by American pilot Hubert Julian that summer whilst not actually owned by him.

The entry was withdrawn at the beginning of September but even in his own extensive coverage of the events, Courtenay makes no mention of his entry nor any reason for its subsequent cancellation. Engaged as press representative by Jim and Amy Mollison, this task would in any case have completely negated his own personal Race ambitions.

25. Wedell-Williams 303 Landplane. NR67Y. (USA)

Entered in the Speed Race by the Wedell-Williams Aero Service Corporation, but with no nominated crew, Jimmy Wedell was expected to be the pilot.

Jimmy Wedell was born in Texas City on 31 March 1900. Due to a limited education he was denied both Army and Navy pilot training and at an early age started work as a car mechanic, eventually opening his own garage business. Having scraped together sufficient funds he bought two derelict Thomas Morse biplanes, rebuilt them and after an hour's instruction by a one-time barnstorming pilot went on to teach himself to fly.

Jimmy Wedell made contact with Harry Williams, a wealthy planter and lumberman, sold him an aeroplane, taught him to fly it and as a result received backing of $2,000,000 to design and build racing aircraft. The Wedell-Williams Aero Service Inc. (1928) operated alongside the Delgado Trade School in New Orleans and Wedell became firm friends with Byron Armstrong who founded the School's aviation programme in 1931.

The 303 Landplane NR67Y was designed specifically for the MacRobertson Races and it was planned that students from the Delgado School should build the fuselage and tail assembly whilst the company staff constructed the wing and the remainder of the aircraft. By the summer of 1934 only a partially finished fuselage had been made when on 24 June Jimmy Wedell was killed in a night take-off accident at Patterson Field, New Orleans, whilst instructing a student in the company's American-built DH.60GM Moth NC924M. Harry Williams cancelled all further work on the 303 which was never completed and the entry to the MacRobertson was cancelled at the beginning of September.

At the time of his death, Jimmy Wedell was recognized as the world's most successful designer of racing aircraft and he held more records than any other pilot.

26. Airspeed AS.5A Courier. G-ACVF. (GB)

This entry made by Sir Alan Cobham was to have been flown by Squadron Leader David E Stodart and his cousin, Sergeant Pilot Kenneth G Stodart. Described as an 'overseas' version of the Courier powered by a 277hp Cheetah V engine, the specification of the AS.5B model had a top speed of 165mph and a range of about 1,500 miles.

Sales agent for the Courier, Aircraft Exchange and Mart Ltd., took delivery of their first demonstrator aircraft in mid September 1933 and immediately received a number of serious enquiries from home and abroad. Contemporary press comment confirmed the fact that the Courier was the first British aircraft to be put up in competition with American machines of the 5-6 seat cabin class, and had a better performance than any of them. "Airspeed Ltd. look like having a very good chance of making a fortune in the next few years, for they can supply what is wanted all over the world: a luxurious saloon car with wings and a high speed."

The entry was not taken up as the manufacturer claimed the aircraft was required elsewhere for 'commercial reasons' but according to Cobham, who cancelled the nomination, withdrawal was due to the machine not having been 'correctly described' on the application form which identified the type simply as an AS.5 and fitted with a Cheetah V engine. David Stodart was furious and protested to the sub-committee, writing to confirm that the technical details were correct and that he personally had paid for G-ACVF to be prepared to the declared specification. The aircraft was not actually issued with a Certificate of Airworthiness until August 1936 when she was defined as an AS.5B/A.

The real reason for withdrawal is more likely to have been a matter of hard economics. Following the restructuring of the Airspeed Company in July 1934, after the Race entry deadline, the directors' opinion was that any advertising value gained by participation in the Races would be small (unless they won) and that no company money should be committed to the venture.

As compensation the Stodarts were invited to fly Race No. 14, the Lynx powered AS.5A Courier G-ACJL, originally scheduled for Aircraft Exchange and Mart's director Alan Naish.

27. DH.83 Fox Moth. VT-AEJ. (India)

The entry was made by V L Chandi in the Handicap Race, nominating A M Murad as pilot with Chandi travelling as passenger.

Murad had recently taken a job with Indian Railways and his new employer was unable to arrange his leave to cover the Race schedule. His sponsor subsequently abandoned any further involvement and the entry was withdrawn. The aircraft had been acquired from the local Air Taxi Service by the Madras Flying Club in September 1934 and was written-off in October 1935.

28. Lockheed 8B Altair. VH-USB. *Lady Southern Cross.* (Australia)

An entry by Sir Charles Kingsford Smith in both Speed and Handicap Races, to be flown by himself with P G (Gordon) Taylor as co-pilot and navigator.

Australian pilot Charles Kingsford Smith (Smithy) had been described by some of those in a position to know as 'the greatest trans-world flyer of them all' or more simply, 'the world's greatest aviator'. During

No. 28. Sir Charles Kingsford Smith's decision to take part in the Races was too late for him to buy a DH.88 Comet and he acquired a Lockheed Altair from the USA, naming the aeroplane 'ANZAC' which was changed after protests.
(National Library of Australia. Canberra.)

Having no valid certification on arrival by ship in Australia, and following opposition to the name, Kingsford Smith's Altair was unloaded at Anderson's Park and flown the short distance to Sydney with the name covered over with brown paper.
(National Library of Australia. Canberra.)

No. 28. Sir Charles Kingsford Smith's Lockheed Altair VH-USB was converted from Sirius NR118W in 1934. In Australia she was renamed 'Lady Southern Cross' but due to delays in certification and technical problems was withdrawn on the eve of the start of the Races. (Ed Coates Collection.)

the First World War he had joined the Australian Army but transferred to the Royal Flying Corps in 1916 with whom he saw action in France and at Gallipoli, was wounded and received the Military Cross. Flying a Bristol Tourer in June 1927 he set a record for the 7,500 mile circumnavigation of Australia and a year later completed the first flight across the Pacific from San Francisco to Brisbane in the Fokker Trimotor *Southern Cross*.

In August 1928, again flying the *Southern Cross* he made the first non-stop flight across Australia and the following month across the Tasman Sea, achieving another first for the return flight in October. The aeroplane made a record time to England in 1929 arriving at Croydon Airport on 8 July and in June 1930 flew from Ireland across the Atlantic to complete the first East-West connection of London and New York.

Further record flights between England and Australia were achieved in 1930, 1931 and 1933; the last, in a Percival Gull, was solo, unlike all the others when a crew complement of between two and four was normal.

P G (Gordon) Taylor had accompanied Kingsford Smith as co-pilot on a number of flights across the Tasman and, in addition to his flying skills, was recognised as one of the finest navigators in the world, specialising in the scientific study of the art.

Driven by an almost obsessive need to do 'the right thing' and support the Empire's aviation industry by buying a British aeroplane, Smithy's somewhat limited first choice was a de Havilland DH.88 Comet, but he believed he needed something faster and turned towards America with interest focused on a new Northrop Gamma. Ambitious plans were considered for the Gamma to carry sufficient fuel for only two stops en-route and to complete the course in 48 hours. This reasoning may have been partly influenced by Kingsford Smith's ban from Turkey but, unlike Jim Mollison's prohibition which was due to an illegal entry into the country, Smithy had voiced criticism of the Turkish Government. Following much diplomatic activity the ban was lifted before the Races although Mollison's exclusion remained in effect. However, the price of a new Gamma proved to be outside the budget and when published, the Race Regulations required a minimum of five intermediate landings anyway.

By the time he was persuaded a Comet would be appropriate to the task, all three aircraft laid down at Stag Lane had been sold. The de Havilland Company investigated every possible consequence of accepting a commission for an additional aeroplane, but found they were unable to do so under their original offers of guarantee and the prospective and prestigious order was turned down with very genuine regret. They did, however, offer to supply a Comet (probably the additional airframe that was constructed mainly to provide a source of spare major component parts) fitted with fixed pitch propellers. Smithy recognized that this was totally unacceptable if he was to have any chance in the Speed Race, and rejected the offer.

Many conflicting stories regarding Sir Charles' intentions had appeared in the press, and mindful that his backer was Sir MacPherson Robertson himself, Kingsford Smith set the record straight with a statement that was published in the London *Daily Telegraph* on 26 April:

"My earnest desire is to fly a British machine, but I consider that British manufacturers are at present handicapped by the absence of any demand to construct high-speed long-range types, whereas America, with greater distances and a demand for speed, has for many years developed a definite type of machine suitable, in my opinion, for the Race.

I was interested in a de Havilland machine, but was unable to secure a controllable-pitch propeller. I also consider that it is rather late successfully to develop and try out this machine. I am deeply disappointed at not being able to follow my natural preference, but I am anxious to do my best for Australia."

Turning again to the USA, he purchased a second-hand Lockheed Altair for $25,000, half the cash having been donated by Sir MacPherson Robertson in a patriotic gesture to assist Australia's best known airman. Sir MacPherson had been prepared to pay the whole of the purchase price of a Comet but Kingsford Smith's dithering resulted in a lost opportunity and now he felt guilty enough to accept from Sir MacPherson only half of the asking price for the Altair, raising the remainder from friends and supporters.

The Altair originally had been built as a Sirius for George Hutchinson's 1930 New York-Paris attempt, a tandem seat, low-wing, single-engine wooden monoplane with a fixed undercarriage, but the aircraft had crashed at Los Angeles due to 'handling difficulties'. Following repairs she was used between 1932 and 1934 by film actor Douglas Fairbanks and director Victor Fleming. After sale to Smithy, she was converted to an Altair and fitted with a new wooden wing incorporating flaps and a retracting undercarriage, a supercharged 550hp Pratt and Whitney Wasp engine and Hamilton variable pitch propeller. In her rebuilt form she was credited with a 205 mph cruise at 15,000ft over a still-air range of 3,500 miles. On 27 June, dockers at San Francisco Harbor were reported to have lifted their strike to load the aircraft onto the deck of the SS *Mariposa* bound for Sydney, Australia.

The blue and silver Altair was named *ANZAC* but in Australia there were serious objections to the name being used 'for commercial purposes', principally from the Returned Soldiers League, and the name, sign-written on the fuselage, was covered over with brown paper until she was renamed *Lady Southern Cross* in honour of Smithy's wife, Mary.

It was proposed to fly the Altair to England via the nominated Race Control Points, effecting a rehearsal in reverse, whilst attempting to break the Australia-England record at the same time. Race pundits were alleging that Smithy had left instructions for the MacRobertson Gold Cup to be engraved omitting the hyphen from his name! However, the aircraft had been delivered from the USA in such haste that the US Department of Commerce had been unable to inspect and test the machine at its revised weights. During the conversion process, fuel capacity had been increased from 150 gallons to 418 gallons.

The Altair arrived in Australia on 17 July with no valid certification and was classified as a 'banned import' resulting in major problems with Customs. Eventually offloaded onto a barge and landed at Anderson's Park, she was immediately flown the short distance from there to Mascot Aerodrome. The authorities in Australia had warned Kingsford Smith repeatedly that they would not issue a Certificate of Airworthiness until the manufacturer had provided stress data for operations at the greatly increased weights proposed by Smithy, and this was not immediately forthcoming. Equally, the US Department of Commerce refused to issue any approvals on the grounds that the aeroplane had been exported in such haste that it was not available for survey by their inspectors, and the stalemate resulted in the absurd allegations that the situation was the invention of the British and the Americans to protect their own interests and to prevent Australia's knight from taking part in the Races. The truth was that this was a problem entirely of his own making.

The promise of a Special Permit issued by the Australian Department of Civil Aviation following a safety inspection was rejected on the grounds that it would not be acceptable to the sub-committee as the Altair would still not conform to the Regulations. After heated debate, the Lockheed Company provided additional structural and performance data, the Department of Commerce advised the Australian Government that they were now satisfied and would not need to inspect the aircraft and the Royal Aero Club offered all possible assistance with documentation required in support of the entry.

Following full-load performance tests at the RAAF base at Richmond, the Altair established a number of records in August and September flying between many of Australia's principal cities. On 28 September 1934 the Australian Defence Ministry issued a Certificate of Airworthiness in the 'Special/Racing' category which included elements of the American approval, the most limiting of which was a restriction on weight due to Lockheed's assessment that the undercarriage shock absorbers were 'inadequate' to cope with the great increases. The only solution was to restrict fuel uplift to less than 300 gallons and the Australian authorities sealed the Altair's auxiliary tanks to ensure compliance. Although the aircraft was now completely outclassed, Smith telephoned the Royal Aero Club in London asking for an immediate assurance that as certificated the aircraft would be accepted under the Regulations. The return call was received in the early hours of Saturday 29 September and at 6.00am, Kingsford Smith and Taylor left Mascot Aerodrome en-route to Darwin and England.

During a stop at Cloncurry in north west Queensland, Smithy asked his old friend Harry Purvis to make a routine inspection during which he discovered a series of dangerous cracks along rivet lines in the engine cowling which may have been caused by hitting a bird in flight but were more likely to have developed as a result of vibration. There was no question of continuing to London and the aircraft returned at low speed to Sydney where the eminent aircraft designer and engineer Commander Lawrence Wackett pronounced that the damage was very serious and ordered a new cowling to be built.

It was estimated that the aircraft would be ready by 6 October but in view of the lack of rest he and Taylor would get before the start of the Races in England, even if they made an immediate departure, Smithy decided to withdraw. Forever the gentleman, on 4 October he cabled the Royal Aero Club in London:

"Deeply regret on account of delays caused by difficulties in repairs that I am unable to participate in the Centenary Air Race. Please accept this as a formal withdrawal, coupled with sincerest good wishes for the winner and the safe carrying out of the most spectacular air race in the history of aviation."

The decision was greeted by the Australian public with disappointment, disbelief and outright hostility born of frustration and the crew received a torrent of critical publicity and abuse. In England, Harold Perrin promised one of his benevolent extensions to the deadline for arrival and Jim Mollison sent a message urging the crew to reconsider. In self-defence, and with an unauthorised fuel load of 514 gallons on board, the crew set off on 20 October, 12 hours before the official start of the Races in England, for a successful flight across the Pacific to California.

29. Bellanca 28-70 Monoplane. EI-AAZ. *The Irish Swoop.* (Eire)

The aircraft was entered in Speed and Handicap Races by the Dublin based charity, Irish Hospitals Trust Ltd., but withdrew from the Handicap Race on 5 September. There was a mild protest at the entry being listed as 'British' when the official sponsor and senior pilot were Irish.

Nominated as the crew were Colonel J E Fitzmaurice, a former officer of the Irish Free State Army Air Corps, and Scotsman Eric Watt Bonar.

Fitzmaurice fought with the British Army in France between 1915 and June 1918 when he transferred to the Royal Flying Corps, but the war had ended before he was fully trained. In 1922 he transferred to the Irish Army.

In April 1928, at the age of 30, James Fitzmaurice acted as co-pilot in the three-man crew of the Junkers W.33 Monoplane *Bremen* which made the first East to West crossing of the Atlantic from Baldonnel Airport south west of Dublin to Greenly Island, just within the boundary of the Province of Quebec, Canada. He had planned a New York to Dublin flight in June 1931 with Captain J P Saul in a Sikorsky amphibian carrying 500lb of mail and another in 1932 with Bernt Balchen, a Norwegian-American pilot he had met in Canada. Balchen was one of the most experienced Polar pilots (both North and South) in the world. In the event, neither flight took place. In 1933, Fitzmaurice travelled to Berlin to negotiate purchase of an aircraft for the MacRobertson Races from a German manufacturer, but was unsuccessful.

Eric 'Jock' Bonar had served in minesweepers during the First World War and

No. 29. The Bellanca 28-70 EI-AAZ which was flown from Germany to Eastleigh Aerodrome, Southampton, by Eric Bonar. In this picture the engine cowling is attached. The wrapped structure at right identified as 'Irish Swoop Airplane Depot' was a hospitality unit which was to serve as a headquarters at the start. (National Library of Australia. Canberra.)

joined the RAF as a pilot in 1922 flying Gloster Grebes with No. 25 (F) Squadron, later becoming an instructor to No. 5 Flying Training School, Sealand. On 24 May 1932, at Barton, he rescued the crew from a burning Siskin for which act of bravery he was awarded the Medal of the Civil Division of the Order of the British Empire.

Bonar began his civilian career as a joyriding pilot with Berkshire Aviation Tours and continued with Northern Air Transport Ltd. Both he and Fitzmaurice had worked with the British Hospitals Air Pageant, a travelling air circus, since 1931, 'Jock' Bonar as the pilot of a DH.83 Fox Moth and James Fitzmaurice as a colourful compère. Fitzmaurice had usually travelled with Bonar as a passenger in the Fox Moth when the circus moved on.

When the Races were announced, the pair left the Pageant to concentrate on raising a MacRobertson entry. 'Fitz' secured the backing of the Irish Hospitals Trust and Bonar believed that a new design on the boards of American Giuseppi Bellanca in Wilmington, Delaware, was ideal. The aircraft offered the potential of a 3,000 mile range, cruising at an estimated 220mph at 15,000ft, depending on the final choice of engine, but some observers believed the aircraft to be grossly overpriced if the sum of $30,000, rumoured to have been paid by the Trust, was correct.

In support of the contemporary view that British engines were best, the team approached Sir John Siddeley in Coventry to discuss the prospects of acquiring an Armstrong Siddeley Tiger engine, only to be met with outright rudeness and rejection on the grounds that Sir John would never dream of having one of his engines installed in such a crazy aeroplane.

By good fortune, the US Navy Department had been testing the 700hp Pratt and Whitney Twin Wasp R-1830 radial engine. Three engines were involved but the trials had been successfully completed using only two, and after negotiation the third was released to Bellanca for their use. The engine could easily accommodate a new Hamilton Standard two-pitch propeller and was much welcomed by Chief Engineer Al Mooney who thought the combination was ideal for the 28-70 Monoplane.

To survey part of the Race route, Bonar flew by Imperial Airways from London via Baghdad and Allahabad to Singapore in July. Fitzmaurice had intended to fly KLM to Batavia but instead travelled to the USA to raise all possible publicity on behalf of his backers. He was later joined by Bonar to supervise completion of the Monoplane which was almost finished by the middle of August.

Fitzmaurice was convinced that the aircraft would cause a sensation as it was fitted with a retractable undercarriage! Designed to land at only 60mph with a full load and without the assistance of slotted wings or flaps he was certain it was amongst the fastest in the world.

The aeroplane was designed and built by 16 men in 12 weeks from the contract signature. The welded steel tube fuselage was faired to the wire-braced monoplane wing which housed the inwardly retracting undercarriage legs. Bonar told reporters that the empty weight of the aircraft was only 994lb, her range 3,220 miles at 235mph and she could achieve a speed of 265mph.

In her paint scheme of green wings and white fuselage representing both Irish Free State and the Vatican, she was rolled out at Wilmington, Delaware, and flown by company test pilot J Allen on 1 September 1934. Fitted with a temporary cowling Bonar flew the aircraft to New York where she was loaded aboard the SS *Bremen* (remarked upon as an amazing coincidence) which sailed on 29 September bound for Southampton. The Race-cowling was to follow by another ship as soon as it was ready.

Due to gale force winds in the English Channel the aircraft could not be offloaded at a British port and was carried through to Bremerhaven where she arrived on 5 October. In Germany the crew was afforded hospitality by Herman Goering, a personal friend of 'Fitz' and, when difficulties were encountered in transferring the Bellanca from the ship to a pontoon, Goering arranged for Germany's largest floating crane, the *Lang Heindrick,* to be made available to lift the aircraft and carry it to the point ashore nearest to the aerodrome. This accomplished, *The Irish Swoop* was manhandled across anti-tank ditches only to discover that the narrow gates of the aerodrome barred further progress. Goering instructed that they should be demolished with explosives. The obstacles removed, the aircraft was carried onto Bremen Aerodrome for assembly and test.

The subsequent flight to England by Eric Bonar started on 9 October and was, according to his own account, without the engine cowling. However, the Bellanca was reported by the press to have landed at Amsterdam 'with a cracked cowling' but eventually flew on to Eastleigh Airport, Southampton, possibly without it. During an interview for a British aviation magazine some years later, Bonar made no mention of the diversion but remembered he was greeted on arrival at Eastleigh by the board of the Irish Hospitals Trust, a British ground crew supplied by Bill Rollason, an old friend from flying circus days, Henry Schwortz, an American engineering foreman loaned for the occasion, and R J Mitchell, Chief Designer for the Supermarine Company.

Had *The Irish Swoop* been offloaded from

the *Bremen* onto a lighter in Cowes Roads as intended, Fitzmaurice was planning to import the aircraft as 'personal property', presumably to deviate around a Customs regulation, but Irish Hospitals Trust insisted that the aircraft belonged to them. Although it had proved impossible to offload *The Irish Swoop*, all the spares carried on the ship were landed at Southampton 'in error' where they remained awaiting the return of their host.

A more substantial engine cowling and baffle system arrived from the USA in good time on board the SS *Aquitania* and were fitted at Eastleigh after which the aircraft was flown to Croydon for Race preparation by Bill Rollason's company. But there were issues with the new cowling, which for the time being remained at Croydon, and the undercarriage doors which were to be fitted before the start and were to cause their own complications later.

30. Northrop Gamma 2-G. NX13761. (USA)

The entry was made by Jacqueline Cochran in both the Speed and Handicap Races.

Jacqueline Cochran chose her surname from a telephone directory when in her teens. She was a fostered child working long hours in a Florida cotton mill until she decided she could improve her prospects. She taught herself to be a beautician then trained for three years to become a nurse. Realising that was not her true vocation she returned to the beauty business, made a fortune from marketing cosmetics, fell in love with aviation and at the age of 22 learned to fly at Roosevelt Field, Long Island in just over two weeks.

No. 29. The crew of The Irish Swoop, *Colonel James Fitzmaurice and Eric 'Jock' Bonar, had worked together with the British Hospitals Air Pageant since 1931.*
(National Library of Australia. Canberra.)

No. 30. Northrop Gamma 2-G entered by Miss Jacqueline Cochran. This photograph submitted with the entry form carried a handwritten message to indicate that the radial engine would be changed and a second seat added to the competition aircraft, NX13761, before the Races.
(Royal Aero Club Archive.)

Northrop Gamma NX13761 in her Race configuration with two cockpits and a Super Conqueror engine installed. Wesley Smith is seated in the front cockpit.
(John W Underwood Collection.)

When the MacRobertson Races were first announced, Miss Cochran was intending to fly with US Navy pilot Ted Marshall, who had taught her to fly, but he was killed while on duty in Hawaii. Determined to become proficient at instrument flying, Jackie Cochran contracted TWA airmail pilot Wesley Smith as her navigation instructor for four months during which time they flew together the length and breadth of the country.

Smith was made redundant after the US Government cancelled air mail contracts with the civil airlines in favour of mail carried by the Army Air Force. As a direct result, TWA cancelled an order with the

Northrop Corporation for the supply of Gamma long-range mail-planes. Jack Northrop, a friend of Jackie Cochran, was able to purchase one of the cancelled Gammas on her behalf and convert the mail hold into a fuel tank, increasing the range to 3,000 miles.

The original plan, subject to the Race Regulations, was for Jackie Cochran to complete the Race route from England to Allahabad with Wesley Smith, and for Royal Leonard to take over from Smith as co-pilot for the flight into Melbourne. While Smith was in Australia conducting a pre-Race reconnaissance, Northrop were confidently proclaiming a 240mph cruise speed for the Gamma at 20,000ft, powered by a new supercharged version of the liquid cooled Super Conqueror engine, a 700hp V-12.

Due to continuing failures of Super Conqueror superchargers the Cochran team was advised to re-engine the Gamma with a Cyclone supplied by Curtiss Wright, a power unit of proven reliability but slightly less horsepower. The sub-committee declined to approve the change due to late notification and Jackie Cochran realised that she was committed to maintaining faith in the Conqueror as it was spares for this engine she had arranged to be shipped to strategic points along the route.

Three days before the Gamma was due to be loaded on board ship at New York, Leonard and Cochran left Los Angeles at midnight, headed east, but the supercharger collapsed over Arizona and with a cockpit filling with fumes, the aircraft was forced landed at an emergency airfield.

Jackie Cochran continued by commercial airline to New York to make arrangements to receive the Gamma while the Curtiss-Wright Company sent engineers to Arizona. Satisfied that all was well, Leonard took off with an engineer in the rear cockpit but the supercharger failed again, and after dropping parachute flares, the Gamma was crash landed in New Mexico.

No. 31. Miles Falcon G-ACTM during a test flight from Woodley Aerodrome carrying the 'B Conditions' markings 'U3'.
(The Aeroplane.)

Harold Brook barely qualified as a Race entrant due his very limited pre-start flying experience.
(Flight.)

Ella Lay being congratulated by members of her family after qualifying for her pilot's 'A' licence at Brooklands on 17 March 1934.
(Via Nick Spencer.)

Jacqueline Cochran subsequently took over Race No. 46, the Granville Gee Bee Monoplane, released by Clyde Pangborn after he had failed to secure financial backing for his own entry and teamed with Roscoe Turner.

31. Miles M.3A Falcon. G-ACTM. (GB)

Entered by H L Brook in the Handicap Race with himself as pilot and an expectation of two passengers.

The prototype Miles Falcon G-ACTM was described as a 'special' Miles Hawk, a low-wing, all-wooden aircraft powered by a 130hp Gipsy Major engine. She was Miles' first true cabin type and capable of a top speed of 145mph. Following submission of his entry papers Harold Brook advised the Royal Aero Club that at the end of May construction of the aircraft had not begun but he was confident that it would be completed in time. Registered to Harold Brook on 10 June 1934, her first flight was from Woodley in the hands of Fred Miles only eight days before the start of the Race. As a prototype civil aircraft the Falcon was obliged to undertake airworthiness trials at the Aircraft and Armament Experimental Establishment (A&AEE) at Martlesham Heath before her long range fuel tanks were installed. The Martlesham test pilots were impressed by her good handling characteristics but less so by poor brakes, no emergency escape hatch in the roof and the absence of a 'no smoking' placard in the cabin.

Brook had declared his ambition to fly to Australia even before he had qualified for a licence. After some gliding experience and less than five hours' dual instruction on Moths with the York County Aviation Club, he was sent solo and qualified for his 'A' Licence at the end of August 1933. At 5.20am on 28 March 1934, with 43 hours flight time in his log book, the 37 year old Yorkshireman had taken off from Lympne in Jim Mollison's old but re-engined Puss Moth G-ABXY, to survey the MacRobertson route and make an attempt on Kingsford Smith's England-Australia record. Just over six hours later, at 12,000ft over the Cevenne Mountains when flying in freezing fog, heavy ice accumulation caused the aircraft to lose height, and it crash landed into deep snow on the side of Mt. Loziere. The aircraft was wrecked but Harold Brook was relatively unhurt and certainly undeterred. The Gipsy Major engine and instruments from the Puss Moth were salvaged, returned to England and fitted to the new Falcon, G-ACTM. When he presented himself to the scrutineers at the start he had barely accumulated the 100 hours' solo flying time required under the Rules.

As the result of advertising for cost-sharing passengers Brook was contacted before the Race by a lady who wanted to fly to Australia to visit relatives. Miss E M Lay, a 60 hour 'A' licence holder was welcomed in the belief that the trip was to be nothing more than routine.

32. Lockheed 9D Orion. NR14209. (USA)

Entered by the Lyon Flight Expedition Co. Inc. in both Speed and Handicap Races, the aircraft was to be flown by R F Lape and Captain Harry Lyon.

The Orion, a low-wing, all-wood monoplane with retracting undercarriage and variable pitch propeller, powered by a single Pratt and Whitney Wasp radial engine, was the entry of Harry Lyon who had commanded a US Navy cruiser during the First World War, afterwards flying as navigator on the *Southern Cross* during the first ever Transpacific flight from the USA to Australia in 1928. Following submission of the entry papers, nothing further was heard and none of the 36 Orions built appears to have been allocated to the Lyon Company. At the time of the Race the entrants were

No. 33. The clipped wing Monocoupe Special NR501W was entered by Jack Wright with the support of the Utica Civil Flight Committee although funds were always in short supply. This photograph of the aircraft submitted with the Race entry forms bears a racing number of earlier vintage displayed on the fuselage due to the limited space on the fin or rudder.
(Royal Aero Club Archive.)

The Monocoupe Special arrived in England on board the liner SS Olympic, *sharing space with Race No. 46, the Gee Bee NR14307. Both aircraft were carried by lighter to the Hamble River and landed at the adjacent aerodrome.*
(National Library of Australia, Canberra.)

quoting registration NR14209, but according to Lockheed production lists these numbers were allocated to an Altair.

33. Lambert Monocoupe 110 Special. NR501W. *Baby Ruth*. (USA)

The entry was made by owner/pilot John H. 'Jack' Wright and John Polando in the Handicap Race.

Monocoupe NR501W was the first aircraft of the type to be modified to what was known as the '110 Special' for owner John Livingstone. Fitted with a 145hp Warner Super Scarab engine, pressure cowling and additional strut fairings Livingstone had been successful in the 1931 racing season but with a wingspan later reduced from 32ft to a little over 23ft, the little 'Clipwing' Monocoupe was described as 'hot'. She was sold in 1933 to 'Utica Jack' Wright who immediately reduced the surface area of the empennage and set up a number of record times when racing later that year. Due to the reduced fin and rudder area which already displayed the aircraft registration, on 3 September Wright asked for permission to paint the racing number on the sides of the fuselage, contrary to Race Regulations. These were quoted back to him in the London sub-committee's official refusal of his request but common sense prevailed when photographs revealed the nature of the problem.

Prompted by Wright, the Utica district

of New York formed a Civic Flight Committee which at a very late stage was still attempting to raise funds to support his entry. Their efforts were finally successful and Wright was confirmed as their nominee. Harold Perrin was obliged to raise the matter of entry fees and the funds were wired to London.

The Civic Flight Committee commissioned an extravagant letter heading upon which, amongst much else, it was claimed that 'Jack Wright's Flight for Utica' was to be part of 'The World's Greatest Air Derby'.

Yorkshire born Jack Wright, a one-time professional basketball player and machine gunner with the American forces in the First World War, was wounded at Fismes in September 1918 and spent his convalescence at Issoudon in France learning to fly! After a time spent developing an insurance business and raising a family in New York he took a serious and successful interest in air racing for which he could later boast receipt of 131 trophies. Asked why he had entered the MacRobertson in the Handicap category only, Wright declared that he did not want to fly at night.

Wright's friend, 33 year old ex-car mechanic John Polando who had learned to fly in 1927, was nominated as co-pilot. He was no stranger to long distance and endurance flying. With Russell Boardman starting on 28 July 1931, flying a Bellanca Skyrocket, *The Cape Cod*, they had travelled 5,011 miles Great Circle distance non-stop from Floyd Bennett Field, New York to Constantinople in just over 49 hours, beating the existing world's record for a non-refuelled flight in a theoretical 'straight line'.

The Monocoupe, named after an American confectionery bar (the expression 'candy bar' was new to many European journalists) arrived at Southampton on board the SS *Olympic* at 12.00 noon on 12 October. The Monocoupe had shared the

No. 34. Charles Scott and Tom Campbell Black met at a London cocktail party in 1933 and agreed to enter the Race as a team. They were introduced to Arthur Edwards by Francis St. Barbe and engaged to fly a DH.88 Comet, an aircraft purchased off the drawing board by Mr Edwards following the release of information by the de Havilland Aircraft Company.
(Flight.)

Arthur Edwards, Chairman and Managing Director of Grosvenor House, who made a private entry in the Races although his aircraft carried the name of his business.
(Miss Annie Edwards.)

voyage with Jackie Cochran's Gee Bee, and both aircraft were unloaded onto a lighter and landed from the Hamble River which ran alongside Hamble Aerodrome.

A minimum of pre-flight preparation was necessary and the little red and white aeroplane had been erected and flown to Heston by 3 o'clock in the afternoon. That evening the crew was being entertained at the Royal Aero Club's pre-Race banquet in London.

34. DH88 Comet. G-ACSS. *Grosvenor House.* (GB)

The aircraft was entered by A O Edwards in both Speed and Handicap Races to be flown by Charles Scott and Tom Campbell Black.

Charles Scott had joined the RAF as a pilot in 1922 where he had acquired a reputation for his aerobatic virtuosity with No. 32(F) Squadron flying Sopwith Snipes and Gloster Grebes. He left the Service in 1926 within which he had held both Heavy and Cruiserweight boxing titles and, having qualified for his 'B' commercial licence, emigrated to Australia to seek work with the fledgling airline companies, acquiring an intimate knowledge of the territory as a result. He had held England-Australia flight records three times previously in 1931 and 1932, having drawn inspiration, he declared, from the achievements of Bert Hinkler and Amy Johnson.

Tom Campbell Black was born in Brighton, Sussex, in December 1899, where his father was town mayor, and learned to fly in 1917. He was a fighter pilot with First World War service in both the RNAS and RAF. During the 'twenties he and his brother Frank owned and managed a coffee plantation in East Africa, neighbours to Lord John Carberry who owned a DH.51.

In 1929, Campbell Black joined the newly formed Wilson Airways based in Nairobi flying a DH.60 Moth. Scheduled services across the country were introduced as the company expanded and Campbell Black was appointed managing director. During the course of buying new aircraft, and through charter flights, he had flown from England to Nairobi 13 times and had piloted HRH the Prince of Wales on safari in East Africa. In 1932 he resigned his post to return to England, taking up an appointment as personal pilot to shipping magnate and racehorse breeder Lord Marmaduke Furness.

Campbell Black had met Scott at a cocktail party at the Royal Aero Club in London a year before the scheduled start of the MacRobertson and both agreed to enter the Race, but only as a team and if a suitable sponsor could be found. In January 1934 they came close to signing a contract with a financial backer but he withdrew on the grounds that his associates believed the Races were 'a crazy adventure'. Within days of this disappointment Scott was called to Stag Lane for an interview with Francis St. Barbe, Business Manager of the de Havilland Aircraft Company, who that same day introduced him to Arthur Edwards, an entrepreneur and speculative property developer in London. Within 20 minutes of their meeting a deal had been concluded.

Edwards had made a successful proposal to develop an hotel and complex of flats on the site of the Duke of Westminster's former house in the fashionable and hitherto exclusively residential Park Lane to cater 'specifically for the American market'. Designed by L. Rome Guthrie later assisted by Sir Edwin Lutyens and built between 1926 and 1930, Grosvenor House, with 448 bedrooms and suites and the largest ballroom in Europe, became one of the famous landmark buildings of the capital and like most modern architecture was not without its critics. Arthur Octavius Edwards was appointed first Chairman and Managing Director of the Grosvenor House development but it was in a private capacity that he engaged the services of both Scott and Campbell Black following his order off the drawing board of a de Havilland Comet.

35. Fairey Fox 1. G-ACXO. *Marobe Goldfields New Guinea.* (New Guinea)

Considered more an Australian entry, the Fairey Fox was backed by the New Guinea Centenary Flight Syndicate, to be flown in both Speed and Handicap Races by Raymond Parer and Godfrey Hemsworth.

Parer, an Australian, left the RAF in 1919 to make the second England-Australia flight with Lieutenant John McIntosh, a trip which lasted eight months, and was so full of mishap that Ray Parer earned for himself the nickname of *'Re-Pairer'*. In 1926 he took a DH.4 to New Guinea to begin commercial air services, and remained there, working principally for mining companies.

Hemsworth, 24 years old and from Sydney, was a pilot with Parer's New Guinea based Pacific Aerial Transport Company at the time of the Race, having joined the busi-

No. 35. Ray Parer, no stranger to the England-Australia route, entered an ex-military Fairey Fox and submitted this standard shot with his entry forms.
(Royal Aero Club Archive.)

Godfrey Hemsworth and Ray Parer in New Guinea from where financial backing for their Race entry largely originated. In some quarters it was thought that Hemsworth's first name might be considered 'cissy' and in all Race communications it was decided he should be referred to as 'Geoffrey' instead.
(National Library of Australia, Canberra.)

ness immediately after gaining his commercial pilot's licence with the Australian Aero Club in 1931. He was known subsequently in New Guinea as 'Geoff' because Godfrey, his real Christian name, was considered by some to be 'cissy'.

Parer heard about the Races during a wireless broadcast, showed immediate interest, and began to search for an affordable, suitable aircraft. His first preference was a Lockheed machine but the necessity to raise £7,000 was reason enough to look elsewhere. He was surprised to find an advertisement in *Flight* magazine offering for sale a Fairey Fox, an ex-RAF light bomber.

Only 28 Fairey Fox aircraft were constructed, powered initially by the Felix engine, a Curtiss D12 licence-built by the Fairey Company in England. The type was re-engined in RAF service with the Rolls-Royce Falcon, apart from six Felix powered machines which were sold. Three of these subsequently appeared on the British civil register, one of which was G-ACXO, built in 1926 and until 1931 in service with No. 12 Squadron at Andover. Acquired by Hon. Mrs Victor Bruce, the aircraft was advertised for sale, including a spare engine, at a price of £200. The sales patter described the antique as 'the only British 'plane with a chance in the MacRobertson Air Race'.

Parer raised £1,400 from friends in New Guinea, persuaded 'Geoff' Hemsworth to join the adventure and together left for England on 26 June 1934. The pair arrived on 6 August and found the Fox to be in 'poor condition' at Hanworth, where in spite of telegraphed instructions, modifications designed by W S Shackleton, also acting as Parer's agent, were still being incorporated by National Flying Services. These included fitting a tailored fairing over the cut-out in the top wing in an attempt to improve lift area with particular relevance to the published Handicap formula, fairings around all the rigging wire fork joints, smaller cock-

No. 36. In April 1931 Glen Kidston and Owen Cathcart-Jones flew Lockheed Vega G-ABGK to South Africa where later in the year Kidston was killed in a Puss Moth crash. The aircraft was shipped back to England and offered for sale. (Simon Kidston.)

Australian pilot Horrie Miller purchased the second-hand Vega which was not subject to Australian import regulations and entered her in the Races with James Woods, shown in his trademark broad-brimmed hat and clenching his pipe. Miller was later replaced as co-pilot by another Australian, Donald Bennett. (The Aeroplane.)

pit openings and racing style, thin profile Palmer wheels and tyres. The cost of the modifications effectively doubled the purchase price.

With some disbelief, Parer discovered that as surplus military equipment, the Fairey Fox was almost certainly ineligible for a civilian Certificate of Airworthiness and immediately entered into long discussions with the Air Ministry who promised consideration of any formal application, but only if the maximum projected fuel load was reduced to a level which the Race crew believed was already less than desirable for reasons of safety alone.

But the pilots insisted that additional fuel tanks must be installed, one between the cockpits and a second faired-in below the fuselage, to provide their minimum operational requirements and raising the endurance to between six and seven hours. After further discussion, the Air Ministry withdrew its objections and the aircraft was issued with a full Certificate. She was registered to the Syndicate on 20 August 1934 quoting her home base as New Guinea.

36. Lockheed DL-1A Vega Special. G-ABGK. *Puck.* (Australia)

An entry by H C Miller of Adelaide for both Speed and Handicap Races, to be flown by himself and James Woods.

Horace 'Horrie' Miller was Managing Director of the MacRobertson-Miller Aviation Co. of Perth, Western Australia. His co-director, David Robertson, was Sir MacPherson Robertson's younger brother.

Scottish born Captain James 'Jimmie' Woods, 41 years old, once a commercial pilot in New Zealand, later with West Australian Airways and more recently with Miller's company, was an ex-RFC pilot with 11,000 hours who had flown a DH.60 Moth from Australia to England in 1933.

Miller and Woods visited Lockheed's Burbank plant in California in July 1934 seeking to purchase a new Vega aircraft for commercial operations in Australia, but due to an import embargo, Miller's order subsequently was cancelled. In August 1934 Woods was hosted by the Douglas Company at Santa Monica and later flew across the USA in a Boeing 247 but, subsequently, Miller found a British-registered Vega offered for sale in England which was not subject to the embargo and purchased her.

Mindful of the problems surrounding the interpretation of the 'substantial' satisfaction of the ICAN requirements, clarification

Race prepared at Heston, the Lockheed Vega was delayed in reaching the start due to a sticking oleo. Donald Bennett believed Woods was spending too much time socialising and not enough on the technical matters to hand. (Richard Riding.)

was sought even though the aircraft was fully certificated in Great Britain. By request, the US Chamber of Commerce in Washington cabled the Air Ministry on 13 October:

"Lockheed serial 155 probably complies substantially ICAN requirements at 5,000lb gross but regret can certify only that airplane complies US requirements at 4,500lb gross."

The fuselage of the all-wood Lockheed Vega incorporated double-curvature ply and sections were built on concrete moulds. G-ABGK, a six passenger, cantilever-wing monoplane powered by a 420hp Pratt and Whitney Wasp SC1 engine, previously had been the property of Lieutenant Commander Glen Kidston RN. The aircraft had been purchased new in September 1930 and was registered G-ABFE but, following her arrival in England, she was re-registered G-ABGK on 3 January 1931 to reflect her owner's initials and was test flown from Croydon at the end of that month. In April, Kidston and Owen Cathcart-Jones flew her in record time from the historic old aerodrome at Netheravon on Salisbury Plain to Cape Town. Kidston, one of the so-called 'Bentley Boys' and, with Woolf 'Babe' Barnato, winner of the 1930 Le Mans race, had plans to build Vegas under licence in Europe but the scheme was abandoned. In May 1931 he was killed when DH.80A Puss Moth ZS-ACC, in which he was travelling as a passenger, broke up in flight over the Drakensberg Mountains in South Africa. The Vega was shipped home to Hamble where she was test flown by Cathcart-Jones on 5 December 1931, positioned to Hanworth the same day and flown occasionally by him under the administrative umbrella of a Kidston family Trust.

The Vega was at Hanworth when Jimmy Woods arrived in England by ship from New York on 6 August. He tested her on 15 August and the following day flew to Rotterdam for the Fokker company, Lockheed's European Agent, to carry out a general overhaul and installation of extra tankage. A variable pitch Hamilton propeller ordered from the USA was fitted by KLM at Waalhaven. The aircraft was then positioned to Heston on 14 September for repainting and final preparation and where British Customs demanded a 20 per cent duty payable on the value of the newly imported propeller. Without argument Woods put the money into bond until somebody had the good sense to understand that the aircraft would be staying in Australia after she next left England so the creation of further paperwork was rather futile. At Heston it was announced that Miller would not after all be flying in the aircraft due to business commitments, and the co-pilot was yet to be nominated.

Owen Cathcart-Jones was considered as a natural replacement to fly with James Woods but he elected for a seat in a DH.88 Comet and Flying Officer Donald Bennett, an Australian serving as an RAF flying boat instructor with 210 Squadron at Calshot, and a Race aspirant in his own right, was offered the position acting as navigator.

Bennett, a qualified engineer, had been promised the loan of a Rolls-Royce engine for the Race but was unable to secure a suitable airframe into which it could be installed. He had prepared himself by studying for and passing examinations for a First Class Navigator's Licence, a rare qualification demanding exceptional ability. He would have preferred a piloting position but failing to secure one accepted the offer to team with Woods whom, in spite of his experience, Bennett regarded as a rank amateur.

The name '*Puck*' was chosen for the aeroplane in memory of the late Hugh '*Puck*' Grosvenor, aide-de-camp to the Governor of South Australia and a personal friend of Horrie Miller. In 1929, '*Puck*' had flown a DH.60 Moth on a 6,000 mile circumnavigation of Australia.

37. Cessna AW. NC7107. (USA)

An entry by David Wehman and Paul Clough was received for the Handicap Race, to be flown solo by Paul Clough.

Piloted by P Cramer, the aircraft had previously flown 10,000 miles from Kansas to New York via Alaska and Siberia in 1929. Following receipt of the entry David Wehman withdrew his name as a 'nominator' on 5 September and nothing further was heard.

38. Short S.16 Scion. (New Zealand)

The aircraft was entered into the Handicap Race by R C Wallace with S S Kirsten nominated as pilot.

By August 1934, Short Bros. had completed only two Scion aircraft, a high-wing cantilever monoplane powered by two Pobjoy Niagara engines. The company expressed surprise when advised of the entry, having received no orders or instructions from the nominee.

Meanwhile, Kirsten, an ex-RAF officer, had reconnoitred the route as far as Baghdad and later enrolled Flying Officer E Windsor as his Race co-pilot. Furthermore, he had advised the organisers that his wife and another woman friend would be carried as passengers.

Efforts were made by Short Bros. to provide a Scion for the Race, competing under their own name, but the idea was abandoned when they declared that no suitable crew could be found. Neither Wallace nor Kirsten entered into any further communication with Shorts or the Race organisers.

39. Bernard 84 GR Monoplane. F-AKEX. (France)

An entry for both Speed and Handicap Races, the nomination was made jointly by

André de Roussy de Sales and Jean Lacombe who were to act as crew.

F-AKEX was built and flown as a Bernard 80 and was modified into an 81 GR with a Hispano 12Nbr engine. The aircraft had a complicated and rather unedifying history, which included wing vibrations that were denied by its design team. Its final near-transformation into the 84 GR indicated a proposed engine change to a more powerful 900hp Gnôme-et-Rhône double radial, but the almost bankrupt Bernard company could not finance the development, and it is not confirmed whether the new engine was ever mounted in the airframe. Before the start, the entry was withdrawn.

40. Comper Kite. (Portugal)

An entry in the Handicap Race to be flown solo by Carlos Cudell Goetz.

Only one Comper Kite, G-ACME, was ever built, powered by a seven cylinder Pobjoy Niagara radial engine developing 98hp, first flying in 1934 and taking part in that year's King's Cup Air Race. The Goetz entry was withdrawn at the beginning of September and the Kite was scrapped at Heston the following year.

41. DH.60G Moth Major. G-ACUR. (GB)

An entry in the Handicap Race by W J Cearns who nominated Sydney P Jackson as pilot.

Sydney Jackson, a Coventry-based speedway rider and described as the best motorcycle polo player in England, was a member of the Auxiliary Air Force and had previously competed in a number of domestic air races. On 19 October Jackson left Croydon in Avro Avian G-ABIE, bound for Australia, but the aircraft was written off in a crash in Italy four days later and Jackson was injured.

42. Fokker XVIII. (Holland)

An entry by KLM in the Handicap Race with either Smirnoff or Prijns nominated as commander.

Although a different aircraft type from the Fokker XXII specified for Race No. 23, registration options for the Fokker XVIII were declared to be any one of PH-AJO to PH-AJS, showing some duplication with data supplied previously. A passenger-load of four was nominated but like Race Nos. 8 and 23, the aircraft was withdrawn on 19 September due to late delivery from Fokker and all support was diverted to the new Douglas DC-2, Race No. 44.

43. Harkness and Hillier Monoplane. (Australia)

Sponsored by All-Australian (British) Aeroplane Entry Ltd., under the patronage of Don Harkness' Drummoyne-based motor racing and garage business, the two-seater aircraft was entered for both Speed and Handicap Races in which the nominated pilot was Flying Officer D Saville.

Saville, a fellow garage proprietor from Sydney and a former RAF pilot, was a reserve member of the British Schneider Trophy team which flew at Venice in 1927. A co-pilot was not nominated.

The twin Cirrus Hermes engined, two seater Monoplane, a flying wing with empennage carried by two triangular section welded steel tube booms, had been designed by T D J Leech and L G R 'Jack' Jones, both lecturers in aeronautical engineering in Sydney, and was quoted to have a top speed of 200mph, a cruising speed of 180mph and a range of 2,000 miles. Open to public gaze, the project's slow progress at the premises of Grace Bros at Broadway, Sydney, was attributed to lack of financial

No. 40. The sole Comper Kite G-ACME was entered in the Handicap Race by Carlos Cudell Goetz, corresponding from Portugal, but the entry was withdrawn in September and the aircraft scrapped the following year.
(Richard Riding.)

No. 42. Like all other KLM entries specifying Fokker aircraft, the Type XVIII allocated Race No. 42 was withdrawn in September.
(Royal Aero Club Archive.)

backing. The sponsors had tried all manner of fund-raising exercises including an auction, dances and sale of certificates for the 'Bob-in-Fund' priced at one shilling. Anybody accumulating ten certificates was eligible for an illustrated souvenir book 'when published' covering the design, construction and financing of the aircraft. By early October, too late for active participation, the aircraft was reported to be substantially complete at Mascot Aerodrome, Sydney, although it had not flown. Having missed the Race, the entire project was abandoned.

44. Douglas DC-2. PH-AJU. *Uiver*. (Holland)

Entered in both Speed and Handicap Races by the Dutch airline KLM with nominated pilots Captain Koene Parmentier and Captain Jan Moll, Flight Engineer Bouwe Prins and Wireless Operator Cornelis Van Brugge.

Within a week of a report emanating from the USA placing doubt on the availability of the aircraft, KLM issued a statement confirming that their new machine would certainly be entering the Races and, in addition to the nominated crew, would carry fare paying passengers and mail.

A decision to fit extra fuel tanks, remove cabin equipment and compete only in the Speed Race, operating via the five Control Points, was reversed when, due to late delivery, the company withdrew their Fokker XXXVI (Race No. 8) which was to have operated in the passenger role. They now considered that a standard passenger, freight and mail operation with the DC-2 would be a far better advertisement for the airline and the practicability of them operating regular services from Europe via the East

No 43. A group of schoolchildren look on as work on the starboard wing of the Harkness and Hillier Monoplane continues at the premises of Grace Brothers at Broadway in Sydney. One of the two Cirrus engines is mounted on a stand.
(State Library of New South Wales.)

Under the bunting at Grace Brothers, May Bradford and Jack Jones check the welded structure of the tailplane of the All-Australian Aeroplane entry. The aircraft was never completed to an airworthy standard.
(State Library of New South Wales.)

An artist's impression of how the All-Australian Aeroplane might have looked in flight. Having missed the Races the entire project was abandoned.
(National Library of Australia, Canberra.)

Indies to Australia, a commercial prospect which they were anxious to promote.

Both KLM captains, Parmentier aged 30 and Moll at 34, had accumulated considerable experience of the majority of the Race route. Their knowledge coupled with the speed of their new American aircraft caused observers to believe they were a Race winning combination. That view diminished slightly when KLM confirmed their intention of carrying commercial payload although it was recognised that a good load carried in an efficient aeroplane was of direct benefit in the calculation of the Handicap allowance.

Moll was born in Surabaya and had learned to fly with the Netherlands' Indies Air Force. He was to act as navigator in the Race. Parmentier, trained by the Dutch Air Force and acclaimed as a leading specialist in instrument flying joined KLM in 1929 and flew the DC-2 during acceptance trials in California, then positioned the aircraft from Santa Monica to New York. There she was dismantled and loaded on board the *SS Statendam* bound for Rotterdam, where she arrived on 12 September, and was then delivered to the Fokker works in Amsterdam for re-assembly.

The Race crew familiarised themselves with the aircraft whilst operating on the regular Amsterdam-Berlin service, and arrived at Croydon Airport on the scheduled KLM service from Amsterdam prior to positioning to Martlesham Heath for weighing. Such scheduling created a psychological advantage which was fully recognised by the press and their fellow Race contestants.

Three passengers were listed for the flight to Australia: Fraulein Thea Rasche, a German pilot who was a competitor in the 1929 Powder Puff Derby in the USA and acting as correspondent for a consortium of European aviation journals; J J Gilissen, a director of Arnold Gilissen and Co., bankers

No. 44. The first DC-2 acquired by KLM, PH-AJU, was test flown in California before being dismantled and carried by ship to Holland.

The KLM crew selected for the Races due to their experience on the majority of the route: Cornelis Van Brugge; Captain Koene Parmentier; Captain Jan Moll and Bouwe Prins.
(The Aeroplane.)

The KLM DC-2, PH-AJU 'Uiver', operated a scheduled commercial service into Croydon Airport when positioning to England for the start of the Races.
(The Aeroplane.)

from Amsterdam, and Roelof Jan Domenie, General Manager of Banco Holandes Unido, based in Rio de Janeiro, Brazil.

45. Vance Viking. NR12700. (USA)

An entry made by Lieutenant Murray B Dilley Jr. in the Handicap Race with Monty G Mason nominated as co-pilot/navigator.

Lieutenant Dilley, a pilot granted special leave from the US Army Air Corps, bought the prototype Viking and all future manufacturing rights after the death of the designer, Clair Vance, in a mailplane crash.

The Viking, normally a single seat aircraft powered by a 660hp Wasp engine, was a flying wing, balanced by twin tailplanes mounted between a pair of booms. The aircraft had been conceived as a high altitude freighter in which payload was carried within the wooden wing. The crew sat in a centre section built from welded chrome-moly tube with dural cladding.

Dilley's first major change in design was to revise the fuel system increasing capacity to about 1,200 US gallons, distributed between 14 separate tanks all within the wing and claiming to give the aircraft a still-air range of 7,200 miles. After the MacRobertson he had hoped to make attempts on distance and endurance records but the aircraft failed to appear in England and nothing further was heard of the project.

46. Granville Brothers R 6H Monoplane. NR14307. *(Quod Erat Demonstrandum.)* (USA)

Entry in the Speed Race was made by Clyde E Pangborn nominating Marion E Grevenberg as co-pilot, later amended to Colonel Roscoe Turner.

Pangborn was unable to raise the financial backing to support this entry for which he had quoted the registration N14324, a Cunningham Hall GA21, and was subsequently invited to form a crew on the Boeing 247D, Race No. 5, for which Roscoe Turner was arranging a charter. Race No. 46 was re-allocated to Jacqueline Cochran and Wesley Smith.

Following the loss of her own aircraft, the Northrop Gamma, Race No. 30, Jackie Cochran agreed to buy the R 6H Monoplane from the Granville Brothers with financial assistance from her friend Mabel Willebrandt, only if the aeroplane could be delivered to New York in time to be shipped to England. In addition she insisted she be paid a royalty on all similar aircraft sold as the result of a military contract then under negotiation with the Chilean Government.

No. 45. The Vance Viking was conceived as a high-altitude freighter carrying cargo within the structure of the wooden wing. For the Races the wing was modified to accept 14 separate fuel tanks permitting a range of over 7,000 miles. The entry was not pursued and the aircraft never arrived in England.
(The Aeroplane.)

No 46. The Granville R6H Monoplane NR14307 arrived at Southampton on board the SS Olympic *on 12 October. Essential work which was due to have been completed during the voyage remained unfinished as the members of the travelling engineering party all suffered from sea-sickness.*
(National Library of Australia, Canberra.)

Loaded onto a lighter in Southampton Docks shared with No. 33, the Monocoupe Special, the vessel was towed to the Hamble River and the aircraft landed at the old Avro aerodrome.
(National Library of Australia, Canberra.)

No.47. The British Klemm Eagle was a high-performance sporting and touring monoplane with a retracting undercarriage and normally powered by a Gipsy Major engine. G-ACVU was specially built for the Races and incorporated additional fuel tanks in the cabin and a 200hp Gipsy Six engine.
(Richard Riding.)

Clad in his Royal Air Force flying overalls with commodious knee pockets, Flight Lieutenant Geoffrey Shaw flew Wapitis with the RAF Reserve at Thornaby.
(Flight.)

Powered by a 700hp Pratt and Whitney Hornet engine and with a fuel capacity of 340 gallons, the Monoplane's metal fuselage was mated to a wooden wing incorporating flaps and supporting a heavily trousered, fixed undercarriage. The aircraft was delivered as promised but still required attention. Two mechanics sailed with her on board the SS *Olympic* to work during the voyage accompanied by a certifying engineer from the Department of Commerce, provided due to the influence of Mabel Willebrandt. Due to stormy conditions encountered during the Atlantic crossing much of the work to install landing lights, flares and wireless equipment salvaged from the abandoned Gamma was not completed as members of the engineering team were struck down with sea-sickness.

On arrival in England the aircraft was offloaded together with Jack Wright's Lambert Monocoupe which had shared the voyage. Both aircraft were landed at Hamble Aerodrome on Friday 12 October from a lighter on the adjacent river in front of a strong press contingent and immediately were prepared for flight by engineers from the Supermarine works at Woolston. Unaware of the change of plans due to the dramas enacted in the USA, uninformed members of the press still believed they were witnessing the arrival of the Northrop Gamma and reported as such.

Wesley Smith positioned the aircraft to Heston which was where for the first time Jackie Cochran actually saw the Monoplane when she and Smith arrived on Tuesday 16 October before flying to meet the Race scrutineers at the start where, in addition to several other names, the press corps quickly dubbed the aircraft the '*Heebie-Geebie*'.

47. British Klemm BK.1 Eagle. G-ACVU. *The Spirit of W. Shaw & Co. Ltd.* (GB)

The aircraft was entered in the Handicap Race by Flight Lieutenant Geoffrey Shaw on behalf of his family's Middlesbrough based firm, the Wellington Cast Steel Foundry.

In August, Harold Perrin contacted the British Klemm Aeroplane Company to enquire after the state of G-ACVU as he said he had read reports that the aircraft was unlikely to be ready in time for the Race. Somewhat bemused, the company replied asking him to identify his sources as the British Klemm Company was a member of a Press Cutting Association and they were aware of no such references.

The Eagle was a two/three seat, low-wing, cabin monoplane normally powered by a 130hp Gipsy Major engine. G-ACVU, the third production aircraft, was specially built for Geoffrey Shaw's entry into the MacRobertson Races and was flying from the manufacturer's base at Hanworth in September, fitted with the larger Gipsy Six engine, providing an additional 70hp. An extra fuel tank in the cabin was filled through a long neck accessible only when the port-side cabin door was opened fully forward and provided a range increased to about 1,000 miles.

Before the Races Geoffrey Shaw resigned his commission in the Royal Air Force having been at one time personal pilot to Sir Sefton Brancker, late Director of Civil Aviation. He had learned to fly on Moths at Cramlington in 1929 and currently held a commission with No. 608 (North Riding) Squadron, Auxiliary Air Force, a day bomber unit based at Thornaby and equipped with Wapitis. The Eagle was registered to him on Friday 13 July 1934 quoting Thornaby-on-Tees as home base.

48. Hosler B. NR14Y. (USA)

An entry into the Speed and Handicap Races by designer and pilot Russell A Hosler.

Russell Hosler built his first aeroplane in 1919 at the age of 17 and, without any form of instruction, taught himself to fly on it. He test flew aircraft for Woodson and Sikorsky, operated a flying school at Huntingdon, Indiana, and a charter service in New York. Also he was a regular competitor in the National Air Races.

Design and construction of the Hosler B at Huntingdon was of his own admission "going slow in every way," mostly due to a lack of finance. Little was known about the machine except that it was a high-wing type with retractable undercarriage, powered by a 500hp Curtiss-Wright D12 engine. At the beginning of October rumours reached the sub-committee that the entry had been withdrawn, but nothing more was heard and the aircraft did not arrive in England. It is believed the 'B' was abandoned prior to completion and all resources were transferred to another racing project, the Hosler Fury, an aircraft not completed until 1941 and destined to make only one flight in ground-effect from the ice covered Lake Wawasec near Syracuse, Indiana, before America became embroiled in the Second World War and all future development was abandoned.

49. Lockheed 8D Altair. NR13W. (USA)

The entry was made by Miss Ruth R Nichols for the Handicap Race.

Ruth Nichols was a well qualified aviatrix who held several world records for long distance and speed. She had qualified for a commercial pilot's licence and ground engineer's certificate and had occupied positions as company saleswoman, airline traffic manager, racing pilot and magazine editor.

In August, Miss Nichols confessed that she would not be competing as she had decided not to purchase the Altair, an aircraft owned by film actor Clarence Chamberlin, and that she would not be joining the crew of another Race entrant.

50. Vultee V.1A. NR13770. (USA)

This Speed Race entry was made by New York, London, Moscow Air Lines Inc. of New York City, with Lieutenant Colonel G R Hutchinson nominated as pilot, Peter Redpath as navigator/co-pilot, and Donald H Vance as wireless operator.

The Company was already planning transatlantic air freight services linking Europe to North America but the capacity of the Vultee, an aircraft powered with a single 735hp Wright Cyclone engine, was already in doubt.

George Hutchinson was an experienced airline operator, a pilot, lecturer and writer of aviation adventure stories who had in 1932 flown a twin-engined Sikorsky amphibian to Europe with a party of eight on board including his whole family. Redpath, a Canadian, was an airline pilot and lecturer in navigation, and Vance had spent five years as a radio operator in the US Navy.

The entrant had corresponded with aviation magazines since the earliest publication of Race details, promising further information on preparation of the Vultee, but from June he had remained silent although it was believed that petrol supplies had been laid down in Ireland in preparation for a refuelling stop during a transatlantic delivery flight.

Nothing further was heard. In spite of packages being gathered for shipment from Croydon the transatlantic air freight service was quietly forgotten, the aircraft did not arrive in England, and the Race entry lapsed. However, the Vultee did make a double crossing of the North Atlantic in 1936 under the name of *Lady Peace*.

51. Savoia Marchetti SM.79P. I-MAGO. (Italy)

The entry was made in both Speed and Handicap Races by Societa Idrovolanti Alto Italia (Savoia) with nominated pilots Adriano Bacula and Alessandro Passoleva.

Very little was known about the aircraft except that it was a new design powered by a Piaggio Stella nine cylinder radial engine developing 560hp. The SM.79P was confidently expected to appear in England well in time for the start but did not do so, and nothing further was heard.

It later transpired that the aircraft had flown for the first time only on 20 October but the entrants had long since lost enthusiasm for the Races and when Francis Lombardi withdrew the CAB PL.3 (Race No. 61), Bacula and Passoleva finally decided they would not take part.

52. Airspeed AS.5B Courier. G-ACLF. (GB)

The aircraft was entered for the Handicap Race by aircraft brokers R K Dundas Ltd., with Australian born Mrs Keith Miller as pilot and Nell Ferguson as co-pilot, the only all-female crew to have been notified.

Jessie 'Chubbie' Miller previously had flown from England to Australia with Bill Lancaster. They left Croydon on 14 October 1927 in Avro Avian G-EBTU, *Red Rose*, and arrived in Darwin on 19 March 1928, a trip which at the time was the world's second slowest at 158 days. Almost immediately afterwards Mrs Miller moved to the USA where she raced, set records and flew as a test and demonstration pilot. In 1929 she was a participant in the Powder Puff Derby and the following year flew an Alexander Eaglerock Bullet, one of few pilots to do so with any degree of success. In 1931 she owned a Laird Racer.

Frances 'Nell' Ferguson was 28, a qualified pilot and a charter member of the '99s' in company with 'Chubbie' Miller, almost certainly the association that brought them together as Race nominees.

G-ACLF was sent on a sales tour of India in December 1933 which resulted in the sale of the only other 'B' Cheetah V-powered version of the AS.5 Courier, G-ACVG. Following restructuring of the Airspeed Company in July 1934, and a resolve by the directors not to commit any company money to the venture, most Airspeed machines including the Courier G-ACLF were withdrawn from the Races for what were described as 'commercial' reasons. The aircraft was sold to North Eastern Airways Ltd. at Croydon in February 1937 and was impressed into RAF service in 1940. She was scrapped in 1943.

53. Lockheed 9D Orion Special. NR14222. *Auto-da-Fe*. (USA)

The Orion was entered in the Speed Race by Miss Laura Ingalls nominating herself as pilot.

Laura Ingalls had established a reputation as a speed and long-distance record breaking pilot, mostly operating outside the glare of the usual publicity which surrounded such adventures. The entry was cabled to London from Argentina but nothing further was heard until just before the start when the specially prepared Orion, powered by a 550hp Wasp and fitted with additional fuel capacity, was officially withdrawn.

54. Blériot III/6. F-ANJS. (France)

The aircraft was entered by the Blériot Company (Blériot Aéronautique Société Anonyme) in both Speed and Handicap Races with Charles Quatremare and Louis Massot nominated as pilots but with the withdrawal of Race No. 13, the Wibault 366, the crew from that aircraft, Captain Edouard Corniglion-Molinier and Captain Leon Challe, transferred to the Blériot, Race No. 54.

The model III/6 was a low-wing, strut braced, four seat cabin machine powered by a supercharged Gnôme-et-Rhône Mistral Major engine developing 725hp. Although the basic design was several years old, the airframe had been modified very considerably with a new wing, retractable undercarriage and additional fuel tanks increasing capacity to 353 gallons, sufficient for a range of about 1,650 miles. Flight testing at Villacoublay near Paris indicated a cruising speed of about 219mph and a top speed of 230mph.

55. Waugh and Everson Evo III Monoplane. (New Zealand)

The aircraft was entered in the Handicap Race by I Waugh and E Everson with

No. 55. The Evo III Monoplane was considered a joke entry on account of her engine specification but she was flown in New Zealand where certification before the deadline for despatch to England was denied.
(NZ Sport and Vintage Aviation Society via Bruce Skinner.)

No. 58. The Airspeed Viceroy G-ACMU was modified to carry additional wireless equipment and fuel and was powered by a pair of Cheetah Mk VI engines. In her standard form the undercarriage was non-retractable.
(The Aeroplane.)

Captain T. Neville Stack teamed with Sydney Turner of Airspeed sales agent Aircraft Exchange and Mart Ltd. to enter an AS.8 Viceroy purchased on easy terms.
(Flight.)

Ernest Everson nominated as pilot.

The powerplant was listed as a 30hp Bristol Cherub III, a two-cylinder engine dating from the late 'twenties. Speculation was rife that this could only be a joke entry considering the specification was clearly that of an ultra-light aeroplane. In fact, the Monoplane was far from a joke. Built by the Everson brothers, Ernie, Arthur and Ron, assisted by Ivan Waugh and John Burns, the aircraft known as the 'Evo III' was the third machine designed by the team, a twin-engined, high-wing single-seater with a span of 36ft, wing area of 216 sq. ft and an estimated endurance of 30 hours. Both of the previous two designs had crashed on their maiden flights.

Construction of the aircraft was completed at Waikato but she was still uncertificated by the time she should have been shipped to England.

56. Airspeed AS.7 Envoy. G-ACVJ. (GB)

Entered in the Handicap Race by Lady Gladys Cobham, ex-RAF officers Flight Lieutenant H C Johnson and Flight Lieutenant G A V Tyson were nominated as pilots.

Hugh Johnson was chief pilot of Alan Cobham's travelling National Aviation Day display and Geoffrey Tyson was a fellow pilot. The entry under the name of Lady Cobham was made on a wave of optimism but, following restructuring of the Airspeed Company in July 1934, G-ACVJ was withdrawn.

The AS.7 was described as a 'military version' of the basic AS.6 Envoy, but the project was shelved at the design stage. As an AS.6A(I) Envoy G-ACVJ was flown by Airspeed agent R K Dundas Ltd. on a demonstration tour of India in January 1935. She was sold to Commercial Air Hire Ltd. at Croydon in July 1936 and delivered to Spain the following month.

57. Beech A-17FS Staggerwing. NR12569. (USA)

An entry in the Speed Race by Mrs Louise Thaden, a one-time sales lady with Walter Beech at the Travelair Company, flying with her husband Major Herbert V Thaden, although there was some speculation that the Thadens were actually planning to fly separately entered aircraft.

The Beech A-17, a streamlined back-stagger cabin biplane powered by a 600hp Wright Cyclone, cruised at 212mph in its standard form and was capable of a top speed of 235mph. Cruising range was over 700 miles on normal tankage with a service ceiling reported to be in excess of 30,000ft.

Mrs Thaden later withdrew to be replaced by Captain Frank Hawkes, a move which attracted popular support in the aviation press, followed by considerable disappointment when the entry was cancelled on the grounds that the aircraft's range was considered not to be sufficient.

58. Airspeed AS.8 Viceroy. G-ACMU. (GB)

Entered in both Speed and Handicap Races by Captain T Neville Stack, to be flown by himself and S L Turner.

Neville Stack, an ex-Royal Flying Corps pilot, had been flying for civil organisations since the end of the First World War, and already held distance records. In 1926 he had flown a DH.60 Moth to India accompanied by Bernard Leete in a second Moth.

Sydney Lewis Turner was a director of the Hanworth-based Airspeed sales agent, Aircraft Exchange and Mart Ltd. and had flown 500 hours in a little more than three years.

Not listed in any official paperwork and probably accepted as 'payload' a third crew member, R G McArthur, is known to have been carried as radio operator.

Basically an AS.6 Envoy, in its modified racing configuration and fitted with a unique pair of 315hp Cheetah Mk VI engines, the aircraft was re-designated AS.8. Stack had originally approached Airspeed for an AS.6 fitted with Rolls-Royce Kestrel engines, but was shaken by the asking price of £15,000. Unable to raise financial backing he agreed to pay a deposit of £1,000 and purchase the AS.8 on 'deferred terms' agreed by Airspeed.

Like several other competitors seeking to maximise commercial aspects of their involvement, Stack and Turner employed a business manager, B S S Rockey of the appropriately named 'Avideals' quoting an address in Park Lane, London.

Registered in the joint names of the two entrants on 20 August 1934. the crimson-and-white painted aircraft showed great promise with an endurance of six hours thanks to an additional 270 gallon fuel tank fitted on the port side of the cabin. The cruising speed of 190mph, and top speed of 210mph were known not to be sufficiently good to achieve a high placing in the Handicap Race.

Some of the newly installed electrical systems had not been proved in operation and were believed to be not entirely reliable. At the beginning of October it was reported that the AS.8 had been fitted with a new Plessey A.C.44 longwave transmitter and receiver arranged to run from the lighting battery and engine driven generator. When used for telephonic communication the set had a range of 200 miles. Described as being of the latest and most modern designs, the apparatus weighed-in at over 62lb.

Apart from the DH.88 Comets, the AS.8 was promoted as the only other British aircraft specifically designed for the Race, although in truth it was based on the layout of on existing type.

Late delivery of engines and the automatic pilot delayed the Viceroy's maiden flight until 19 September. To save valuable time, an RAF test pilot from Martlesham Heath flew to Portsmouth to complete the certification trials at the manufacturer's airfield when it was revealed that too late for major rectification, engine performance was down and the tare weight above estimates.

59. DH.84 Dragon. G-ACJM. (GB)

The entry was submitted by Alan S Butler in the Handicap Section but with no nominated pilot.

Alan Butler was Chairman of the de Havilland Aircraft Company and a very experienced racing and sporting pilot who would almost certainly have flown the machine himself. He had intended to use the long range DH.84 Dragon G-ACJM which had arrived at Heston on 9 August 1934, having been flown directly from Canada by Len Reid and Jimmy Ayling. Only three days after arrival in England, G-ACJM hit the boundary fence when landing at Hamble and although thought not to have been badly damaged, the aircraft was never repaired.

The entry was withdrawn at the beginning of September, possibly as a result of the de Havilland Company's programme for the three DH.88 Comets which was draining manpower and administrative resources to a serious degree and causing immense concern within the business. However, it is more likely that Alan Butler was still recuperating from an operation and grounded on doctor's orders, a dictate that prevented him from attending any of the pre-Race social events arranged by the Royal Aero Club in London.

60. DH.89 Dragon Six (Rapide). ZK-ACO. *Tainui.* **(New Zealand)**

The aircraft was declared as the 'official' New Zealand entry nominated by Oliver Nicholson, President of the New Zealand Centenary Air Race Committee, to be crewed by Squadron Leader James Hewett

No. 60. Funds to support an official New Zealand entry for the Races were raised by a number of methods including a draw for big cash prizes. This order form is for the draw scheduled for 25 September, less than a month before the start in England.
(National Library of Australia, Canberra.)

To be delivered for the development of air services in New Zealand, Dragon Six ZK-ACO was painted tangerine, green and black and was collected from Hatfield by her three-man crew just before the start of the Races.
(The Aeroplane.)

No. 62. The forms for Fairey Fox 1 G-ACXX were received from New Zealand. The accompanying photograph was of a military Fox with the previous serial of G-ACXX, J8424, added by hand.
(Royal Aero Club Archive.)

and Flying Officer Cyril Kay with Frank Stewart acting as radio operator. Each member of the crew signed a formal contract with the Committee on 11 July 1934.

Assured of financial backing after approaches to the New Zealand Government, the prospective entrants contacted de Havilland to enquire after the purchase of a Comet but were too late and instead were offered a brand new DH.89 Dragon Six (Rapide), suitable for entry in the Handicap Race. The type was accepted on the grounds that the aircraft could be sold for civil transport services in New Zealand and instructions were cabled to the de Havilland Aircraft Company at Hatfield who were requested to submit entry forms for both Speed and Handicap Races.

Various fund raising activities were employed to support the venture including a 'National Appeal' which sold tickets at one shilling each for inclusion in a draw with a top prize of £1,000. There was little time between the date of the draw, 25 September 1934, and the start of the Races in which to deliver profits.

Born at Kihikihi in New Zealand in 1891, James Duff Hewett joined the New Zealand Army in 1911, transferring to the Royal Flying Corps in 1916, and flew as a fighter pilot with No. 23 Squadron over the Western Front, later serving with the RAF in India. He returned home in 1925 to join the New Zealand Territorial Air Force in which capacity as Commanding Officer of No. 1 TAF Squadron by 1934 he had achieved the rank of Squadron Leader. As a commercial pilot he became Managing Director of Falcon Airways, operating from their private airfield at Oraki.

Cyril Eyton Kay was born in Auckland in 1902, joined the RAF and was posted to Egypt where he was taken ill and promptly sent back to England. During a period of leave in 1929 he navigated a Desoutter Monoplane from London to Sydney flying with Harold Piper, and was generally well acquainted with long-distance flying.

The third crew member listed as a 'wireless operator' for the Plessey equipment on board was Frank Stewart who was also charged to take photographs and make a film record of the flight. Although some of the cabin windows were blanked out, small transparencies were let into the lower fuselage aft of the trailing edge of the bottom wings. Specified for drift sightings the features were utilised as ideal camera ports.

The crew arrived in England on 24 August, in plenty of time to become acquainted with the new aircraft at the de Havilland factory at Hatfield where ZK-ACO, named *Tainui* after a canoe famous in Maori mythology, received her Certificate of Airworthiness on 9 October. Three cylindrical fuel tanks, thought to have been salvaged from DH.84 Dragon G-ACJM following an accident at Hamble, housing an additional 240 gallons of fuel, had been fitted in the cabin, increasing range to about 1,500 miles.

61. Caproni Aeronautici Bergamaschi P.L.3. I-TALY. (Italy)

The P.L.3, a low-wing, wooden monoplane with flaps and a retracting undercarriage, was designed and built by Caproni subsidiary CAB specially for both the Speed and Handicap Races. Fitted with a Fiat A59 engine, (a licence-built Pratt and Whitney Hornet) and sponsored by Major Francis Lombardi, the aircraft was to be flown by himself and co-pilot Vittorio Suster.

Both pilots had flown in combat during the First World War and subsequently had gained extensive experience with civil passenger air services and record and endurance flying in Europe, Africa and South America. Lombardi acted as 'semi-official' personal pilot to Mussolini.

The aircraft was designed to carry 455 gallons of fuel distributed amongst four tanks, sufficient for a range of 2,500 miles at a cruising speed of 199mph. The P.L.3 flew in July 1934 but by the beginning of September was reported still at the CAB factory at Ponte San Pietro, near Bergamo, and there were growing doubts that at the estimated weights and loadings, the aircraft would conform to the required ICAN standards. However, other reports stated that the aircraft had performed satisfactorily during tests at Montecelio but Lombardi's enthusiasm for the Races had gradually declined. On 13 October the Royal Aero Club was advised that the P.L.3 positively would not be taking part. Three days later the aircraft was registered to Francis Lombardi, later taken on charge by the Regia Aeronautica and eventually scrapped.

62. Fairey Fox I. G-ACXX. (New Zealand)

Entered by James Baines in both the Speed and Handicap Races, with himself nominated as navigator and Flying Officer Harold Gilman as pilot.

The entry forms were received from New Zealand where both crew members had learned to fly although Baines had been born in Woodford, Essex, and Gilman in Sydney, Australia.

Aged 27, Baines, previously a pilot with the New Zealand Territorial Air Force and an airmail pilot at Hawkes Bay with 2,000 hours experience, arrived in England on 26 March and maintained his skills by flying club aircraft. Gilman had been seconded to the Suffolk Regiment at Aldershot in 1928 but transferred to the Royal Air Force the following year and joined No. 101 (Bomber) Squadron at Andover where he had become Squadron Adjutant in 1933. He was granted special leave from the RAF in which Service he was currently Flight Commander of B Flight with the recently re-formed No.15 (Bomber) Squadron, operating Hawker Harts at Abingdon.

The Felix-powered Fox was purchased from Anderson Aircraft at Hounslow and moved to the National Flying Services' facilities at nearby Hanworth for modification, sharing hangar space with the Fairey Fox G-ACXO, Race No. 35, entered by the New Guinea Syndicate. The most significant change to G-ACXX was the increase in fuel capacity extending range to 1,750 miles. She was registered in the name of James Baines on 8 October 1934 quoting Witney Aerodrome, Oxford, as home base.

G-ACXX was one of only three civil registered Fairey Fox aircraft powered by the Felix engine, and the second to be entered in the Races. The crew shared with Ray Parer and Geoff Hemsworth all the pre-start traumas in respect of modification and certification and willingly co-operated with their potential rivals. After a reporter visited the NAS facility in late September he wrote that in his opinion it was 'very doubtful' that G-ACXX would be in any position to be at the start.

63. DH.88 Comet. G-ACSP. *Black Magic*. (GB)

The entry was made in both Speed and Handicap Races by 'James Mollison and Mrs Mollison', the only married couple flying as a crew. The entry forms quoted their address as Grosvenor House, Park Lane, London.

Arthur Edwards, Chairman and Managing Director of Grosvenor House, had nothing but admiration for Amy Johnson (now Mrs Mollison) following her solo flight to Australia in May 1930 and had offered her

No 63. Amy and a suspicious-looking Jim Mollison at Hatfield in the company of de Havilland Chief Test Pilot Hubert Broad before the maiden flight of their DH.88 Comet aircraft. (National Library of Australia, Canberra.)

Jim and Amy Mollison posing with their DH.88 Comet G-ACSP at Hatfield. Painted gloss black overall on delivery, G-ACSP was the first Comet to fly and subsequently was used for all the flight testing and development that was possible in the short time available.
(Mildenhall Museum.)

and her husband free accommodation at the hotel whenever they wanted. Now there was speculation as to why Edwards had not offered his own Comet for the Mollisons to crew. He was probably only too well aware that the couple's relationship was turbulent and unstable and had decided to pick a crew of his own choosing in order to protect his investment.

Jim and Amy were already famous record breaking pilots in their own right. In February 1934 Amy was in the USA and had toured the Douglas factory at Santa Monica in California from where she reported home that the only chance any non-American competitor would have in winning the Race would be if all the others lost their way.

In 1931 Jim Mollison had landed illegally in Turkey, had been arrested and imprisoned, and as a result was banned even from flying over the country. A major decision before entering the Race was whether to hope the Comet would not falter when crossing 400 miles of Turkish territory and not to consider the awful consequences of a forced landing, or alternatively to plan a route to avoid Turkey altogether, adding several hundred miles to the distance from London to Baghdad. Although for purposes of the Race the British Government had intervened on Mollison's behalf, the Foreign Office had to advise the Royal Aero Club that the issue of a Turkish visa had been very positively refused.

64. Cord Vultee V-lA. NR13771. (Australia)

The entry was sponsored by H W G Penny, a 25 year old Sydney radio announcer, trained to fly by the New South Wales Aero Club. The aircraft was to be flown by himself and Lieutenant Commander George R Pond, a pilot who had been a co-pilot to Charles Kingsford Smith in the *Southern Cross* when they set an un-

No. 64. The Cord Vultee NR13771 was an Australian entry from which regular broadcasts were to be made on behalf of a Sydney radio station. By the end of September the aircraft was still unserviceable in the USA and even with a scrutiny deadline extended by the Royal Aero Club, the aircraft failed to arrive in England. John W Underwood Collection.

official world's endurance record in 1928 for a trimotor and who in May 1934 made a successful eastbound transatlantic flight in a Bellanca.

Acting as reserve pilot and radio operator was Roger Downes, tasked with tending the new Marconi short/medium wave transmitting and receiving telegraphic and telephonic wireless and direction finding apparatus. Warren Penny proposed to use the equipment to make regular commercial broadcasts covering the progress of his entry. Downes, a Sydney based journalist, sailed for Batavia on 27 July from where he intended to take the KLM service to Europe during which he would study the route, meeting the team in London.

The Vultee, powered by a Wright Cyclone, still was not serviceable by the end of September and Penny, already in England, was faced with a transatlantic ferry if the aircraft was to be available to scrutineers by the requisite date. He left for the USA on board the SS *Majestic* with an agreement from the Royal Aero Club that arrival at the starting point could be extended to, but not beyond, 16 October, but the Vultee did not appear and Penny's broadcasting aspirations were thwarted.

Following publication of the list of entries only one major question remained to be answered: from where in England would the Races start?

6

Entrants with Authority

Arthur Hagg, Chief Designer for the de Havilland Aircraft Company Limited, caught in what appears to be a relaxed attitude as final adjustments are applied to the three DH.88 Comets entered for the Races. In reality the situation in the Works at Hatfield was described as feverish as the Race starting-date approached and Hagg forced himself to take a weeks' holiday to indulge in his other great interest, sailing. (BAE Systems.)

BEFORE THE INITIAL excitement generated by the prospect of a great air race had developed into frenzied practical activity, much thought had been devoted to the types of aeroplane which might be most successful, given the conditions imposed by the rule makers and the physical extremes of distance, terrain and climate.

From a provisional entry list of over 70 machines two aircraft types are worthy of particular consideration. Both they and their successors were destined to stamp their very individual marks onto the pages of aviation history. They were the de Havilland DH.88 Comet and the Douglas Commercial Type 2, the DC-2.

de Havilland DH88 Comet.

When the first details of the MacRobertson Races were announced no aeroplane in Europe could comply with the ICAN requirements and still reach Baghdad non-stop, 2,300 miles from London.

The Rules did permit in-flight refuelling but experiments were still in their infancy and a fleet of tanker aircraft would have been required, placed strategically along the 12,000 mile route. That in-flight refuelling was permitted by the Rules probably says more for the Race sub-committee's awareness of the developing systems than any serious prospect of competing aircraft taking practical advantage.

The opinion expressed by most British aircraft companies after careful consideration of the Rules was that total compliance would be difficult, whilst design, construction and testing of a specially tailored machine within the time limits would be very expensive and almost physically impossible.

The de Havilland Aircraft Company took a different view. The directors considered that the Company had the brains, the facilities and the capacity to prepare a design within the Race Regulations, and provide, perhaps, the only chance of a British winner. Geoffrey de Havilland's personal view was that no matter what, his company could not do nothing. "We cannot stand by and let this Race be won without any British effort," he told his fellow directors at one of their informal weekly gatherings. Already desperately short of time, in January 1934 the board agreed that the Company should proceed with the design and construction of an aeroplane which would, without question or doubt, conform to the published Rules and Requirements.

Only ten months separated the decision to go ahead from the start of the Races. The economic climate was not good and in addition to coping with certification of the new DH.86 which had flown on 14 January and the DH.89 Dragon Six, later renamed Dragon Rapide, which flew on 17 April, the workforce of 2,500 was still occupied with the final stages of their graduated move from the Stag Lane factory to the Company's new site in rural Hertfordshire. Indeed, with Stag Lane Aerodrome sold as building land, any new project laid down at the Edgware factory would need to be moved to Hatfield for erection and test flying, a recipe for potential damage, delay and additional expense. But the thought of building a machine which could win the Race was incentive enough. Thorough planning and control would be a major key to ultimate success.

The cost of designing and building a limited run of special aircraft was going to be very high with little hope of recovering any of the development expenses, but the value of publicity and prestige would be incalculable, and rejection of the challenge little worse than ignominious. A heavily subsidised selling price would be essential to ensure the project was as attractive as possible to potential customers, and a figure of £5,000 per aeroplane was agreed by the de Havilland board as a figure which they considered any serious participant might reasonably be expected to pay. At the same time they accepted that the true cost of each individual aircraft was likely to be more than three times that amount.

The de Havilland business had been built up from 1920 by what Captain de Havilland referred to as the 'next step' policy, building for the future on what had been learned in the past. The decisions to design and build a new aeroplane against a very stringent specification, to offer customer guarantees of delivery and performance, and to sell at an anticipated financial loss, were all taken after the most careful consideration, although Geoffrey de Havilland admitted afterwards, "we were forced to gamble."

The Company decided that in order to meet all its promises in the time available and with the schedule of other production running through the factories it would be possible to build only three machines and possibly some essential spares. The maker's designation was DH.88 and the chosen type name, Comet, attributed to a suggestion made by Hubert Broad. Advertisements were placed in the press in January 1934 seeking customers for the special aeroplane:

"The de Havilland Comet is now being designed for the MacRobertson International Air Races. Orders are invited for a limited number of this long-distance type of racing aircraft."

All orders confirmed by 18 February 1934, they declared, would receive the manufacturer's guarantee of delivery date and specification. Much to their relief, by the deadline, contracts had been signed for the purchase of all three aeroplanes: one each for Mr and Mrs James Mollison, Mr Bernard Rubin and Mr Arthur Edwards. The aircraft were allocated Works' numbers 1994, 1995 and 1996 and were registered G-ACSP (21 August 1934), G-ACSR (23 August 1934) and G-ACSS (4 September 1934).

The DH.71 Tiger Moth was built as a flying test-bed for the new de Havilland Gipsy engine and for research into high speed flight on low power. The aircraft produced record-breaking performances in both climb and level speed.
(BAE Systems.)

Unusually for a de Havilland aeroplane the all-metal DH.77, built against an Air Ministry specification for a fast interceptor, was powered by a 325hp Napier Rapier engine and was capable of over 200mph.
(BAE Systems.)

The sales could only reflect confidence in shapes on the drawing board; there was nothing else.

Race Regulations ideally defined a machine with a 3,000 mile range flying at 200 mph. For the aircraft to have any prospect of winning, landings needed to be kept to the minimum with maximum speed between each. There would be provision for two crew members and limited equipment, and the basic concept of a wire braced, single engine, low wing monoplane with fixed undercarriage was quickly evolved, based on previous design work in respect of the DH.71 Tiger Moth racing test-bed, and the DH.77 Interceptor. No technical details were to be released to the customers or the press until after the closing date for Race entries in June 1934, by which time construction, undertaken behind tight security at Stag Lane, should be well advanced.

At their new facilities at Hatfield, de Havilland had built a wooden mock-up based on early ideas before the design philosophy gradually shifted towards a more detailed specification which favoured a twin engine, stressed skin, all-wood low-wing monoplane featuring a one-piece cantilever wing with flaps. The wing section was RAF 34 (modified), creating a thin aerofoil for speed and a good compromise between a high lift/drag ratio and docile low-speed behaviour. It was to be constructed around three light spar webs running tip to tip and covered top and bottom with two layers of thin spruce planking laid diagonally, one layer set at 40 degrees to the other, pinned and glued. It was a method successfully used in the yachting world and well known to de Havilland's chief designer, Arthur Hagg. The method, known in some quarters as 'the lobster claw' would create a torsionally stiff, light structure which was immensely strong. The leading edges and wing tips were to be ply covered. A similar method of planking was used to construct top and bottom sections of the fuselage shell which when complete was simply lowered onto the wing centre section and secured.

When the three Comets were in the early stages of construction by Fred Plumb and his team at Stag Lane, in a discreet corner of the factory from which casual visitors were diplomatically steered away, half scale spars in eight feet sections were rigged from a stanchion supporting the factory roof and loaded with shot bags until they broke. It was already a familiar exercise at the site, mostly supervised by Richard Clarkson, and there was hardly a more convincing demonstration of structural strength.

Technical Director and Chief Engineer C C Walker and engine designer Major Frank Halford were deeply committed from the earliest stages. While the design staff gradually converted their thoughts into working drawings, de Havilland's team had already convinced themselves that engines of low frontal area fitted with variable pitch propellers was the only combination likely to ensure a high cruising speed in addition to acceptable take-off performance, particularly in the heat of the tropics where the take-off run from dusty aerodromes would be considerably improved.

A further innovation was the decision to fit a simple retracting undercarriage (described as 'retractile' in all contemporary correspondence) in order to improve streamlining. The legs retracted backwards into a nacelle behind each engine when the undercarriage fairing became an integral part of the cowling, and vital to the flow of cooling air. Later, the design of the operating mechanism, based on a worm gear and cable and which took fourteen and a half turns of a large hand-wheel on the starboard side of the front cockpit fully to lower or retract, was to be the subject of a legal dispute between the de Havilland Company and Airspeed (1934) Ltd., one of whose directors, Hessell Tiltman, was a former de Havilland employee. Tiltman alleged infringement of his patent, but told friends that the argument was more a matter of principle than anything else.

A scale model of a DH.88 Comet wing was mounted in a special jig, rather than from a hangar wall which was the normal practise, and subjected to a programme of severe load tests to destruction.
(BAE Systems.)

One of the best structural tests which was both practical and worthy of good publicity was to encourage students from the de Havilland Aeronautical Technical School to add their cumulative weight to the situation in preference to the more usual application of shot bags.
(BAE Systems.)

One Comet pilot later wrote that the handwheel could only be turned about one tenth of a revolution at a time and that a cycle to raise or lower the undercarriage took up to five minutes.

Hagg decided a castoring tailskid was preferable to a wheel to assist braking on grass and to eliminate possibilities of shimmy and reduce the prospects of swinging on take-off and ground looping during the landing run. Also it was easier to accommodate a spare skid carried on board. A pair of split flaps was fitted over one fifth span underneath the fuselage, between the two engine nacelles, to assist when landing at small airfields. Although not large in area the pilot had the choice of four preselected positions and the flaps did steepen the glide angle during approach and helped considerably when set at half position for take-off. For landing, the flaps were lowered at 110mph and were described as being 'very stiff, just about taking all one's strength to get them down!' All the fuel was to be carried in the fuselage with the tandem seat cockpit set well back on the centre of gravity. The disadvantage of this layout was that the crew was ensured of almost no external view during the critical phases of taxying, taking-off and landing, and minimal sight of the ground in cruising flight.

Apart from the selection of the most suitable power unit itself, the biggest problem faced by Frank Halford and his engine team of John Brodie and Eric Moult was the choice of a propeller. No British company could offer a suitable variable pitch unit, and de Havilland, who were keen to acquire a manufacturing licence, turned their attention urgently towards the USA. Once the basic philosophy of the DH.88 design had been agreed in February 1934, Frank Hearle, the company's General Manager, together with Chief Designer Arthur Hagg, took passage across the Atlantic, and reached an understanding with Hamilton-Standard for the supply of forged aluminium propellers for the three Comets, plus spares, and in addition an agreement to manufacture Hamilton-Standard propellers under licence should Hamilton's current negotiations with an active and rival European concern, (Fokker in Holland), come to naught.

A number of propeller units were shipped back to Hatfield but it was always known that the American blades were more suited to a big radial engine application and not ideal for the slim, in-line layout proposed for the Comet. When eventually he saw them, Frank Halford described the blade roots as being like telegraph poles.

Halford's engine family included the successful four cylinder Gipsy Major of 130hp and developed from it specifically for the DH.89 Dragon Six (later re-named Dragon Rapide), was the Gipsy Six, a six cylinder in-line engine developing 200hp. This unit was considered entirely suited to installation in the Comet. After 'racing' modifications had been applied, which included an increase in the compression ratio by the simple expedient of shaving the face of the cylinder heads, the installation of high-lift valves and new pistons with higher crowns, the engine produced 223hp at 2,350rpm. Due to a perceived lack of time, the pistons were manufactured by casting rather than the preferred method of forging, a fact which was to cause great anxiety throughout the duration of the Races. The team only learned afterwards that forged pistons would have been well within the capacity of the chosen manufacturer to supply inside the tight schedule!

Halford later cleverly redesigned the rocker system and air-intake supply to the carburettor in order to reduce the frontal area of the cowlings. In this configuration the engine model was known as the Gipsy Six R.

When compared with the basic design, the 'R' engine was estimated to produce 18 percent more power on take-off and, without the complications of supercharging, an additional 60 percent at full throttle when cruising at 10,000ft. Ten Gipsy Six R engines were eventually manufactured by the Engine Division of the de Havilland Aircraft Company at their Stag Lane factory.

Work commenced to adapt the engines to accept the variable pitch Hamilton propellers together with their associated hydraulic pumps, heavy and cumbersome in an aircraft for which every additional ounce of structural weight meant a reduction in fuel uplift and reduced range. The first engine was fitted by the Company's Chief Engine Installation Engineer, John Walker, to Geoffrey de Havilland's own racing DH.85 Leopard Moth, G-ACHD, in which 'the Captain' had been victorious in the King's Cup Air Race the previous summer causing the Company seriously to consider marketing a new model of the Leopard Moth, the DH.85A, fitted with a Gipsy Six in place of the standard Gipsy Major. The aeroplane would have offered serious competition to progressive designs from other manufacturers but following an extensive re-assessment of the world economy, the prospect was, very reluctantly, abandoned.

Apart from the lack of total suitability of the propellers, additional technical concerns arose that might not be resolved before the guaranteed delivery dates, including British certification of the Hamilton installation, and these factors convinced the de Havilland team that a simpler alternative propeller system must be found although plans continued for Hamilton propellers to be fitted on what was to be the flight-test aircraft, Comet No. 1994, the Mollison's G-ACSP.

Frank Hearle suggested that the French Ratier Company, known to have developed an unsophisticated pitch changing system for propellers fitted to 240hp Renault engines in a Caudron monoplane, should be contacted as a matter of extreme urgency and he was promptly despatched to Paris. The Ratier system relied upon a simple sprung pneumatic piston. The propeller blades were set at a pre-determined fine pitch angle for take-off and held in position by an air chamber inflated by a bicycle pump through a Schrader valve. Once airborne and at about 150mph, pressure on a disc fitted in the centre of the spinner, allowed air to be released from the chamber causing the blades to move automatically into a coarse, cruising, pitch. The optimum diameter of the disk was determined solely by trial and error. The penalty for simplicity was that the blades remained in coarse pitch until they could be re-adjusted on the ground and overshooting from a bad landing was, therefore, always to be treated with extreme caution. An added problem was that all taxying after landing was always in coarse pitch.

To accept the most suitable Ratier model, the Type 1306, modifications were required to the splined Gipsy Six R crankshafts and also to the skirts of the spinners which had been manufactured in anticipation of Hamilton units. Setting up the fine adjustments to the pitch change mechanisms caused anguish and frustration on the shop floor but it had to be mastered. It was vital that both propellers altered pitch simultaneously in the air.

Hearle had acquired six Ratier units for immediate installation with one in reserve: numbers 1 and 2 were allocated to Bernard Rubin's G-ACSR, 3 and 6 to Arthur Edwards' G-ACSS and 4 and 5 to the Mollison's G-ACSP. The flying test-bed Leopard Moth was fitted variously with propellers of a single fixed pitch represent-

Captain de Havilland's own DH.85 Leopard Moth G-ACHD was traded-in for a new model and the old aircraft became a running test-bed for the Gipsy Six engine. The aircraft was tethered in woodland on the northern boundary of Hatfield Aerodrome and the engine run for long periods at high power. Note the old fuel barrels and the grossly oversize air intakes. (Richard Riding.)

Largely built at Stag Lane the three DH.88 Comets scheduled for the MacRobertson Races were completed at Hatfield. Note the DH.71 Tiger Moth G-EBRV stored in the roof where she remained until October 1940 when the workshop was destroyed in a bombing raid. (The Aeroplane.)

The first Comet to be completed was G-ACSP and initial test flights in an unpainted condition and carrying the 'B Condition' test markings E.1 were with Hamilton propellers.
(The Aeroplane.)

Once the decision was made to use Ratier propellers it was necessary to modify the spinners to accommodate the pitch-angle change mechanism. In this view the blades are at fine pitch with the pressure-activated disc proud of the nose of the spinner.
(The Aeroplane.)

ing the fine or coarse angles to be experienced in service and after many hours of ground running called up for the issue of engine and propeller approvals, was first flown at Hatfield in May by Hubert Broad and later by several other Company pilots.

Richard Clarkson, aerodynamicist and flight test observer, recalled that activity within the Company was 'feverish'. By June, Arthur Hagg had come under such intense pressure that he forced himself to take a weeks' leave to go sailing off the south coast. All the aircraft were trial assembled at Stag Lane before they were dismantled and sent to Hatfield by road, the 44ft wing proving particularly awkward.

As scheduled, aircraft No. 1994, G-ACSP, was the first Comet to be finished and immediately following further flights in the DH.85A Leopard Moth G-ACHD, in the early morning of 8 September 1934, six weeks before the start of the Races and a month before flying the second aeroplane, G-ACSS, Hubert Broad completed a short test flight around the Hatfield circuit. The aircraft was unpainted and carried no markings apart from the de Havilland company's 'B Conditions' identification 'E.1'. Asked for his impression of the aeroplane Broad declared he was "tickled to death."

An eagerly awaited News Release issued by the de Havilland Aircraft Company from its head office at Stag Lane on 12 September 1934 was headed "The de Havilland Comet, a special long distance racing aircraft designed to conform to the conditions of the MacRobertson International Air Race." It was the first public announcement to provide any details of the Company's new aeroplane:

"The 'Comet' is a twin-engined, two-seater, long range, low-wing cantilever monoplane. Throughout its design the aim of the designers has been to produce an aeroplane which conforms in every respect to the conditions and regulations governing the MacRobertson Race from England to Australia and which will have the maximum chance of success, having regard to the route, the lengths of the legs, the climatic and other conditions likely to be encountered. Purchasers have been guaranteed a top speed of 200mph.

Outstanding features are the remarkable cleanness of design, the exceptionally thin wing section attained by the use of stressed skin construction, the completely retracting undercarriage and the employment of controllable pitch airscrews.

The de Havilland technical staff has concentrated its efforts on the design of the Comet since the announcement in January last of the Company's intention to produce a limited number of the type. Three machines are being built and have been entered in the race by Mr and Mrs J A Mollison, Mr A O Edwards (pilots Mr C W A Scott and Mr T Campbell Black) and Mr Bernard Rubin. Work on the building of the airframes and the special racing Gipsy Six engines commenced some three months ago. Day and night work has been necessary to ensure their completion in time for the event.

Wooden construction, including several novel methods of employing it, to which reference will be made below, has been used almost entirely throughout the machine.

The view of the front cockpit of Bernard Rubin's green Comet illustrates the compact instrument layout, the bowl compass and the large diameter handwheel used to lower and retract the undercarriage. If a warning light was ignored it was possible to leave the undercarriage legs at an intermediate position. Note that the side-opening canopy incorporates the front windscreen.
(BAE Systems.)

DH.88 Comet E.1 (G-ACSP) in primer and fitted with Hamilton propeller blades. The root ends of the blades which can be seen were described as 'telegraph poles' by engine designer Frank Halford.
(BAE Systems.)

The fuselage, which is as nearly a perfect cigar shape it has been possible to achieve, is of semi-monocoque form with removable panels permitting access to tanks, controls, etc. The method of construction adopted is a development of the well-known de Havilland box-type system. The straight parts and single curvature sections are of plywood attached direct to spruce longerons, while where curvature is in two directions the skin is built up of double laminations of spruce planking. The nose and tail fairings are of beaten Elektron sheet. The two pilots are accommodated with dual controls in tandem in a cockpit aft of the wings under a hinged canopy conforming to the general lines of the fuselage. Three fuel tanks, having a total capacity at 258 gallons, are located in the body, two of 128 gallons and 110 gallons capacity being slung by a vibration-damping method forward of the cockpit and a smaller one holding 20 gallons immediately to the rear of the back seat.

Controls are of the normal stick and rudder bar type but interesting features have been incorporated. Ailerons are operated on the de Havilland patent differential system. A type of Frise aerodynamical balance has been used and the surfaces are statically balanced by a suitable distribution of lead in the extreme nose.

Rudder and elevators are balanced by conventional bob weights on projecting lever arms. A mechanism has been inserted in the operating system of each to give a very low gear over the centre part of the control range with rapidly increasing movement towards the limits of the control travel.

The object of the gearing device is to provide lightness of control and to eliminate violent response to fine movements of the controls at high speed.

Wheel brakes are connected differentially through the rudder bar for ground steering. A hand lever is used for arresting the forward movement of the machine on landing and parking.

A large wheel to the right of the front seat operates the undercarriage retracting gear through a system of cables and worm gear. On the left hand side of the cockpit is located a lever operating the air-brake flaps. A single dashboard carrying a very complete instrument equipment is mounted in front of the front seat where it can also be seen by the second pilot.

Provision is made for mounting a combined medium wave receiver and Marconi Robinson homing device behind the instrument panel.

Undoubtedly the most interesting part of the Comet from the structural point of view is the main planes. Stressed skin construction has been carried further than has previously been attempted and a remarkably thin cantilever wing has resulted.

The main planes are built up in one piece attached to and faired into the lower side of the fuselage.

Three members, which have the appearance of conventional spars but which are merely for the purpose of dealing with shear and for transmitting the loads to the covering, extend from tip to tip. A cellular construction is formed by closely spaced discontinuance ribs. The covering has taken the place of the flanges of a conventional spar and, in fact, the sheer members and the skin combine to form one enormously robust spar.

The covering is composed of diagonal double skin planking reinforced where necessary by a third, and even fourth, lamination. The top skin at the point of maximum bending of the wing is 9/16 inch thick tapering to about 1/8 inch at the tips.

The greatest depth of the wing is 11 inches at a point where maximum bending and torsion due to the overhung weight of engine and propeller have to be dealt with. Schrenk type flaps which, when retracted, conform to the curvature of the wing and body, are fitted to the under surface of the inner sections of the wing and the fuselage extending from one engine nacelle to the other. Operation is direct by means of a lever in the cockpit. Their purpose is to steepen the glide and to increase maximum lift when coming in to land.

The undercarriage, which is in two halves, is completely retractable. The wheels enter gaps in the cowling behind the engines. Mudguards, attached to the compression legs, close the slots through which the wheels pass and fair into the general line of the engine nacelle. A special feature of the undercarriage is that it is not critical as to the extent it is lowered and does not depend upon a locking device for safety; in fact, so long as it is anywhere approaching the fully down position a safe landing can be made.

The engine cowlings fit very closely to the engines, are free from unnecessary openings, louvres, etc., and form an almost perfect streamline for engines and undercarriages.

All surfaces are very carefully filleted with the fuselage and most careful attention has been paid to the reduction of resistance. Not a nut or screw head will project on the finished machines and the whole surfaces of bodies, main planes and empennage will be highly polished.

A fully castoring tailskid has been selected instead of a wheel as it offers less resistance and helps to pull the machine up on landing.

Dimensions and weights:
Span	44ft
Length	29ft
Height	9ft
Wing area	212.5 sq. ft.
Total weight	5,250lb.
Horse power	444hp.
Wing loading	24.7lb.sq. ft.
Wing loading after 90% fuel has been used:	16.1lb.sq. ft.
Power loading	11.8lb. per hp.

The Gipsy Six Racing Engine.

The engines used in the Comet machine are standard Gipsy Six motors altered in several particulars to give increased power output, decreased fuel consumption and reduced drag.

By using a modified piston and cylinder head the compression ratio has been raised from the standard figure of 5.25 to 6.5, with a corresponding gain in output. The engine operates satisfactorily at this ratio on standard service fuel to D T D Specification 224. This fuel will be used for the Race.

To take advantage of the Hamilton variable pitch propeller which has been adopted, the normal speed of the engine has been increased to 2,550rpm. The use of this propeller has necessitated a new crankshaft with an appropriate hub fixing at the front end and the provision of a temporary oil supply at 100lb/sq inch in order to rotate the blades in their sockets when the fine-pitch position is required. This high pressure supply is obtained from the usual engine pumps through duplicate oil relief valves and in no way affects the normal lubrication system. The control of the pressure and so of the pitch position, is in the hands of the pilot.

Hubert Broad completed all the development test flying on the unpainted E.1 (G-ACSP) initially fitted with Hamilton propellers. Here the aircraft is seen approaching for a landing at Hatfield Aerodrome. Richard Clarkson was the Flight Test Observer on most occasions. (The Aeroplane.)

Following the first test flight Hubert Broad's views were anxiously solicited by Captain Geoffrey de Havilland, Arthur Hagg, Charles Walker and Frank Halford. Note that the windscreen part of the canopy was a fixed structure on the first airworthy aircraft. (BAE Systems.)

DH.88 Comet G-ACSR at Hatfield on one of the few days available to the Race crew before the aircraft was due to be flown to the start. It was a special day for Alan Butler's two little childen, Carol and David, who were allowed a private inspection.
(Alan Butler.)

A special coupling is provided at the rear of the crankshaft for driving a rotary vacuum pump which is used to operate the Sperry gyro compass.

By certain modifications to the valve rocker gear and its casings, the overall height of the engine has been reduced by 26 mm and the shape of the cowling much improved. An alteration to the induction manifolds, and the use of a smaller collecting scoop for the cooling air have decreased the overall width of the engine and helped towards a material reduction in the frontal area.

The standard arrangement of alternative hot or cold air supply for the carburettors is retained in case adverse atmospheric conditions are encountered during the Race. Warm air, if required, is taken through a flame-trap from the vicinity of the cylinders.

In its modified form, the engine develops on the bench a maximum output of 224 hp at 2,400rpm at sea level. Actually in flight this performance is further improved by the effects of the high forward speed on the carburettor air intake. Flying at an altitude of 10,000ft, the full-throttle output is 160hp at 2,250rpm and the fuel consumption 0.48lb/BHP/hr."

It was inevitable that some modifications to the aeroplanes would be necessary, and an immediate requirement was to attempt to improve forward visibility. Broad had complained about this after sitting in the cockpit on the ground but he had been rebuffed by the Chief Designer who told him that as he had not yet flown the aeroplane he was in no position to assess the situation. The significant nose-up attitude which blocked all forward vision on the approach moved Broad to repeat his request, now based on practical experience. Subsequently, the fixed front windscreen was incorporated into the hinged canopy, the depth increased by one and a half inches and the tailskid increased in length.

A landing light was added, beautifully profiled into the nose, and a rudimentary instrument panel was installed in the rear cockpit which carried a copy of most of the blind flying equipment in the front. All flying controls were duplicated although landing and take-off would be the responsibility of the occupant of the front seat who also had sole command of the undercarriage winding gear. Waller and Cathcart-Jones had arranged for specially constructed air cushions to be fitted to both seats in G-ACSR, padded right up the back and round the shoulders to relieve as far as possible the fatigue of direct bodily contact with the vibrating airframe.

The Mollison's Comet G-ACSP was equipped and instrumented to carry out all the prototype test flying as distinct from the brief production checks needed on the other two aircraft. The acceptance tests completed at Martlesham Heath proved the performance of the Comet at maximum weight to be better than anticipated. Top speed at 10,000ft was 225mph, and during the all-important take-off trials, obstacles were cleared by a wide margin. The value of the variable-pitch propellers was evident from figures released before the Race which showed that with their assistance an extra 1,000lb of load could be carried safely off the ground whilst conforming to all the ICAN requirements. The extra weight represented a fuel uplift sufficient for more than 1,000 miles of additional range. At an all-up weight of 5,500lb stalling speed was just under 80mph, falling to 65mph at 3,500lb.

G-ACSP was fitted with Ratier propellers before what is thought to have been her twelfth flight on 26 September when the necessity for precision when setting the propeller pitch on the ground became apparent. Hubert Broad noted that the port side propeller changed pitch unexpectedly at 110mph but the starboard propeller did not follow suit until a further 50mph had been achieved. With practice in setting and the discovery that not all the springs in the system provided identical tensions and were exchanged, the pitch change was found to operate smoothly if the aircraft was accelerated in level flight and then dived gently through the 130mph mark, at which point the change-over occurred automatically.

The tension in the de Havilland camp hardly relaxed during the short but intense period of development flying carried out by Hubert Broad. After one spectacularly untidy arrival back on earth, Broad told his observer, Richard Clarkson, to get out of the aircraft where it was in the middle of the aerodrome, and join him in a walk back to the bar where they would enjoy a brandy. Broad suggested that their employers should be renamed "Death Boxes Limited" a not inappropriate name he considered, "in view of the fact that most of those engaged on the project were so tired that any conversation with them was like talking to dead men."

In addition to his three solo flights in G-ACSP, Hubert Broad flew an additional 17 trips with Richard Clarkson which included eight climbs to 10,000ft, two to 14,000ft and many lengthy speed and consumption tests. Using either G-ACSR or G-ACSS, Broad flew four additional tests at Hatfield and two at Mildenhall, a total of more than 20 hours development flying in 26 trips which was not considered enough but sufficient, just.

The timetable could hardly have been

On 2 October the Comet G-ACSS was named 'Grosvenor House' in a ceremony at Hatfield Aerodrome. With the aircraft still unfinished and the start of the Races less than three weeks away, it was a necessary interruption to the programme and great publicity. The owner of the aircraft and her sponsor for the Races, Arthur Edwards, was already on his way to Australia via the 'hotels of the Empire'.
(BAE Systems.)

helped when on 2 October the attendance of the unpainted G-ACSS was demanded at a naming ceremony at Hatfield. With Scott and Campbell Black amongst the group of invited guests, the aircraft became *Grosvenor House* after Lady Richard Hinton, wife of the London based Agent General for Victoria, used a long-handled hammer to smash a bottle of champagne against the protected hub of the starboard engine, a launching recorded for posterity by the newsreels and a battery of press cameramen. If Arthur Edwards was seeking publicity for his London business there could have been no better start although he was not there to witness it. In September he had left London on a 27,000 mile circular tour of 'the hotels of the Empire' declaring: "I shall endeavour to be in Melbourne to greet my pilots when they win!"

It was from Melbourne a week before the start of the Races that Edwards cabled back to London: "Tell Scott and Black have feeling they will win but let it be safety first win or lose." Scott replied: "Your plane is a beauty your pilots will do their damndest the rest is on the knees of the gods."

The three Comets finally were made available to their crews only days before they were due at the start. G-ACSS, the second aircraft to be completed first flew on 9 October and G-ACSR three days after that. All the crews assembled at Hatfield and using G-ACSP for tutorial were briefed by Bob Loader on how to set the Ratier propeller pitch, remove engine cowlings and start up. Cathcart-Jones later noted: "Most of the directors of the de Havilland Company were in attendance throughout the whole period during which the machines were being prepared for handing over. One had but to raise the smallest question to be instantly given the expert advice of Mr C C Walker, Arthur Hagg, Major Halford or Captain Broad who was as yet the only pilot who had flown a Comet."

Cathcart-Jones flew G-ACSR on 12 October after a dual session with Hubert Broad, then Waller did the same. Jim Mollison made three circuits in G-ACSP, and Charles Scott accepted G-ACSS the following morning. Scott flew one session and next day reported to Mildenhall. During four 'production' test flights conducted by Hubert Broad, G-ACSS had accumulated only 83 minutes flying time. A similar amount had been logged by Scott when the aircraft finally left for Australia; none of the others managed more than an hour. Of all the Race pilots, only Amy Mollison did not fly a Comet at Hatfield although she did so, against the wishes of her husband, on the Thursday before Sturday's start.

The scheduled delivery of the three Comets had been achieved against a self imposed penalty of prodigious proportions and awful potential consequences. There were many problems and scares, but the special teams which had been established worked themselves to a high pitch of efficiency.

Now everything was up to the various skills and courage of the pilots and the assistance that could be rendered by the de Havilland engineers placed strategically along the route. The design and works staff in England could do little except wait.

The Douglas DC-2.
Donald Douglas, President of the American based aircraft manufacturing company that bore his name, was in his late thirties when he received a letter from Jack Frye, Vice President in charge of Operations for Transcontinental and Western Air (TWA).

Frye's airline was in trouble. More than a year previously, one of its foreign manufactured machines had crashed killing the coach of a famous football team and the company had been the subject of bad publicity ever since. An immediate change of equipment to the Ford Trimotor did little to improve the image; the new machines were neither comfortable nor economical, but they were American, and made of metal, and that made some difference.

Jack Frye had written to all the major American manufacturers detailing a basic specification for what he perceived as an ideal new aircraft for TWA. It was to be an all-metal monoplane with a maximum weight of 14,200lb, fuel for just over 1,000 miles at a speed of 145mph and a passenger capacity of twelve. In addition to saying what he wanted, Frye asked when the first aircraft could be delivered.

Douglas gathered his design team for a conference then sent them away to continue investigations. Ten days later they had an answer. With the use of new engines, propellers and materials, they estimated they could offer TWA an even better aeroplane. Frye had specified three engines; Douglas thought they could improve on performance with just two. The undercarriage would be semi-retractable; there would be wing flaps, variable pitch propellers and comfortable seating inside a heated passenger cabin. The airframe would be built entirely of metal and full use of stressed skin constructional techniques were to be employed. The date was August 1932.

The outline sketches of the proposed new aircraft were shown to senior TWA officials in New York a fortnight after Frye had written his letter. They were most impressed but the aftermath of that fatal crash was still painful and an amendment was noted against the original specification: the machine must be capable of safely continuing the take-off from any airfield on the TWA network after losing the critical engine. Douglas designers could give no positive answer immediately except to say that they believed they could comply with the additional request. It was the age of practical testing. No matter what the results from the slide-rules, new aeroplanes would either perform or they would not.

TWA signed a contract with Douglas in September 1932 and immediately plans were put in hand to build the aircraft at the Santa Monica factory in California. The Douglas Commercial, first model, was designated DC-1 on drawings which were translated on the shop floor into the biggest twin-engine, land-based aircraft built in the USA until that time. On 22 June 1933, the prototype DC-1 was pushed out of the workshops, and following the usual pattern of adjustments was declared fit to fly on 1 July. Carl Cover and Fred Herman stepped aboard to see whether she would.

During the take-off run, one engine cut out then immediately regained power. As Cover lifted the aircraft off the runway both engines cut out then back in again as the pilot instinctively put the nose down. With both engines misfiring, the crew decided on a precautionary landing straight ahead, and the DC-1 was put safely down in a field. Exhaustive tests on the engines revealed that experimental carburettors had been

This photograph of a DC-2 in the colours of Transcontinental and Western Air (TWA) was submitted by KLM with their Race entry forms and shows the early type of elevator without the aerodynamic balance.
(Royal Aero Club archive.)

Late in August DC-2 PH-AJU was flown from her birthplace at Santa Monica, California, to New York where, when this photograph was taken, she was minus wings and most of the tail but with both engines and propellers still attached. With her flying surfaces packed into crates KLM's first DC-2 was made ready for shipment to Rotterdam where she arrived on board the SS Statendam *on 12 September.*
(The Aeroplane.)

Following erection at the Fokker factory at Amsterdam DC-2 PH-AJU was flown to Waalhaven near Rotterdam where she was manoeuvred on the ground by the conscription of local manpower. The aircraft is in striking contrast to the two Fokker FVIIA and one Fokker FVIII visible in the background. (The Aeroplane.)

fitted to the Wright Cyclones. As the nose of the aeroplane was raised, a flap had automatically cut off the fuel supply. Turning the carburettors through 180deg. neatly solved the problem.

The DC-1 was subjected to the most severe in-flight testing that could be inflicted on it by both manufacturer and customer, and apart from some induced problems, there appeared to be no hidden vices. The single engine take-off performance trial carried out at Winslow, Arizona, was faultless, and the aircraft flew the 240 miles to Albuquerque on the power of the port engine only, overtaking a scheduled commercial flight which was operating normally on three.

An immediate order was placed for 20 aircraft, but a revised specification called for a minimum of 14 passengers causing Douglas to lengthen the fuselage and increase the wing span. Such drastic alteration demanded a new company designation, and the production aeroplane was identified as the Douglas Commercial Type 2, the DC-2.

The sole DC-1 eventually was delivered to TWA who used the aircraft for development flying, mostly associated with new navigation equipment. Fuel capacity was 1,600 gallons, and during a three day loan to the National Aeronautics Association, the body most critical of the MacRobertson Regulations, the DC-1 established 19 new records for speed, distance and load carrying.

The first production DC-2 flew on 11 May 1934 powered by a pair of 730hp Wright Cyclone engines. TWA accepted the aeroplane on 14 May and she entered commercial service four days later, less than two years after Jack Frye had first approached American manufacturers. The first of an eventual order for 19 machines for the Dutch airline KLM was nominated for the MacRobertson, and its appearance in England was a profound shock to leaders of the European industry. For the most part, the lessons they should have learned were not fully appreciated or at worst ignored.

To most observers the sheer size of the DC-2 was impressive, and the basic design appeared to be more advanced even than the Boeing 247, which was itself a revelation compared to European standards. In the USA, the DC-2 proved so popular that her operators were continually taking business from lines operating almost every other type of aircraft, including the latest Boeing. Arthur Gouge, Chief Designer of Shorts, made a close inspection of the DC-2 in England and immediately abandoned any future ideas for a new biplane airliner. Instead he concentrated on the S23, an all-metal flying boat with a wingspan of 114ft and a maximum gross take-off weight of over 52,000lb. Inspired by his appreciation of the DC-2 at the start of the Races, the S23 was regarded in some quarters as Gouge's masterpiece.

Like the wooden Comet the all-metal DC-2 used the newly accepted principles of stressed skin construction: a light primary frame covered with a skin which itself carried part of the load. The advantages were obvious when the sleek lines of both aeroplanes were compared with others of more traditional design and construction. The British industry's biggest order ever placed for a commercial transport aircraft had been for eight machines, each of which had been practically hand-built. An order book for 60 DC-2s had permitted Douglas to tool-up for quantity production; whole sections of the airframes could be rapidly manufactured in special presses, mostly eliminating the need for hand-beaten developments. The design permitted use of aluminium sheets supplied at maximum production length which were applied fore-and-aft along the fuselage, and span-wise. Riveted to the basic frame, the result was a series of strong panels with a uniformly smooth surface. Three-bladed Hamilton variable pitch propellers, effective flaps and retracting undercarriages were standard equipment now, expected by the airlines, and refined by the manufacturer to a high standard of working efficiency as the result of growing operational experience.

Dimensions and weights:
Span   62ft
Length   61ft 11¾in
Height   16ft 3¾in
Weight   18,560lb
Horse power   1,750hp
Ceiling   22,450ft
Range   1,000 miles
Cruise speed   200mph

Even before the Races began, the free publicity generated by the mere appearance of the DC-2, already engaged in scheduled services, was more productive than the most expensive advertising campaigns which could have been devised. While some were still cogitating on their various reasons and excuses for not entering the Races, a new commercial machine recognised as a potential world beater was being prepared to show her merits on the airway to Australia.

However, one British columnist decided that the entry of the DC-2 in particular was mere propaganda. "The Americans have made an audacious assumption that such aircraft could expect to compete with the fastest designs on the continent," he sneered. Time would prove his opinion right or otherwise.

7

Reality and Disappointment

Following a thorough survey of all aerodromes in the London area no single one was considered suitable for housing all the anticipated field of entrants. At the end of August the Air Ministry offered the use of the new and still unfinished RAF Station at Beck Row, Mildenhall, Suffolk, some 50 miles from the heart of the capital. (The Aeroplane.)

IN SPITE OF THE prolonged discussions on the technicalities of the Rules and the route and every other facet of shepherding a flock of racing aeroplanes halfway round the world, eight weeks prior to the start of the Race no announcement had been made positively to identify the aerodrome from which it would begin.

To alleviate anticipated pressures on one site, the earliest notifications published by the Royal Aero Club had indicated use of at least two aerodromes in England, hopefully in the London area, the Speed racers starting from one at the Handicap entrants from another, an idea soon abandoned.

The number of prospective entries had been the root cause of the lengthy search for suitable facilities and the sub-committee considered it was absolutely correct in its resolve to ensure adequate provision was made on one site for all Race aircraft. The organisers were not to know that in spite of their anxieties, only one third of the 64 entries, something of a disappointment in itself, would eventually materialise.

Every civil aerodrome in the London area was in contention but almost all were rejected. Four sites in southern England were considered more seriously and each was thoroughly inspected: Hatfield, Harmondsworth, Eastleigh and Gravesend. In the eyes of the sub-committee, no single one proved totally acceptable.

The London sub-committee proposed that Major J S Buchanan should open discussions with the Air Ministry on the prospect of using an RAF aerodrome with the prime candidates for investigation being Upper Heyford in Oxfordshire, Mildenhall in Suffolk and Hendon in north London. On 11nJuly Buchanan reported back that the only available site was RAF Digby in Lincolnshire which would be unoccupied until October.

On 14 August, Flight Lieutenant Christopher Clarkson was invited to inspect the new military aerodrome at Mildenhall in Suffolk which he reported to be 'nearly as big as Portsmouth'. The following day an official from the Air Minis-

By October two 'A Type' sheds were complete at RAF Mildenhall, known as the West, illustrated, and the East, with two additional sheds in course of erection. On 14 October, as the Royal Aero Club took up temporary residence, the site was visited by a DH.82 Tiger Moth, K2583, from No. 24 Communications Squadron based at Hendon.
(National Library of Australia, Canberra.)

try telephoned Harold Perrin with a formal offer of the site and members of the sub-committee were invited to make their own inspection on 22 August. In Suffolk they were warmly welcomed but expressed concern that Race activities would interfere with the continuing process of building up the site for RAF occupation. However, they were advised that if the sub-committee chose Mildenhall, the Royal Air Force would do everything possible to assist.

A plan to use Hatfield temporarily dressed with Bessoneau and Hervieu hangars borrowed from the Air Ministry store at Didcot and erected by RAF personnel, was the strongest alternative. At a meeting with Wilfred Nixon and Frank Hearle from the de Havilland Aircraft Company held in London on 27 August, it was revealed that the company would be in a position to offer limited permanent hangarage in addition to their aerodrome, all at no charge, but a number of essential modifications would be required to local infrastructure. The sub-committee convened at the aerodrome two days later for a tour of inspection and estimates were presented: £300-£400 for the works and £500-£600 for the additional hangars. It was suggested that the first call on gate receipts would be to cover the necessary improvements and alterations to the infrastructure and import of facilities with any remainder divided between the de Havilland Company and the Royal Aero Club.

Although Hatfield was conveniently situated for London the expense of setting up all that was required finally tipped the balance. The sub-committee members were impressed by the size of Mildenhall Aerodrome and the two completed hangars but were concerned at the lack of basic amenities and the general remoteness of the place. However, the Air Ministry's offer was accepted, not without considerable misgivings but equally with enormous relief and the news was released to the world:

"The Royal Aero Club announces that the Air Council has kindly given sanction for the new Royal Air Force aerodrome at Mildenhall, Suffolk, to be used as the base for the Air Race to Australia. The Race will start from this aerodrome on October 20th.

Mildenhall was made available to the Club after careful examination had shown that no civil aerodrome possessed the necessary hangar accommodation to house all the aircraft which have been entered.

The Club will be wholly responsible for the organisation at Mildenhall in connection with the Race and the arrangements which have been made will enable competitors to use the station from early in October.

But for this arrangement, the Royal Air Force would have formally occupied Mildenhall on October 13th.

The hangar accommodation consists of two 'A' Type sheds, each with a floor area of 30,000 sq. ft.

The approximate dimensions of the aerodrome are 1,400 yards by 1,100 yards, with a maximum run of about 1,600 yards.

The Royal Aero Club desires to take this opportunity of publicly expressing its thanks to the Air Council and the Royal Air Force for the special facilities which are being provided so that all competitors may have satisfactory accommodation for their aircraft."

The new RAF aerodrome near the village of Mildenhall was developed as part of Great Britain's defence against the perceived 'continental threat'. The government purchased the land in 1929 and the first building, offices for the resident engineers, was started in 1930 by Fred Hale and Sons and completed in 1931. The site, for which the budget was reported to be £1,000,000, was, in October 1934, still under construction in the hands of W and C French who had taken over Hale's contract, and the Newcastle based company Redpath Brown, who were responsible solely for the erection of the hangars. Work was sufficiently far advanced for the site to be made avail-

Part of the preparation process for the surface of a grass aerodrome in which kit the most modern item is, perhaps, a chequered flag. Note the advertising banners attached to the perimeter fence beyond the East shed.
(Flight.)

Air Ministry regulations prohibited officials from sleeping on the aerodrome site and none of the completed domestic accommodation was ever offered. Those not taking up rooms in hotels within a wide radius made their own local arrangements and caravans were a popular choice although, if used overnight, they had to be parked outside the perimeter fence.
(Flight.)

able for the safe operation of aeroplanes, and the two hangars already completed were considered more than adequate. The Air Ministry offered the aerodrome without charge but insisted on a substantial comprehensive insurance policy to cover any damage to their new property. The Royal Aero Club raised the necessary cover against a premium of £100 and at the same time initiated a legal liability policy as promotors of the Races as far as Koepang at an additional premium of £125. The first RAF personnel under the command of Flight Lieutenant H B S Ballantyne moved in on 15 August 1934. One local resident subsequently produced a photograph of a military aeroplane which had used the same field in 1913 having taken part in army manoeuvres near Thetford.

In the heart of rural Suffolk, Mildenhall was more than 50 miles from London and ten miles beyond Newmarket, where by unfortunate co-incidence a major horse race meeting was scheduled for the week before the start. As a result accommodation for organisers and competitors alike proved to be scarce but Perrin's expanded staff managed to secure rooms in Mildenhall village, Thetford and Bury St. Edmunds which could only be taken for the full week, 14-20 October, at prices varying between eight shillings and sixpence and half a guinea per night. For members of the public wishing to stay close-by to view the arrivals, trials and the start, without local contacts accommodation was very nearly impossible and everybody was scattered over a 20 mile radius. However, there were those who be-

lieved the comparative isolation of Mildenhall would dissuade only but the very keenest enthusiasts from travelling there and that the organisation was less likely to be interfered with if the site was not overwhelmed by crowds. The proximity of the Newmarket races would also allow harassed officials the opportunity to spend an odd afternoon relaxing at the sport of kings.

In another display of munificence, the sub-committee arranged that it would not be necessary, for the duration of the Races, for competitors to have passed an examination for a Wireless Air Operator's Certificate, but could instead apply for a special permit for a fee of five shillings!

The Royal Aero Club set up its temporary headquarters in East Anglia from Friday 12 October. Following what was described as 'a bit of a struggle with the Air Ministry' a Members' sitting room, bar and offices were situated in marquees and enclosures and car parks marked out. Car parking arrangements were contracted to National Car Parks who organised sites opposite the main gate. Entrance charges were set at 1/- for members of the public and 2/- to park a car. Royal Aero Club members were requested to pay the same car park fees and were allowed free entry although charged 5/- for each guest.

Unable to agree terms with the first choice of caterers, Messrs. Carpenters, full services were to be maintained at short notice throughout the week by the NAAFI organisation at a level described as being comparable with anything available at a good class hotel and dinner, served with a smile, on demand until the early hours.

"Mildenhall Aerodrome resembled a luxurious camp, marquees are lined with gaily striped canvas, waiters in evening dress and the glass doors of the Aero Club give the impression of a well-appointed

Supervised by Jimmy Jeffs, a smudge fire was constructed as the only method of determining wind direction before wind stockings were imported and erected. Note the can of petrol, Air Ministry issue dustbin, wood-wool and the tail of Lindsay Everard's DH.85 Leopard Moth G-ACKM in the background.
(Flight.)

An early form of mobile wireless communication was tried at Mildenhall linking the aerodrome with the Information Bureau.
(Flight.)

London Club."

In spite of the sub-committee's best efforts, Race officials were quick to discover that facilities were minimal. Station Headquarters was not formally opened until 16 October when the establishment came under the command of Wing Commander F J Linnell OBE, which at least permitted some degree of liaison at a high local level.

Every item necessary for setting up a temporary but operational airfield with engineering facilities for aeroplanes and provision of the basic comforts and needs of the competitors, officials and spectators, had to be carried to Mildenhall. Having prior knowledge, many of the 60 deputed Club officials faced with the prospect of a long vigil arranged for the hire of caravans which they parked adjacent to other temporary accommodation on the site although it was made clear by the Air Ministry they could not be used for anyone to stay overnight.

The occupying force of civilians had to provide their own windsock; the first makeshift indicator, an adapted potato sack, was superseded by a pair of pillowcases stitched together, and replaced eventually by more functional units supplied free of charge as splendid fluttering advertisements for Pratts Ethyl amongst other petrol grades.

Jimmy Jeffs, the aerodrome controller more used to supervising the Watch Office at Croydon Airport, was seen presiding over a smudge fire whose smoke trail gave a good clue to wind-drift. During an early tour of his temporary domain Jeffs discovered a hole a yard square in the middle of the aerodrome's otherwise immaculate grass surface which was immediately marked with obstruction flags until it could be filled in. Signal flares and lamps had to be imported, and an altruistic member of the Royal Aero Club loaned portable floodlights for which his company had a sales agency.

The site possessed the most modern and efficient fuel storage facilities, but these were not made available to the sub-committee, and the contracted refuellers were obliged to provide tankers which were more practical and offered mobile advertisements for their products.

To assist competitors with engineering tasks at Mildenhall, ten senior students from the College of Aeronautical Engineering (Chelsea and Brooklands) had been selected and accommodated at a school in the nearby village of Barton Mills. Under the direction of their leader, Walter Clennell, they were allocated to work in each of the two hangars where they were on duty from just after mid-day on Saturday 13 October and in which it was discovered the electric lighting had not yet been connected.

The General Post Office established a battery of telephones and teleprinters in a marquee dedicated to the Press so that news could be rapidly and efficiently received and distributed and a Post Office counter was arranged in what was to be an RAF stores building in connection with which the General Post Office issued a circular on 10 October:

"A temporary post office and telephone exchange will be open from 13-20 October (or later in the event of the Races being postponed from day to day). The classes of business to be transacted and the hours during which the various services will be available are as follows:

Money orders, postal orders and the sale of stamps, 8.00am to 1.00pm and 2.00pm to 7.00pm.

Telegraphs 8.00am to 9.00pm with probable extensions on 19/20 October.

Telephones: 24 hours."

Several competitors were to ask Mr Meadwood, the Postmaster seconded from Bury St Edmunds, to witness their wills. A special insurance pool was organised, open until four days prior to the start when, for example, a £10,000 aeroplane could be covered for a premium of £1,000, one third of which was recoverable in the wake of a trouble-free flight to Australia or damage-free retirement along the route. Similar percentage premiums were required for pilot's life cover, although no rebates were promised on this business!

Arrangements had been made for the rapid deployment of search parties should the need arise. At Mildenhall competitors

Michael Hansen in the uniform of a Lieutenant in the Danish Army Air Corps. He was no stranger to long flights in a small aeroplane.

The first competitors to arrive at Mildenhall were Daniel Jensen and Michael Hansen, engineer/navigator and pilot respectively of the Desoutter OY-DOD. Both officers in the Danish Army Air Service, their Race involvement was as civilians although both appear to have a wings badge on the lapel of their suit jackets.
(The Aeroplane.)

were asked to confirm how soon they wished a search to be commenced after they had been reported missing and a request from officials in the Dutch East Indies to clarify who would be liable for the cost of any search was answered by Harold Perrin without hesitation: 'the contestants'.

On that first Friday morning sub-committee members and Race officials toured their province, such as it was, followed by a posse of pressmen soon to be fed a constant stream of information from Messrs Robertson, Bowyer and Russell acting as 'Press Marshals' on behalf of the Society of British Aircraft Constructors and the Air Ministry. Taking a deep breath the Royal Aero Club declared itself ready!

In addition to the names of the Race sub-committee the general public could learn the identities of the British Race officials at Mildenhall by the purchase of an official programme for sixpence. Chaired by Lord Gorrell, the Race Stewards were all Lieutenant Colonels: Shelmerdine, McClean, Bristow and Moore-Brabazon. Major Alan Goodfellow was appointed Clerk of the Course with Flight Lieutenant Swinbourne of the Royal Australian Air Force as his deputy. Timekeepers were A G Reynolds, Colonels F Lindsay Lloyd and A H Loughborough while Handicappers were Captain Wilfred Dancy and Fred Rowarth. Flight Lieutenant Christopher Clarkson was appointed Chief Marshal, Jimmy Jeffs Aerodrome Control Officer and H B Howard Clerk of Weights and Measures. The Information and Service Bureau was organised by Ivor McClure of the Automobile Association's Aviation Department and Harold Perrin inevitably became Secretary of the meeting, assisted by Terence Bird, who had been employed by the Club specifically to fulfil this function on a three month contract from 23 July, following recommendation from the Automobile Association, at a salary of £1. 10s. 0d per day..

Alan Goodfellow noted the responsibilities of his team:

"The Chief Marshal is in charge of the aircraft and is responsible to receive them on their arrival, allot them to their various positions in the hangars, see that they are presented to the technical checking officials at the allotted places and times, guard them against injury and interference throughout the week and finally present them to the starter on the starting line in the right order and at the right times. He has under him a staff of six marshals, the AA Scout Squad and a party of young men from the College of Aeronautical Engineering to handle the aircraft. His deputy is responsible for checking up the equipment which must be carried by the various aircraft.

The Clerk of the Scales and Measures is responsible for all the technical checks of the aircraft such as weights, measurements, fuel contents etc. required by the Rules both in connection with the Speed and the Handicap Races. He has under him a number of officials of the AID for carrying out the actual checks together with assistants who are responsible for filling in the special Race logbooks of each competitor as and when the checks are completed."

In order to show 'an appreciation of sportsmanship' the Royal Aero Club organised a banquet on 16 October to which were invited as many of the Race competitors who already were in England. From early September, members of the public had been invited to contribute towards a 'Hospitality Fund' and in common with the accepted practice of the day, contribution lists were circulated to the Press who published details in full, identifying the donors of fifty guineas or five shillings.

Due to the proximity of the Race starting date it was decided that the banquet should be brought forward and eventually it took place at Grosvenor House in Park Lane, London, on Friday 12 October, a sumptuous white-tie affair, to which Royal Aero Club members were permitted to take guests 'including ladies' at a cost of one guinea per ticket. Arthur Edwards invited all crews and passengers taking part to be his guests. Five hundred diners sat down and nine competing pilots replied to the toast in their own individual manner, and mostly without the aid of a reluctant microphone. The general opinion was that it had been a successful evening and an important diplomatic prelude to departure.

At Mildenhall Aerodrome next day news soon came that Race No. 61, the Italian Bergamaschi machine was cancelled and that *The Irish Swoop* was at Southampton with a damaged cowling. Apart from an itinerant Moth which flew around the aerodrome without landing, believed to be Brian Lewis' demonstrator Leopard Moth G-ACLM cruising from Heston, what appeared to be a gloomy aeroplane-less day had drifted on until teatime leading one journalist to file his first Race account so:

"With the elaborate marquees flapping and billowing, two most capacious and entirely empty hangars, and hordes of officials, white-overalled mechanics, and yawning waiters, Mildenhall Aerodrome had taken on a dismal aspect by the time dusk was falling last Saturday, the first day of the reception. It was something like a

REALITY AND DISAPPOINTMENT

Derby Day without a racehorse.

Nobody really expected many machines, but everybody hoped that some would come along. After all, Sunday might provide virtually impossible flying weather and the Royal Aero Club might or might not give the entire entry a time extension.

Slowly the quite appreciable number of hedge watchers drifted away, and an evening haze blotted out the trees on the flat horizon. The few remaining spectators wondered how many machines would turn up on the following day."

Having filed his copy, how irritating that Michael Hansen should appear suddenly overhead in his red and black Desoutter with the title of his sponsoring Danish newspaper *Berlinske Tidende* displayed prominently on the fuselage sides. Hansen and his navigator/engineer Daniel Jensen had left Copenhagen at 9.00am but headwinds and a Customs stop at Lympne had delayed them. Only moments later Jack Wright and Johnny Polando in their Monocoupe approached from exactly the opposite direction, and flew round within the aerodrome perimeter at hangar height in a highly banked racing turn. It was certainly a non-standard circuit and the high speed landing was safe enough in the near darkness.

The day ended with the Clerk of the Course addressing all his officials and stressing "the importance of carrying out all duties in the right spirit whatever may be the difficulties and discomfort and of giving every competitor the utmost possible help which the Rules permit."

In the published Regulations, Sunday 14 October was the day declared for reception of Race aircraft with a cut-off time of 4.30 pm dictated by daylight. Each aircraft and crew were then to be subjected to 24 different checks by officials, all of whom were going to be very hard pressed. Alan Goodfellow was alarmed, therefore, when advised that without consultation, 24 hour extensions had been granted to five aircraft, 48 hours to six others and 72 hours to three more.

Sunday dawned cold and blustery with a northerly wind and officials were made aware that aircraft would be landing towards the hangars and enclosures. First arrival was Flight Lieutenant Geoffrey Shaw in the Eagle, thought to be one of the best equipped machines in the Race. In an effort to wring maximum publicity on behalf of his own sponsoring family business, if anybody should be in doubt, under the aircraft's fairly unimaginative name painted on the engine cowlings: '*The Spirit of W. Shaw & Co. Ltd.*' was added '*Wellington Cast Steel Foundry, Middlesborough, England*'.

Shaw was followed by Waller and Cathcart-Jones in the green Comet who just missed an aerobatic display by Jack Wright in the Monocoupe. Shortly after, the red Comet came in. Charles Scott, designated handling pilot for the Race, admitted to a BBC correspondent that before flying G-ACSS at Hatfield he had never handled a twin-engined aircraft. His landing was

Jack 'Utica' Wright and John Polando pose for the press after their arrival at Mildenhall. Wright's preparation for a Race to Australia seems to have been very limited and he had to borrow money and a set of maps.
(Flight.)

The clipped wing Monocoupe Special arrived late on Saturday and demonstrated racing turns at hangar height within the perimeter of the aerodrome. She had been landed at Hamble Aerodrome fully rigged the previous day ensuring that only minor checks were required to certificate her as airworthy.
(The Aeroplane.)

With the aid of wing-walkers and guides provided by men of the Royal Air Force and students from the Chelsea College, Flight Lieutenant Geoffrey Shaw taxies in after his arrival on a cold and blustery Sunday morning.
(Flight.)

Under the command of Owen Cathcart-Jones, green Comet G-ACSR arrived on Sunday morning, shortly after an aerobatic display by Jack Wright in the Monocoupe.
(Flight.)

Red Comet G-ACSS was flown to Mildenhall from Hatfield by Charles Scott who, by agreement, was to occupy the front seat on every sector to Australia. Grosvenor House enjoyed an escort to her hangar space provided by an RAF motorcycle outrider.
(Richard Riding.)

Cyril Davies and Clifford Hill arrived at Mildenhall in their Fairey IIIF at lunchtime on Sunday. Race preparation of the aircraft had been so hurried to meet the arrival deadline that the racing number had not been applied to the tail.
(The Aeroplane.)

ing a good landing. As the weather was brightening the Viceroy touched down, then the New Zealand Hawk, the last machine to appear that day.

Owing to the late handover by her makers, McGregor and Walker were able to undertake only one proving flight on their Miles Hawk: Woodley-Shoreham-Lympne-Mildenhall-Woodley before returning to register in Suffolk. To achieve any sort of result in the Handicap Race the crew had decided to operate at full throttle all the way, accepting the inevitable increase in fuel consumption, although Frank Halford, designer of the Gipsy Major engine, was unable to confirm an exact consumption during protracted operation at 2,400 rpm. With a maximum fuel capacity of only 55 gallons, such knowledge was all-important in calculating range and the laying-down of supplies at specific points en-route.

Upon presentation for technical scrutiny the Hawk crew was dismayed to discover that when fully loaded their aeroplane was 30lb over the limit. With no possibility of reducing the fuel load, 30lb of aircraft or crew were to be cast off. Wheel brakes were removed, heavy windscreen frames replaced with celluloid baffles, seat cushions abandoned, and a worn tail skid substituted. Four pounds were saved by a major revision of maps and folders, and the crew dispensed with overcoats and flying suits, limiting themselves to the barest minimum kit. Luggage was confined to a single set of shaving gear and a pair of toothbrushes.

Some commercial sponsorship might well have been lost by the removal of the braking system for at a stage obviously too late for change the crew of *Manawatu* was

nicely judged in contrast to the green G-ACSR which 'dropped a wing and wobbled a bit!'

Davies and Hill in their two seat Fairey IIIF G-AABY landed at lunch time, just before Lindsay Everard who flew from his private aerodrome at Ratcliffe in his DH.84 Dragon G-ACEK (described as a 'Rapide Dragon' by C G Grey) to take up duties and who brought the proper windsocks and more smoke indicators.

Melrose in his silver and black Puss Moth VH-UQO was followed during mid-afternoon by the third Comet G-ACSP, painted gloss black, a scheme which Jim Mollison believed best suited a high-speed racing aeroplane and in which he bounced and overshot twice in coarse pitch before mak-

handsomely credited for the use of Bendix Brakes in pre-Race promotional material published by their manufacturer.

From Sunday the general public was admitted to enclosures sited on either end of the two completed hangars at a cost of one shilling a head, although the Royal Aero Club charged its members five times that for the privilege of sitting together in their own reserved area between them. It was not long before the Club's Chief Marshal, Christopher Clarkson, was reporting that crowd control had already broken down and that four competitors were complaining of damage to their aircraft. Fencing for the enclosures was described as 'neither as high nor as strong as could be desired' and hundreds of spectators had trampled it down in weak spots and gained access to the hangars. An official had found it necessary to chalk a message in poor French on the doors of the West shed: 'Défense D'entres'. It was rather accusing.

For visitors there was never any assurance of seeing competitors flying their aircraft apart from an occasional test. No suggestion was ever made that the public would be allowed near the aircraft, a point not completely understood by many folk lured by the immense publicity given by the newspapers and who had undertaken long journeys to Suffolk, some walking cross-country from well outside the county boundary.

Jimmy Melrose set a fast pace as he taxied over the concrete, newly laid around the area of the technical site. He had flown to England from Australia in his DH.80A Puss Moth expecting to race home in a new DH.85 Leopard Moth on delivery, but the aircraft was not ready.
(The Aeroplane.)

Jimmy Melrose, at 21 years of age, was known as 'The Young Australian' and built up a substantial following from young women. His aircraft carried his name and address on the starboard engine cowling and the name of his sponsor, his mother, on the port side.
(Flight.)

Regarded as strong favourites to win the Speed Race, the arrival of Jim and Amy Mollison in their black Comet G-ACSP attracted a deal of attention. In contrast to many others the crew wore heavy flying suits and were equipped with seat parachutes. Preparation time at Hatfield had been so short that the aircraft arrived without her name which was signwritten on the nose during the next few days.
(The Aeroplane.)

The prohibition on casual visiting aircraft landing at Mildenhall lest they coincided with Race machines arriving or on test, prompted the business instincts of Aircraft Exchange and Mart Ltd. to rent a fifty acre field close to the site where they provided fuel and oil and an engineering service. In addition to a fixed landing fee of half a crown, the company added a passenger charge of one shilling per head, with an additional sixpence per person for transport to the appropriate enclosure. In spite of the Air Ministry Notice which seemed to give authority for use of this field by prior permission of the Control Officer at Mildenhall, Alan Goodfellow, Clerk of the Course, insisted that the site was within the banned area and, therefore, illegal.

Meanwhile, the local grocer was receiving telephone calls from pilots who wished

Following a gloomy Sunday morning the weather brightened during the afternoon when the scarlet and white Airspeed Viceroy arrived. The terms of her handicap almost guaranteed that even before the start she would not be in the running for a prize.
(The Aeroplane.)

With a full escort of Chelsea College students, Henry Walker taxies the Miles Hawk to her accommodation in the East hangar. Due to administrative errors at Woodley and late completion of the aircraft, the New Zealand crew were able to fly only one meaningful test-sortie prior to departure for Australia.
(Flight.)

Chelsea College students, aided by Scouts from the Aviation Department of the Automobile Association, position Comet G-ACSS into the West hangar and past the notice chalked on the door in an attempt at French.
(Michael Ramsden.)

to take advantage of the temporary landing ground. The Notice had carried the wrong telephone number!

Together, the two completed 'A Type' sheds were capable of accommodating all the expected aircraft by size and number and allocation of space at Mildenhall was relatively simple. Aircraft proposing to use Shell and Stanavo (Standard-Vacuum Oil Co.) products were directed to the West shed and those contracted for Pratt's Ethyl and Essolube to the East. From the earliest stages the Race organisers had made it clear that refuelling, in addition to arrangements for hangarage, crew accommodation and expenses, engineering support and repairs were the total responsibility of the entrants.

Before the Races both Shell and Stanavo had established offices in London to field enquiries and to provide information and assistance. Shell's Aviation Manager, Captain H. 'Jerry' Shaw, was despatched on an Australasian tour to organise matters for both the forthcoming Empire Air Mail route and the MacRobertson. During time in Queensland, he was intrigued to discover that the 40 mile Daly Waters to Birdum scheduled Qantas service was operated by a two seat DH.60 Moth. With great efficiency and at great expense fuel companies arranged for ample stocks of petrol and oil at all 27 nominated Checking Points along the route. The cost of fuel uplifted was to be covered by the competitors but lubricants were promised free of charge. It was believed that some American entrants had declined the invitation to race because of anticipated fuel bills in addition to all else.

Several permanent buildings on the Mildenhall Aerodrome site which had been complete for some time were declared out of bounds and the organisers were obliged to erect marquees for their own domestic arrangements. Competitors and officials soon generated an atmosphere reminiscent of an RAF mess and some good parties brewed up in the bar. The imported piano in fine tune, was augmented by Neville Stack and his ukulele, a combination which for years had been much in demand at social evenings and guest nights. The revelry

In addition to the immense effort put into organising the Races from London and Melbourne, it was the fuel companies who committed massive resources at all the stopping points. At Mildenhall, denied access to the RAF's permanent system, their bowsers were busy all week as competitors required fuel for weight analysis, leak checks, capacity calibration and test flying.
(The Aeroplane.)

Still without her racing number, Fairey IIIF G-AABY is subjected to some serious engineering investigation against a background formed by part of the extensive tented headquarters set up by the Royal Aero Club.
(Richard Riding.)

was much appreciated by many of the residents of this unreal world temporarily established in the middle of nowhere.

By racing number, a pigeon hole mail box system for competing crews was installed in the so-named sitting room, soon to receive shoals of letters and telegrams from around the world. Most of the 64 slots were to remain permanently empty, whilst other mail arriving would be 'returned to sender' or despatched elsewhere before the week was out. Into each box, to augment the Race Handbooks which carried complete instructions on how to get logbooks signed and procedures for clearing Customs at each airfield, the organisers also slipped copies of reports on changed situations along the route, official Notices and Bulletins, and there were plenty of those. Nobody could argue that crews were not being kept sufficiently well informed.

Ivor McClure's Service and Information Bureau provided a marshal to look after each competitor on an individual basis. An extraordinarily good idea conceived by the organisers required that each official carried a notebook complete with carbon paper. Details of requests from competitors or other officials were entered and initialled by all concerned. The system was to prove its worth later when facts were disputed.

Whilst Jack Jarvis and Mac McIsaacs weighed competitors' baggage, H B Howard, Clerk of the Scales, was responsible for check weighing all aircraft that could be accommodated on the 10,000lb limit Avery pads provided at Mildenhall. Six units had been procured for the week which, the sub-committee was assured, would enable 12 aircraft to be accurately checked per day but the weighing pads were far too slim, tailored for small aeroplanes, and large diameter wheels fitted with low pressure 'doughnut' tyres could only be accommodated with great difficulty. Early realisation of the impending problem had caused the sub-committee to advise each of the aircraft likely to be affected. It was suggested to KLM that the DC-2 should be weighed by Imperial Airways on arrival at Croydon on Tuesday 16 October but ten days before she was due this idea was cancelled and the DC-2, Boeing 247, Pander and Viceroy all were re-directed to Martlesham Heath near Ipswich where the permanent facilities of the Aircraft and Armament Experimental Establishment were better able to deal with aircraft of their size. Special arrangements then were necessary for the circulation of aerodrome information for, as a military facility, Martlesham Heath's details were not generally available to foreign visitors.

Following an agreed schedule, technical scrutineers checked each aircraft, entered details in the Race log books and elicited the 'L, W, P and A' constituents of the Handicap Formula. Seals were applied to engines and airframes and, where applicable, to fuel tanks. To assist the Royal Aero Club's own marshals seven officials from the Aeronautical Inspection Department were drafted in to double-check weighing reports, instruments and confirm the measurement of wing-spans, an important ingredient in the Handicap formula. Perrin was advised they would expect payment. When the KLM DC-2 was thoroughly checked, every item on board corresponded precisely with the requirements and was documented as such. It was quite a revelation to some competitors, but Parmentier shrugged off the compliments on the grounds that it was their everyday business, and merely routine.

Without Jimmy Melrose on board the Australian Puss Moth was found to be marginally over the authorised weight

Students from the Chelsea College showing some determination when adjusting the Avery scales prior to weighing Comet G-ACSS. (Gerald Lambert Collection. By Courtesy of the Suffolk Record Office. Bury St. Edmunds.)

The width of the weighing pads was insufficient to accommodate the wheels of the big twins all of which were re-directed to Martlesham Heath. (Gerald Lambert for Flight.)

optically on a rotating cylinder and the reflected light converted into a radio signal. It was reconstituted in Melbourne by the reverse procedure. Because of the distance involved and the problems of interference from conventional signals, the historic transmission was made by on/off signals, the basic concept of digital broadcasting.

The Rules required entrants to carry sufficient supplies to sustain the life of each person on board for three days. The water ration was optimised at 12 pints per person per day, but emergency food packs were left to individuals on the basis that their contents must also satisfy the scrutineers. If properly packed and sealed the emergency rations were permitted to be designated as payload. The Danes Hansen and Jensen revealed that their mandatory survival pack which originally had been defined as 'standard RAF Desert Survival' would include bottles of whisky and cognac, cigarettes and a supply of ginger nut biscuits. Cyril Davies requested that water in the radiator of his Fairey IIIF G-AABY might be counted as emergency supplies and the sub-committee agreed.

Harold Brook arrived in his Falcon on Monday 15 October having taken delivery only the previous Saturday, and introduced his passenger, Miss E M Lay who was to share expenses. On 4 September Brook had confirmed with the Race Committee that he still hoped to carry two passengers and was reminded by Harold Perrin that according to the Rules of the Handicap Race, having declared two passengers, a pilot and 400lb of additional payload would be recorded in all documentation.

The New Zealand Dragon Six ZK-ACO flew in from Hatfield painted uniquely with a tangerine fuselage and cowling stripes, silver flying surfaces and bottle green engine nacelles and interplane struts, the crew declaring it was their first flight in the type.

Lord Nuffield's entry, the AS.6 Envoy G-ACVI, Race No. 3, en-route to Martlesham Heath for weighing, was reported to have been damaged when she

limit, and all non-essential items were discarded including the carpet. A substantial fruit cake baked by the pilot's grandmother and destined for Australia, remained between the entrant and eligibility. The Clerk of the Scales confessed that his 'official heart' was melted and the Puss Moth was cleared to take-off overweight by one fruit cake. But Melrose ate half the offending confection before his departure, preserving honour all round.

A photograph of Jimmy Melrose crouching under his aeroplane in the hangar at Mildenhall, surrounded by baggage and equipment was the first ever to be transmitted from England to Australia by short-wave radio. The transmission took place on 17 October using the Amalgamated Wireless (Australasia) Ltd. radio-picturegram service. In London, the image was scanned

The crew of the New Zealand Dragon Six, ZK-ACO, James Hewett, Frank Stewart and Cyril Kay, shortly after arrival at Mildehall on Monday, photographed wearing their competitors' badges stamped with their racing number.
(Gerald Lambert for Flight.)

On 17 October 1934 this picture of Jimmy Melrose with some of the emergency equipment carried on board his DH.80A Puss Moth was the first ever to be transmitted from England to Australia by short wave radio.

Lockheed Vega G-ABGK arrived on Monday having been granted a concession due to her undercarriage problems at Heston. She was photographed during one of the periods when visitors were excluded from the hangars.
(Richard Riding.)

forced landed with an oil pressure problem at Conington Aerodrome near Peterborough. Later in the day her second pilot, Flight Lieutenant David Anderson arrived at Mildenhall in an RAF Tomtit to explain the position and request an extension but in the event the aircraft, still unserviceable on Wednesday, was withdrawn.

One of Harold Perrin's extensions to the nominated arrival time had been granted to Lockheed Vega G-ABGK, Race No. 36, which was delayed at Heston with a sticking main oleo, a persistent problem which Woods believed could be overcome by careful landings. According to Donald Bennett the determined professional pilot, Jimmy Woods treated the whole enterprise less than seriously and when he should have been attending to technical situations on the aircraft he was more likely to be found socialising in London. As a crew, Woods and Bennett was not an ideal pairing.

Although the undercarriage problems remained largely unresolved the Vega arrived at Mildenhall at last light. Jimmy Jeffs put out his portable floodlights in readiness, the units throwing a wide beam over a distance of 1,000 yards, but ignoring the lights Woods, a consummately skilful pilot, slipped the aircraft in to land alongside the hangars.

If entry to the aerodrome and enclosures was difficult enough, then passes for admission to the hangars were almost impossible to persuade from the Secretary's grasp on the rare occasions he could be found in his office. But controls were necessary. Four aircraft were reported to have been damaged by over-enthusiastic spectators so it was almost inevitable that during the week additional rules and regulations were published, and more special passes issued, which did nothing to enliven the spirits of the press corps who already found themselves working under some restriction. The RAF police guarding the main gate were so diligent that they turned away a newly posted officer they did not recognise.

On Tuesday, together with the Royal Aero Club's security guards and others acting under direction of individual entrants or companies, six uniformed police officers appeared. Each group viewed the others with suspicion but all were impotent when the keys to one of the hangars were lost and nobody could get in through the main doors until a duplicate set was found.

There was some consternation when a Comper Swift appeared, apparently unan-

Against the background of the early skeletal framework of one of the two additional A Sheds under construction, the two-tone blue Airspeed Courier taxies in on Tuesday, exhibiting a high degree of 'panting' on the fuselage side panels.
(The Aeroplane.)

Ray Parer left his sickbed to comply with the limits of his late arrival exemption and flew Fairey Fox G-ACXO from Hanworth, accompanied by a mechanic who had been working on the aircraft.
(Gerald Lambert for Flight.)

nounced, probably due to the pilot having requested landing instructions from the village grocer. The luggage locker, designed around the length and bulk of a full set of golf clubs, was opened to disclose three rubber dinghies and several tins of dope, all addressed to Race No. 58, Neville Stack's scarlet and cream Viceroy, G-ACMU. In contrast, the Stodarts' Courier G-ACJL arrived wearing the sober two-tone blue house colours of Aircraft Exchange and Mart.

Jim Mollison's arrival the previous Sunday afternoon had frightened some spectators when Comet G-ACSP had indulged in much needed landing practice. During an intended overshoot with both propellers in coarse pitch, the machine had barely cleared the far hedge and had staggered into the air in a semi-stalled condition. On Tuesday afternoon Mollison flew two further circuits, dropping a wing on the first approach from which he skilfully recovered, and bouncing in from about six feet on the second. There had been such little time for preparation at Hatfield that the name *Black Magic* had to be signwritten in gold lettering on the long nose during the few hours of grace G-ACSP spent in the West hangar at Mildenhall. The name was the Mollison's own choice and had no connection, as seemed likely at the time, with the popular line of 'dark chocolate' introduced by the Rowntree company in 1933.

Due to the late start on the programme of modifications to their Fairey Fox G-ACXO, Ray Parer had requested extra time prior to presentation at Mildenhall and the sub-committee had granted him 48 hours' grace. When Parer was suddenly struck down with malaria and confined to a hospital bed, Geoff Hemsworth requested an additional concession but this was immediately refused. Parer's answer was to leave his sickbed and fly the Fox to Mildenhall, accompanied by a mechanic from Hanworth, landing very late on Tuesday where he instantly felt much better and was later joined by his co-pilot.

On Tuesday evening the Royal Aero Club hosted a Competitors' Dinner in their marquee which was described as being 'not very much of a success owing to the wintry conditions under canvas'. Everyone was anxious to get away early to somewhere offering warmth.

In strong north easterly wind conditions *The Irish Swoop*, Bellanca Monoplane EI-AAZ, Race No. 29, touched down on Wednesday afternoon to be met by a team of six mechanics who raised the tricolour of the Irish Free State over the camp. She followed the DC-2, Boeing 247 and Pander all arriving from Martlesham Heath. The Pander originally had been scheduled into Martlesham on Saturday 13 October but the late delivery of new tyres ordered from the USA delayed her arrival for Customs clearance at Croydon until mid-day on Tuesday. The crew was reminded to ensure their refuelling contractor could provide sufficient petrol at Martlesham Heath to see the Pander at maximum declared weight on the scales. That having been achieved, the aircraft was late in arriving at Mildenhall due to the slow rate of draining fuel down to an acceptable level.

The Irish Swoop had arrived without its racing standard cowling. At Croydon, Rollason Aircraft Services had been unable to rectify a structural problem which had manifested itself and a Pratt and Whitney technician later was flown to Mildenhall from London in a Puss Moth to offer assistance.

By evening there was concern for the

REALITY AND DISAPPOINTMENT

Smith-Cochran Granville, *Q.E.D.* Mildenhall had been notified of the aircraft's departure from Heston at 4.30pm, but long after its ETA there was still no sign of it. Beyond last light the aircraft finally arrived, stalled from an estimated 50ft onto a brilliantly floodlit patch of grass and taxied in showing no signs of distress or damage. There was no immediate explanation for the late arrival, and conscious that Australia was a great deal further from Mildenhall than their present distance from Heston, Wesley Smith was asked with some caution whether he had been lost.

It was almost certain that a pilot of Smith's experience could not have been off course during the relatively short sector from Heston, even in a machine from which the visibility was considered to be little better than appalling. Smith disclosed that after take-off, Miss Cochran asked whether photographers might greet them at Mildenhall. Assured that they would, Smith was ordered to stay airborne until after dusk, in the hope that no photographs would be taken. The lady, it appeared, did not wish to be pictured wearing that day's outfit.

One pressman described the scene:

"As we all surrounded her machine waiting impatiently for her to dismount, she showed she was in no hurry. First, she fiddled about in her cockpit, then she doffed her flying helmet, combed her hair, powdered her face and generally tidied up. After some minutes of this she descended. As soon as we closed in on her for a few words she drew a typed document from her bag with a 'here is what my attorney says I may say.' She brandished it for all to read and stalked off."

The crew of the Pander Postjager who were already familiar with the Race route as far as the Dutch East Indies: Radio Officer Pronk, more usually employed by Radio Holland, Gerritt Geysendorffer and Dirk 'Dick' Asjes.
(Gerald Lambert for Flight.)

The Pander arrived with the other twins in the rain on Wednesday, after being weighed at Martlesham Heath, and immediately became the centre of interest for the patient members of the press corps and newsreel crews anxious to record every detail.
(Gerald Lambert for Flight.)

The Irish Swoop flew up from Southampton in the rain on Wednesday afternoon to join the party which had developed around the hospitality centre erected by her sponsors.
(The Aeroplane.)

The day after their arrival, Wesley Smith and Jackie Cochran posed for photographs by their Gee Bee. Miss Cochran was anxious to explain that the couple's swift exit the previous evening was due to a belief that the aircraft had been damaged on landing and she wished to divert press attention.

Work scheduled to have been completed on the Gee Bee during the voyage from New York to Southampton remained unfinished as members of the engineering party were struck down by severe sea-sickness. She was the only Race aircraft not test flown between arrival and the start of the Race on Saturday morning.
(Gerald Lambert for Flight.)

"There was tremendous publicity surrounding the contestants. I had never experienced anything like it. It was a waste of time and I did not like the scene there in London one bit!"

She explained that her rapid departure from the aircraft in the gathering gloom following her initial reluctance to disembark was because she and Smith were convinced that the heavy landing had probably broken a wing, and she did not want the newsmen to have the opportunity of looking too closely. Miss Cochran added that she soon discovered that Gee Bees always tended to land in the same violent manner and because the undercarriage legs were so close together, it was almost impossible to keep the aircraft straight on the runway. Later she expressed further views:

"Gee Bee stands for Granville Brothers," she said, "a Springfield, Massachusetts, airplane company which made fast, unstable, dangerous planes. The nearly cute name is a sham. They were killers. There were few pilots who flew Gee Bees and then lived to talk about it. Jimmy Doolittle was one. I was another."

Whether Smith was content with his aeroplane or there were other less public reasons, the Granville alone of all the gathered competitors made no further attempts to fly prior to the start of the Race the following Saturday. In later years Jackie Cochran admitted that the ride from Heston to Mildenhall had been 'unsatisfactory' as no second seat had yet been installed and she had been forced to sit on what she described as a 'cracker box', a biscuit tin, in the rear cockpit. She described the aeroplane not only as the fastest in the Race but also the coldest and noisiest. With no form of intercom, communication be-

The words on the card read: "I have no comment to make. Please consult my lawyers."

Not primarily interested in the day's fashions or the state of the aeroplane, the press had been following a story line concerning the squabble between Jackie Cochran and Clyde Pangborn in which Pangborn was alleged to be demanding money for the transfer of the Gee Bee entry after he decided to crew with Roscoe Turner on the Boeing 247D. Jackie Cochran refused to pay and suggested that if she had been a man, the matter would have been settled behind the hangar.

Immediately after arrival Miss Cochran, described as wearing 'a parachute over a tweed lounge suit', was spirited away to a secret address which even the best endeavours of the press failed to discover but was believed to be in London. The intrigue did nothing to dispel their self-generated myth concerning this 'mysterious American woman entrant' which Jackie Cochran despised:

REALITY AND DISAPPOINTMENT

tween the two crew was always difficult and resulted in them having to pass written notes as best they could.

Both *Black Magic* and *Grosvenor House* flew briefly on Wednesday and the following day Jim Mollison made four almost faultless landings and spent a lot of time taxying around the aerodrome. Amy Mollison was in awe of the aeroplane and was the only Comet pilot not to have received any dual time at Hatfield. Jim had forbidden her to fly solo, but against his wishes and after a public argument, she made preparations to fly circuits under the gaze of the world's press. Jim had been tempted to increase the insurance on the machine in which he and his wife shared the financial risk, a fact of which Jim was reminded by Amy's father, Will Johnson, but he declined as he considered it 'unnecessary'.

Amy's first landing was good but on the second something was seen to become detached in the area of the undercarriage. Unaware of the problem she taxied back for another take-off but a car was sent out to stop her and investigation revealed that a fairing had broken loose, a minor snag that was soon rectified by de Havilland's resident working party.

Less happy was the experience of Cathcart-Jones who landed the green Comet G-ACSR at dusk on Thursday with the undercarriage in an intermediate position, having deliberately raised it by four turns of the big handwheel during circuit

DH.88 Comet G-ACSP, still without her signwritten name 'Black Magic' being assisted to swing into wind for a training flight on Thursday with Jim Mollison in the front seat. Later in the day his wife Amy flew her first solo circuits in the aeroplane, the only one of the six Comet pilots in the Race not to have flown the type at Hatfield.
(BAE Systems.)

Wednesday was the only really wet day at Mildenhall during the week spent on final preparations. Green Comet G-ACSR received attention to her tail skid when parked on the new compass base whilst her suitably dressed crew surveyed the scene.

Refuelling the front tank of G-ACSR on Thursday for a training flight by Owen Cathcart-Jones which ended in near disaster. Only a phenomenal effort by de Havilland employees and friends in the British and French aircraft industries allowed repairs to be completed in time for the start.
(Gerald Lambert for Flight.)

Fairey Fox G-ACXX was the last of the competitors to arrive and touched down late on Thursday evening, delayed due to continuing technical problems and fog in southern England.
(The Gerald Lambert Collection. Reproduced by kind permission of the Suffolk Record Office, Bury St. Edmunds.)

When Harold Gilman and James Baines checked in with their Fairey Fox they were advised that due to their late arrival it would be impossible to inspect their aeroplane in accord with the provisions of the Handicap Race, but they could still take part in the Speed event. On the understanding that they had absolutely no chance of winning a prize, the crew elected to race-on for the sheer adventure.
(Gerald Lambert for Flight.)

flying to facilitate engine cooling. About 200 yards after touchdown both undercarriage frames promptly folded back and the aircraft slithered along on her belly accompanied by a trail of sparks, suffering damage to the radius rods, engine cowlings and both propellers, but little else.

Normally, a warning light in the Comet's cockpit would have notified the pilot that the undercarriage was not fully extended via a system that was connected to the throttles, operating when the levers were closed. Due to the nature of the circuit work being undertaken the aircraft was 'rumbling' on approach, using bursts of power, a technique which effectively and unwittingly overrode the protection of the warning system.

It was lucky there was no fire as the only fire engine on the airfield expired half way to the site of the accident but a reserve RAF tender from Bircham Newton had already been promised for stand-by duties on Friday and Saturday. de Havilland test pilot Hubert Broad flew back to Hatfield in the company's DH.84 Dragon G-ACPX, positioned for communications duties, and quickly returned with replacement airframe parts, complete undercarriage assemblies and the only spare propeller blade.

On the previous Monday Broad had been refused permission to land the Dragon at Mildenhall and put down at Newmarket Racecourse where a landing fee was demanded. Much to his embarrassment, Broad discovered he had no money with him and had to borrow the coins to make a telephone call to Hatfield, explain his predicament, and get someone in authority there to seek official approval for him to continue on to Mildenhall.

The last of the entrants destined to be part of this historic pageant were Baines and Gilman in the Fairey Fox G-ACXX, Race No. 62, much delayed by technical problems which had been further aggravated by fog in southern England. It was ironic that they should have been last. Once Mildenhall had been announced as the starting aerodrome, James Baines had made frequent practice approaches to land at the site in rented Club aeroplanes. Now he had arrived in near darkness to discover a disabled aircraft lying in the middle of the airfield marked by a surrounding circle of cars each with its headlights switched on.

When the Fox crew checked in they were advised that due to their late arrival it would not be possible fully to inspect the aeroplane in accord with the provisions of the Handicap Race, but they would be allowed to compete in the Speed Race. Having come so far the two friends could do nothing but accept their new situation in the knowledge that they now stood absolutely no chance of winning a prize.

REALITY AND DISAPPOINTMENT

A busy scene at Mildenhall as final preparations were completed. The picture distills an atmosphere of expectancy and anticipation from the man on his bicycle to the animated policeman, the radio commentators to the de Havilland Company's communications Dragon and the enclosure packed with cars and spectators. Gerald Lambert for Flight.

Utilising the Handicap formula originated in Melbourne, Fred Roworth and Wilfred Dancy, official Handicappers to the Royal Aero Club, had produced 'V', the all-important 'airspeed' element of the equation. Division of the route distance by the calculated airspeed resulted in a Handicap flight time in hours, minutes and seconds appropriate to each aircraft declared for the Handicap Race. The results were published on the eve of departure:

	Entrant	Speed according to Formula (mph)	Time allowance over scratch	Flying time (12,314 miles) according to Formula
7	Desoutter Mk II	113.18	42.02.24	108.48.00
16	DH80A Puss Moth	114.30	40.58.12	107.43.48
47	Klemm Eagle	115.44	39.54.36	106.40.12
31	Miles Falcon	119.29	36.27.36	103.13.12
2	Miles Hawk	120.57	35.22.12	102.07.48
60	DH89 Dragon Rapide	140.08	21.09.00	87.54.36
14	Airspeed Courier AS.5	140.54	20.51.36	87.37.12
35	Fairey Fox	143.43	19.05.24	85.51.00
15	Fairey IIIF	146.68	17.10.48	83.56.24
33	Lambert Monocoupe	154.02	13.11.24	79.57.00
44	Douglas DC-2	168.03	6.31.12	73.16.48
36	Lockheed Vega	177.71	2.31.48	69.17.24
19	DH88 Comet	182.81	0.36.00	67.21.36
34	DH88 Comet	182.84	0.35.24	67.21.00
63	DH88 Comet	182.83	0.35.24	67.21.00
58	Airspeed Viceroy AS.8	184.44	0.00.00	66.45.36

8

Final Adjustments

Pilot Roy Tuckett and cameraman John Chapman flew from London to Koepang in February 1934 in a specially modified DH.80A Puss Moth, G-ABPF, to film part of the Race route. The legend carried on the aircraft's engine cowlings simply reads: 'London to Melbourne Air Race. Film Expedition'. Supported by the Vacuum Oil Company the result of their combined endeavours was screened in the Old Town Hall, Mildenhall, on Thursday 18 October, 'for one night only'.
(The Aeroplane.)

ON THURSDAY EVENING many of the MacRobertson Race crews went to the cinema. In February, Roy Tuckett and John Chapman had flown from London to Koepang in the specially prepared DH.80A Puss Moth G-ABPF and had secured film coverage of many of the principal features of the route. During their return flight to London they sold the aeroplane in Singapore, but time was not a factor. Through the good offices of the Vacuum Oil Company who had supported the expedition, crews who had never before ventured towards Australia were able to view in comfort something of what lay ahead.

At Mildenhall in 1934 films were screened in the Old Town Hall in Andrews Street and on 18 October, 'for one night only', at 5.00pm and 8.00pm, John Chapman's one hour film was shown together with the crime thriller '*Channel Crossing*' starring Matheson Lang. Although primarily the film was for the benefit of the competitors the public was allowed admission (seats at 1/6, 1/3 and 1/-) as invited by the posters which were displayed around the district:

"At the request of the Royal Aero Club, a film of the entire route of the Air Race from Mildenhall to Melbourne will be shown for the instruction and guidance of Competitors in the Race and to show beforehand the difficulties, dangers and air conditions which may be expected to be encountered whilst flying over the various oceans, mountains, deserts etc. en-route.

By the courtesy of the Royal Aero Club we are permitted to admit the public to such accommodation as will not be required by the competing pilots and staff. This is strictly limited, but seats may be reserved at the Town Hall during Wednesday and Thursday."

Confirmation was finally received that the Blériot III/6, Race No. 54, reportedly delayed through weather, had suffered a broken undercarriage bracing link at Villacoublay the previous Sunday whilst in the very act of taxying prior to take-off for Mildenhall. The story was covered in *The Aeroplane* in the unique style of C G Grey:

"We were all sorry when we heard that the Blériot entry has broken its undercart. It is such an old friend. First of all it appeared at the Berlin Show with a fixed undercarriage about four or five years ago, then it turned up at the Paris Show two years ago; that time we seem to remember that it had a retractable undercarriage of sorts. And now the poor thing seems to have no undercarriage at all?"

Nothing had been heard from either of the Italian entrants, the Savoia Marchetti SM.79P and Caproni P.L.3, Race numbers 51 and 61, and no extension had been requested. The assumption that neither would appear was reinforced by a statement published by the Italian Government alleging British competitors held an advantage as most of the compulsory stopping points were in British territory.

Meanwhile, nobody could offer anything but praise for the engineers from the de Havilland Service Department. Rounded up from their homes near Hatfield, they loaded two hire cars with spares and tools and left for Suffolk at 9.00pm on Thursday. Working continuously for 21 hours on G-ACSR, they fitted the new undercarriage components and repaired torn cowlings. Fortunately the engines themselves appeared to be undamaged but were subjected to a series of comprehensive checks.

It was fairly common knowledge that de Havilland could raise only the one spare Ratier propeller blade, and this already had been collected from Hatfield by Hubert

At the end of a training flight on Thursday evening, Owen Cathcart-Jones landed Comet G-ACSR with the undercarriage partially retracted. A working party assembled by the de Havilland Aircraft Company worked continuously for 21 hours on site in addition to the progress being made elsewhere.
Alan Butler.

FINAL ADJUSTMENTS

To repair G-ACSR it was necessary to fit new undercarriage frames and to repair torn cowlings. The structure of the airframe and the engines were undamaged but, most seriously, the propellers were bent.
(Gerald Lambert for Flight.)

During retraction tests on Friday afternoon, Mildenhall received a visit from the Secretary of State for Air, Lord Londonderry, who was hosted on a tour of the site by Lindsay Everard. His visit might have been a fact finding mission on the state of the aerodrome as much as anything as it was Lord Londonderry who pursued the policy of increasing Great Britain's bomber force during his four year period of office from 1931.
(The Aeroplane.)

Broad. At Mildenhall Percy Bryan from the Hatfield Drawing Office raised an instant repair scheme and thanks to the good offices of Captain A G Lamplugh of British Insurance, on site as a Royal Aero Club steward and enrolled to act as a gate attendant, the less badly damaged blade was sent to the Fairey Aviation Company works at Heath Row Aerodrome, Harmondsworth, a commercial competitor in the metal airscrew market, where as a gesture of sportsmanship it was quickly straightened. The Fairey company did not hold the necessary approvals to sign off the work and the repaired propeller could not be released for service. It seemed likely that the whole of the rescue exercise might founder until de Havilland's Chief Inspector, Harry Povey, asked his Fairey counterpart if he would be prepared to accept the signature of Ratier's Chief Engineer? Yes, he would! The enthusiastic M. Dreptin was still in England having taught the Hatfield engineers the intricacies of setting the pitch changing mechanism and, following confirmation of the arrangements on Thursday evening he reported to Heath Row at seven o'clock the next morning. In association with his hosts he was able thoroughly to inspect and approve the repair and sign the necessary release documentation. The blade was rushed back to Suffolk and Cathcart-Jones was airborne at 4.45pm that day testing the aeroplane in preparation for the next morning's departure.

The de Havilland Company's top people found accommodation at The Angel in Bury St. Edmunds where they insisted M. Dreptin should join them and where some of the wives attempted to teach him the basics of English. Staying in a town with such a name was reported to have caused some initial difficulties.

The Service Van from the Dagenite Company was kept well employed at Mildenhall too. Technical representatives, checking on the status of their installed equipment, discovered amongst other things that switches in some aircraft had been left live overnight, causing batteries to run down. In some cases during Race preparation, new batteries had been installed which were found not to have been charged first.

The jets in the carburettors of each of the Comets' Gipsy Six engines were changed as Eric Moult, Frank Halford and Eric Mitchell compared race distances against results obtained from the minimal in-flight consumption trials which had been possible in the available time. They judged that without the change none of the Comets could fly non-stop to Baghdad. Engines were run at night to judge mixture from the condition of the exhaust flames, then the jets were changed again. On each training and familiarisation flight the crews were asked to note oil and cylinder head temperatures and a special watch was maintained on oil consumption. On one occasion a pool of spilled petrol caught fire beneath a Comet on start-up and Richard Clarkson, observer to Hubert Broad on all the Comet's early test flights, beat at the gathering flames with his notebook whilst others joined in with their caps and jackets.

The three Comets had been weighed un-

Having completed the repairs to G-ACSR and reassembled all that was necessary, final adjustments to engines and the newly installed propellers were completed in time for engine runs and a test flight late on Friday. Engine designer Frank Halford (on the right, less hat) was on hand to offer his advice and assistance if required by the thoroughly capable de Havilland team.
(The Aeroplane.)

Roscoe Turner inside the fuselage of the Boeing 247 which had been fitted with six additional fuel tanks. To comply with weight limitations it was necessary for a 195 gallon tank to be sealed at Mildenhall which imposed restrictions on the sector lengths that could be flown.
(Boeing via Carl Cleveland.)

der supervision at Hatfield with all their tanks filled with petrol. The same machines refuelled at Mildenhall all weighed differently, the green Comet G-ACSR being some 100lb over the limit. Bob Loader from the de Havilland Publicity Department was so frustrated that he seized one of the portable scales and weighed himself, protesting that the indicated weight was 20lb in error. Discussions with Anglo-American Oil, the contracted fuel supplier, resulted immediately in the preparation and delivery by road of aviation spirit with a lower specific gravity, which partly solved the problem. It was soon realised however, that most of the overload had been caused by the late installation of larger capacity oil tanks and metal covers over the magnetos. Harold Young, Resident Technical Officer at Hatfield, contacted the Air Ministry and arranged for an agreed increase of 50lb in the all-up weight.

Before the Race Campbell Black ('CB' to his friends) had proposed marriage to the noted actress Florence Desmond, who had perfected a broad Yorkshire accent which she used in a sketch on stage to parody Amy Johnson, now Mrs Mollison. 'Dessie' had promised to give 'CB' an answer to his proposal after the Race which appeared rather insensitive when Campbell Black was already subject to emotional strain.

Scott and 'CB' worked through long sessions of propeller hand-swinging practice, developing the routine which could be all-important along the route, away from the expert advice readily available in England. They flew well as a team from the beginning, a fact noted by Jim Mollison who disliked them both but the feeling was mutual. Scott and Mollison were old rivals on the record breaking scene and Scott had, according to Amy, snubbed her after her arrival in Australia in 1930. One incident noted at Mildenhall was when a particle of dust or grit got into CB's eye and he was looking for assistance. Jim Mollison is alleged to have taken a clean white handkerchief out of his pocket, blown his nose on it with some flamboyance, and handed it to Campbell Black who promptly dropped it on the ground and walked away.

Apart from the odd personal rivalry there was no mass outbreak of hostility which the press might have been anticipating and hoping to exploit. The atmosphere was rather the opposite with competitors helping one another. The Clerk of the Course discovered that amongst the aircraft there was a woeful lack of emergency smoke flares but the limited supplies had been 'passed round' to ensure that at the time of its official inspection any specific aircraft was well supplied in accordance with the Rules.

During official scrutiny the Boeing 247, Race No. 5, was found to have a large '57' painted on the front fuselage. Although Race No. 57, Louise Thaden's Beech A17 Staggerwing had been withdrawn, the matter was taken up with Roscoe Turner who pleaded for the markings to be retained on the grounds that "Mr Heinz has paid a lot of money for them to be there!" In fact, the H J Heinz food company defaulted on what Turner thought was a firm agreement and paid nothing.

With maximum fuel and stowage of a week's supply of emergency rations including Thermos bottles filled with water, the Boeing was almost at the gross weight limit authorised under the Rules before the crew was added. Turner's observation was characteristically wry:

"Our ship is fine and the weight is OK. She's absolutely grand and perfectly ready to fly the trip, but she'll have to fly without us!"

At first nobody could explain the apparent difference between theoretical and actual weights until it was decided that calculations made at American fuel weights would show a result lighter than at British weights due to differences in the refined spirit. The gap was estimated to be about 0.2lb per Imperial gallon with temperature differences to be considered too. It was a practical reminder that the business of fuelling large transport aeroplanes in different parts of the world was more critical than most people ever realised.

To comply with the Rules, Turner had to accept that one of the 195 gallon long-range fuselage tanks was to be sealed closed. He was dismayed as he declared later to a newspaper reporter on return to his home town:

"This meant that instead of cutting a straight course to Australia via the five Control Points we had to go zigzagging down to the other side of the world in 1,000 to 1,500 mile hops. Our maps were useless. We had to arrange stops at additional points along the route!"

When the Viceroy was inspected it too was well over the weight limit and the only prospect was to reduce fuel loads very considerably limiting maximum range to about 1,000 miles, a hopeless situation for a Speed Race entrant. To make matters even worse for Stack and his crew, when the handicaps were announced on Friday the Viceroy was declared the scratch entry, although several other types were known to be faster, and the Handicap Formula was again the subject for scrutiny and criticism.

In view of the problems foreseen with the possibly inadvertent overloading of fuel after departure from the closely monitored situation prevailing at Mildenhall, the Race Stewards issued a late note to all competitors reminding them of their en-route responsibilities:

1. Fuel and oil will be checked at each Control Point after refuelling tanks unless all tanks can be filled to capacity without exceeding the permissible weight.

2. It will be in the interests of all competitors to provide suitable means for checking the contents of all their tanks.

3. It will accelerate the checking at the Control Points if competitors limit their tank capacity so that the maximum permissible weight is not exceeded when the tanks are full.

Jack Wright's Monocoupe *Baby Ruth,* one of the smallest machines in the Race, was also obliged to carry less than maximum fuel to stay within the limitations of her American Certificate of Airworthiness.

Under the mathematics of the Handicap formula an additional fuel tank in the cabin did not sit well with a very small wing area.

It was noted that when Davies and Hill in their Fairey IIIF left North Weald for Mildenhall the aircraft's gun firing mechanism was still attached to the control column and no compass was fitted although the extra tankage had been installed. Now Davies was advised that due to all-up weight limitations he could not use the long range tank, reducing the maximum range to 600 miles and leaving the crew to reflect on the fact that the Timor Sea crossing alone was 520 miles. Davies believed his figures had been established by the RAF Research Department at Farnborough, and accused Club officials of creating a 'danger factor' rather than the sought-after 'safety factor' which was at the forefront of the Race ethos.

Fuel was of concern to the New Zealand team in their Dragon Six too, but for different reasons altogether. Their aircraft had been modified to carry extra petrol by the installation of three additional tanks slung from the cabin roof, and the only means of reaching the single seat in the cockpit was by wriggling along the floor, through a shallow alleyway underneath the suspended containers.

Jim and Amy Mollison may have offered some advice as their transatlantic flight the previous year had been in a DH.84 Dragon rigged in an identical fashion. The aircraft had crashed in Connecticut almost within sight of their New York destination, but there had been no fire for the crew to worry about as all the tanks were empty.

On 9 October 1934 when King Alexander of Yugoslavia was starting a State Visit to France he was shot by a lone gunman who jumped onto his open top car. The assassination was captured on film by a Fox-Movietone crew and edited into a six minute newsreel. The film was banned from being shown in France by the government, quickly followed by those of Holland, Germany and Hungary. Before the start of the Race there were rumours that the Mollisons had either asked for or had been offered £30,000 to fly a copy of the film across the Atlantic to New York, foregoing the Race to Australia. As a result Amy Mollison received an uncomplimentary letter in her mailbox at Mildenhall. Embellished with Jim's colourful comments it was posted on the pilots' notice board until it was removed by order of the Race sub-committee.

Volumetric space as much as weight was at a premium in all the smaller aircraft, especially for entrants such as Melrose and Hansen who were obliged to stow all mandatory equipment alongside and around themselves within a severely limiting cabin. The philosophy adapted by Cathcart-Jones and Waller was that all items which would not be required in the air were screwed to the cockpit walls or stowed inside the long nose. They admitted that their emergency stores also were packed inside the front fuselage where they were beyond the reach of the crew and in direct line for demolition in the event of a serious accident.

Large scale maps, cut, pasted and folded, were carried in addition to the route strip maps, and provided the safeguard of wider coverage. En-route refreshment was limited to flasks of black coffee, barley sugar and malted milk tablets. The Comet's crew dressed in fur-lined flying suits, gloves and boots, woollen scarves and caps. Discomfort endured on the ground during transit, particularly from the Middle East onwards, they declared, was more than compensated for by the protection enjoyed at their high cruising levels, where frost was certain to form on the inside of the canopy.

Those who examined the Danish Desoutter were astonished to see that the only Race modification appeared to be installation of an extra 25 gallon fuel tank in the cabin which left engineer/navigator Daniel Jensen with practically no space to sit, let alone work. In addition to the tank whose capacity effectively doubled the standard range, Jensen was surrounded by tins of oil, a spare Heine propeller, promotional material and by comparison with others, masses of baggage.

Richard Fairey was not impressed when he inspected his company's two civilianised Foxes and was sceptical that either would ever complete the course. There were constant problems with the radiator on Parer's aeroplane which caused a deal of anguish for the Chelsea College students throughout their stay. On Thursday evening it was discovered that the radiator had corroded through from the inside and was now leaking. With great urgency a replacement was located and fitted but next day it too had sprung a leak and the students worked all night to get it serviceable by 6.30am.

The keenness, application and good humour of the students was evident to many and drew favourable comment from those in a position to appreciate what they were achieving. And they were happy to take on any chore. One of Harold Llewellyn's first tasks was to put up coat hooks in the competitors' cloakrooms. Unhappy with the untidy and useless rope barriers laid out on the floors of the hangars, the students painted white lines instead, said to be equal to anything seen at Wimbledon. 'No Smoking' signs were made and hung from the hangar roof beams some 30ft from the ground, achieved without the assistance of ladders. After helping to weigh and tag aircraft and loads, Llewellyn and his fellow students were called upon to help change the wheels on the DC-2, taking the brake drums out of the old wheels and re-fitting them in the new.

"The Flight Engineer, Bouwe Prins, supervised our work and I well remember his great concern that it was correctly done, and his attention to the locking wire of the set-screws," Llewellyn wrote in his diary. Later, while changing an inner tube on Melrose's Puss Moth, Llewellyn became aware of an audience, and looking up was greeted by Their Majesties King George V and Queen Mary.

Friday 19 October was a day of good weather which saw many competitors airborne on final check flights. The Comets remained in their hangar where undercarriages were still being modified or repaired although just before darkness fell, G-ACSR was wheeled out for a short but satisfactory circuit, joined briefly by Jimmy Woods in the Vega. Being manhandled in the opposite direction was the Falcon which had lost her engine in the middle of the airfield, an embarrassment which, according to helpers, caused Harold Brook to lose his temper.

The day was punctuated with support aircraft movements, including a Chance-

The KLM DC-2 which boldly displayed her racing number and the identity of her operator under her nose and belly. The brake drums were changed on Wednesday night, the engineers working with the aid of floodlights as the hangar lighting circuits were still waiting to be connected.
(Mildenhall Museum.)

The New Guinea entry, Fairey Fox G-ACXO, was fitted with a new radiator on Thursday and engine runs were completed alongside the Royal Aero Club's tented headquarters with a member of the ground crew holding down the tail. By Friday morning the new unit was leaking too and the engineers worked all night to have the aircraft ready for the start on Saturday morning.
(Richard Riding.)

Amongst a miscellany of aircraft bringing 'official visitors' on Friday, causing many frustrating diversions, was a Royal Air Force Armstrong Whitworth Atlas.
(Ted Vaisey.)

FINAL ADJUSTMENTS

Following the renewal of her brake drums, the KLM DC-2 was test flown on Friday and was photographed as she taxied past the Hawker Hart, K3074, of No. 24 (Communications) Squadron which had carried the Secretary of State for Air, Lord Londonderry, from Hendon. (Ted Vaisey.)

A major team effort was required when refuelling some of the Race entrants and it was often essential that a member of the crew or an engineer was on hand to identify the location of tank caps and confirmation of capacities. Lockheed Vega G-ABGK was test flown to check on the state of her landing gear after the problems encountered at Heston. (Gerald Lambert for Flight.)

Vought flown by the American Naval Air Attaché, an RAF Atlas, a civilian Stinson and the KLM Fokker VIII *'Snip'* which arrived from Amsterdam carrying both its designer and Freiherr von Bismarck, German Ambassador, sundry spares, the passengers for Australia and the special airmail solicited for carriage in the DC-2. The company's Chairman, Dr Albert Plesman, founder of KLM, was there to receive the 22,000 letters which were to be franked with a special hand-stamp in the East Indies or Australia. Income from their transportation was expected to equal the DC-2's total fuel bill, if not to exceed it.

One of the passengers, Roelof Domenie, had travelled by ship and train from Brazil to The Hague where on 12 October he had bought a return ticket from London to Melbourne at a cost of 4,500 Dutch guilders.

No official General Post Office airmail was contracted for carriage on any of the British entries, and none of the competitors was allowed to solicit mail against financial reward due to the GPO monopoly. British philatelists joined others from across the world in accepting KLM's offer to carry individually produced covers on board the DC-2.

A system of envelopes within envelopes ensured that covers posted to an address in Holland would be delivered to Mildenhall, carried as official Dutch airmail to Australia, and returned to the senders as soon as practicable. KLM also published its own official cover. Postmarked 10 October 1934 in Holland, the plan was to backstamp each in Melbourne and for 50 covers to be forwarded to New Zealand, but by sea.

In view of the lack of any GPO initiative, Major Alan Goodfellow, Clerk of the Course and a keen aerophilatelist himself, approached the sub-committee to seek approval for a scheme of his own. He originated a batch of envelopes which were flown by Railway Air Services from Manchester to London and then taken by road to Mildenhall, where the special Post Office set up temporarily on the aerodrome kindly back-stamped each one as proof of receipt. On the reverse side of the envelopes, Goodfellow wrote the address of his aunt, Lady M E Spencer, who lived at South Yarra, near Melbourne, and affixing the necessary penny-halfpenny airmail postage stamp, slipped two covers into each of the Race logbooks. It was hoped the receiving officials in Melbourne would see what was required and post them on to Lady Spencer, when each was to be franked with an historic date-stamp.

Many other entrants carried 'unofficial' postal covers and in view of the GPO's monopoly all faced potential prosecution. Several had covers hand-stamped in the village of Melbourn, near Cambridge, with the intention of adding Melbourne's postmark at the conclusion of a successful journey. The crew of the New Zealand Dragon Six ZK-ACO carried 100 covers, each with the correct airmail rate affixed. The covers were posted in Mildenhall on 19 October and addressed to 'Flying Officer C E Kay, Mildenhall Aerodrome, Post Office' from where they were collected and loaded on board their transport.

Turner and Pangborn carried 1,000 covers in the Boeing 247, posted in London W1 on 11 October and addressed to 10 Bury Street, St James'. Once in Australia the covers were to be delivered to Colonel Roscoe Turner, care of the Centenary Council, Melbourne. Waller and Cathcart-Jones carried mail franked in Cambridge on 19 October and Melrose's envelopes were cancelled in

Royston on the same day but no private mail was carried on either the Dutch Pander or by Ray Parer and Geoff Hemsworth or, surprisingly, by the Mollisons.

On Friday, Lord Londonderry, Secretary of State for Air, arrived in a Hawker Hart of 24 Squadron, escorted from Hendon by another Hart from 601 Squadron, Auxiliary Air Force, carrying Air Marshal Sir Robert Brooke-Popham, AOC Fighting Area. Lord Londonderry was met by Mr Lindsay Everard MP, Chairman of the Royal Aero Club's Organising sub-committee and one of the country's most progressive private owners, and escorted around the site in the company of Race Steward Lieutenant Colonel Francis Shelmerdine, otherwise Director of Civil Aviation. The officials coped with the interruption to an already frenetic schedule but it was only a prelude. Then in a moment of embarrassment, one of the Harts was damaged as it was being wheeled into a hangar. Repairs were effected overnight and she was declared serviceable again before dawn.

Confirmation was received that HRH The Prince of Wales would be arriving just after 2.00pm in his DH.84 Dragon G-ACGG, the aircraft painted in the distinctive red and blue colours of the Guard's Flying Club, then a telephone call from Newmarket announced that the King and Queen, who had just left the racecourse, were intending to visit Mildenhall which was on their route home to Sandringham and would be arriving in 15 minutes. Senior

HRH The Prince of Wales owned de Havilland Gipsy Moth, Puss Moth and Fox Moth aeroplanes before graduating onto his first twin, DH.84 Dragon, G-ACGG. He arrived in the aeroplane at Mildenhall on Friday afternoon for a pre-planned visit.

Whilst some of the hangar erection gangs employed by Redpath Brown took a break from their duties, remaining on their perilous perches, Charles Scott explained the pitch control mechanism for the Ratier propeller system fitted to Comet G-ACSS, to HRH The Prince of Wales. (Gerald Lambert for Flight.)

Their Royal Highnesses King George V and Queen Mary arrived at Mildenhall on a practically no-notice visit on Friday afternoon. Lindsay Everard hosted HM The King who disliked aeroplanes, and Harold Perrin is pictured with HM The Queen. Jim and Amy Mollison are standing alongside Comet G-ACSP which has now been signwritten with her name 'Black Magic'.
(The Aeroplane.)

HM The King expressed concern for the welfare of 'the young Australian' James Melrose and what appeared to be a small and frail flying machine. At 21 years of age Jimmy Melrose was already a seasoned aviator with practical experience of long-distance flying and was planning to fly the route only by day in a series of appropriately modest sectors.
(Mildenhall Museum.)

After the departure of the Royal party on Friday afternoon, Jack Wright took off in his Monocoupe and flew a series of aerobatic manoeuvres which included 'knife edge', a high speed exercise unknown to most of those watching.
(Gerald Lambert for Flight.)

Committee and Club Officials were hurriedly re-assigned to guide the Royal party and Chief Marshal Christopher Clarkson was deputed to host Her Majesty The Queen.

Having never been inside an aeroplane Her Majesty was invited to inspect the DC-2 and was most amused when Roscoe Turner, forever the showman, insisted that the King should climb aboard to see inside the Boeing, and together they promptly disappeared through the hatch. Turning to Clarkson, Her Majesty declared with some amusement that the King absolutely hated aeroplanes and had not been near one for years. The visit was commemorated by Turner's presentation to His Majesty of a model of the aircraft.

Their Majesties may have put a seal of approval on the 'unofficial' carriage of air mails when they gave Jim and Amy Mollison a letter to be delivered to HRH Prince Henry, Duke of Gloucester, their third son who was due to arrive in Melbourne aboard the cruiser HMS *Sussex* prior to presentation of the Race prizes scheduled for 10 November. The Royal party was not to know that the crew of *Black Magic* had declined to carry other mail but by asking the favour had displayed great faith in their prospects as 'hot favourites' for a safe and early arrival in Australia.

After meeting Jimmy Melrose, His Majesty showed deep concern for the safety of the young pilot, and questioned the advisability of permitting such a tiny aircraft packed with equipment, to fly to Australia. Melrose and his Puss Moth made a deep impression on The King who repeated his fears to Harold Perrin as the party prepared to leave.

In anticipation of the expected visit of HRH The Prince of Wales, all the Race aircraft had been paraded on the new concrete between the two A Sheds and, walking ahead of his parents and the Royal party, was soon familiarising himself with cockpits that shortly were to reflect hope, glory, joy and disappointment, not to mention some degree of danger. After a long inspection Roscoe Turner presented the Prince too with a model of the Boeing 247D. Later, the Yorkshire born Jack Wright, described by the British press as loquacious and 'the embodiment of the brash American' announced that he shared the same birthdate with the Prince, offering a long-handled file, asked whether His Royal Highness would "use this to sign your monicker on the nose of my plane?" The invitation was diplomatically declined on the grounds that not only did Royalty not autograph anything, but all other competitors might make the same request. "Oh well! Just spit on it or something" Wright is alleged to have suggested. To those standing close by, and with some contrition, he remarked, "Gee, I hope I didn't embarrass His Royal Highness!" As everybody had roared with laughter the embarrassment appeared to be all Jack Wright's! To dissipate his disappointment, he took off in the red and white Monocoupe and performed what some observers called 'a series of breathtaking aerobatics', including a new manoeuvre which later became known as 'knife-edge' flight.

The intensity of interest in proceedings

A view inside the East hangar, an 'A Shed' with a floor area of 30,000sq.ft. The two completed hangars offered for use at Mildenhall were considered appropriate for a field which numbered over 60 aeroplanes at one time. With only a third of that number actually taking part each entrant enjoyed the luxury of space and the capacity for almost unhindered ground manoeuvring.
(National Library of Australia, Canberra.)

at Mildenhall had heightened as the week progressed, taking the organisers completely by surprise. Blue arm-banded guides attempted to smooth ruffled feathers by organising small parties for conducted tours of the hangars although they appear to have been poorly briefed and lost credibility and confidence in the eyes of an educated public. The never ending clamour for official passes reached its height on Friday during the visit by the Royal party, but on that occasion the hangars and buildings in use by the Club were sealed off, and to the fury of diplomat and general, influential friend and captain of industry, all but the surprise visitors were excluded.

According to a frustrated Alan Goodfellow, the work of the marshals had piled up and much of Thursday night was spent in organising a programme of work for the following day in the hope that all competitors could be sent off to bed at 6.00pm with confidence that all was in order for Saturday's start. He made a note in his diary:

"It was a very kindly and much appreciated act on the part of the Royal family to come to Mildenhall and, had the competitors arrived in time with their machines in proper order, Friday afternoon would undoubtedly have been the ideal time for the visit. Unfortunately, as in every other air race, things were left to the last minute with the result that instead of finding everything peaceful and prepared their Majesties found (and left behind) a desperate last minute scramble to be ready. As an example the American Wesley Smith actually arrived at the aerodrome at 4.30pm today for the purpose of weighing in his personal kit. Fortunately he is only in the Speed Race or he would certainly have had to be disqualified."

Inevitably, there was disappointment and frustration amongst visitors but the following was perhaps stronger than most letters on the subject published by the aviation press:

"Sir:
In MacRobertson Week I flew to Mildenhall, and after battling against appalling weather conditions, I landed at a dangerously undulating 'Official' landing field, for which I paid the sum of three shillings and six pence (not grudged). Transportation by car was provided to the aerodrome and back at the rate of five shillings, a total distance of 2.5 miles.

On arrival at the aerodrome gates I was besieged by Commissionaires for another five shillings and this for the colossal privilege of being allowed to enter the aerodrome; in other words, the Royal Aero Club's hundred or so square yards of concrete paving provided for the exclusive use of air-minded people, nicely situated behind the hangars and roped off at every conceivable point.

As for obtaining lunch, or even a drink in the tents provided for that purpose, the idea was preposterous. No! Interested visiting airmen, who, after flying in such weather probably deserved lunch and liquid refreshment far more than any of the Royal Aero Club's 'arrival by car' officials, must go hungry and thirsty. Had it not been for the chance meeting of an influential friend I should most certainly have had to go without refreshments of any description. Indeed, the very air seemed permeated with odious and petty officialdom.

After repeated efforts to gain admission to the hangars to inspect the entries for the Race (the sole object of my visit, since circumstances did not permit of any future opportunity) and subsequent lengthy wandering around in the rain, I was eventually escorted round the hangars with a party by a Royal Aero Club official who displayed absolute ignorance to the elementary questions put to him by members of the tour party. This tour was arranged after I had endeavoured to complain to the Secretary of the Royal Aero Club, who, I was informed, had been detained at the Newmarket Races."

Perhaps the atmosphere is best conveyed by two contemporary reports both published in *The Aeroplane* on 31 October 1934, a week after the Race had been won and everybody was that much wiser. The first was penned by journalist Anne Thrope, an invitee from *Woman's Weekly* magazine:

"To my surprise I discovered that there was such a thing as the Service and Information Bureau operating under the direction of Mrs Nigel Norman. In doing that I flatter myself that I was the only journalist who found the real centre of human interest on Mildenhall Aerodrome, including the Royal Aero Club bar.

The competitors' mail had just come in. On a table ten foot square 'Juice', 'Jack' and 'Jane' (as they called each other) were arranging some 200 letters, and about as many telegrams, into neat piles. The room was full of people all talking at once. A lost dog that had been brought in was crying for its master. A rending and crunching noise came from the Mollisons going through their mail on a mounting pile of envelopes and other *debris*. Two men with blue and white armbands were filing letters in a steel rack, each compartment bearing the competitor's racing number. Their voices rang out in a monotonous sing-song,

Geysendorffer six, Hansen seven, Wesley Smith forty six, Molinier, return to sender.

At the far end of the room a group of people were clamouring to be shown round the hangars. At intervals people came in and shouted, 'Has anyone seen Stack?' A distressful competitor was trying to keep his temper and get back the lock of his cabin door, which he said the Bureau had lost. Through the discordant clamour pierced the voice of a tall woman speaking into two telephones at once.

It was the fairest imitation of chaos I had ever seen. I settled myself in a corner to get some real malicious pleasure.

I was nearest to the rape of the lock, so I got an insight into the trouble. It ran something like this. The competitor had lost the keys to the lock of the cabin on his aircraft. The Bureau got him some new keys, but by that time the lock was missing. When the new lock came the keys would not fit it and the lock would not fit the aircraft. Now he wanted the old lock back. He said the Bureau had got it. The Bureau said it hadn't. What I wanted to know was, how was the Bureau going to get out of it. Well, it did.

The blue and white armband who was looking after the competitors was summoned. He produced his notebook. In it there was the carbon copy of a message to the aircraft constructor asking him to forward a new lock *similar* to *the enclosed*. The competitor threw his hand in.

I asked about the notebook, suspecting chicanery. It appeared that each S.I. (Service and Information) man had been issued with a field-service message-book of Army pattern. Every job they did for a competitor was written in the notebook, initialled by the competitor and a carbon copy kept. Purely a protection against complaints that might crop up later. The supply of chaos looked like being limited.

The *bruhaha* at the desk was going on non-stop:

"Oh, Mr Wright, we want you. Don't go away—most definitely not. The Aerodrome Control Officer has refused permission for you to land. The Bishop of Ely? Mr Who?

Two visitors to the West hangar during what appears to be a quiet period. Hangar passes were always at a premium and especially so after aeroplanes were reported to have been damaged. The students from the Chelsea College improved security in company with a number of other organisations brought in for the purpose.
(National Library of Australia, Canberra.)

Oh, Mr Seely. We shall have to broadcast for him. Will you hang on? What's your number in case we get cut off? The passports for yourself and Mr Polando have arrived from the A A. There are no flares at Rambang."

So there were no flares at Rambang! I wanted to know what the Bureau was going to do about *that*. I pulled a grey flannel sleeve.

The S I Bureau dealt with the problem by issuing an Aviation Notice. They were printing these Notices all the time. A copy was placed in the pigeon-hole of each competitor and other copies were posted in the Royal Aero Club bar. The grey flannel sleeve showed me some. 'The following Italian aerodromes are unserviceable....' A Greek aerodrome was also taking a rest. Life-boats would not count as pay-load unless... and so on.

There were Information Bulletins as well. Roy Tuckett's film of the Australian route was being shown in Mildenhall. The Sydney *Morning Herald* offered £50 to the first competitor to reach Melbourne carrying a copy of *Punch*. The O C, RAF, was offering to put mattresses and blankets in the billiard room.

The KLM pilots are bringing their wives—Tell Swinbourne that Foreign Office annexes to passports have not been received for the following—The Rolls left for Cambridge an hour ago; The Talbot will be back from Thetford in about a quarter of an hour—Are you in a desperate hurry?—No, it's only a question of accommodation for them—Stack's wireless operator wants some information—It's in the Stop Press file on the table—All right, Dutch wives to Ely—what?

Thea Rasche has lost her luggage!

Grey flannel had been replaced by Harris tweed. I pulled that. Had there been any trouble with the females in the Race? Particularly not. Miss Cochran was rather trying because she had successfully kept her address a secret and the press—something about dogs in a village street.

For a brief moment only one telephone was in use. But it was working for two. Someone wanted to book a seat in the Douglas airliner—£400—No, single—What's *your number?—We'll ring you back.*

I sat out the enquiry that followed with considerable interest. The Clerk of the Scales said that the Douglas had 600lb load to spare, so it was O.K. by him. The Handicapper had written all over the tablecloth in the restaurant and said that an extra passenger would make the handicap *worse* for the Dutch, not better. Formulas play tricks like that. The Clerk of the Course, who had a temperature of 101 and was drinking quinine in the Control Office, said it was too late for extra passengers, and why did they want to worry about telephone calls for drunken undergraduates, anyway. Which apparently was true.

Commander Harold Perrin came in. The Chief Aircraft Marshal came in. The Assistant Clerk of the Course came in. The Assistant Secretary came in. A pompous-looking little man with 'Steward' on his arm came in. Immediately there was a tremendous row. Someone shouted, 'Aircraft landing.' Everybody ran out except the tall woman, who went on bawling into two telephones.

This had been going on for six days."

The second eye-witness account of the final hours before the start was written under the pseudonym 'Blue Band' indicating perhaps a strong possibility that the author was one of Harold Perrin's armbanded officials with easy access to high places:

"Of course," remarked one of the Newmarket Fillies after I had explained to her for the fifth time that Cathcart-Jones's Comet was not the Blériot which went to ground in Paris, "*all* flying people are mad." A bit hard, I thought, but I remembered the honour of the old Club and refrained from saying that women like her are what drove us so.

As a matter of fact, she was not so far wrong. Mildenhall last week was a nightmare in which was a series of whirlpools going round in ever narrowing circles faster and faster as Saturday drew nearer and into which one plunged madly, occasionally coming up for breath at the Aero Club bar.

Whirlpool number one was officially called 'Secretary's Office', principally because the Secretary was hardly ever there. Harold Perrin ate, slept and entertained Royalty with that truly British calmness in the face of great stress which always astonishes foreigners, especially those who want hangar passes, and that brings us to another whirlpool.

When the week started in comparative calm, letting people see the machines in the hangars was easy. But as the week progressed and the fact became evident that the great British public were taking a far more flattering interest in the Race than the modest organizers had ever imagined, something drastic had to be done, and therefore Perrin's Pretty Parties were organized.

In theory various club officials were told off as hangar guides and they were given twelve hangar passes which they were instructed to keep in their pockets and on no account to part with. The purpose of this mysterious rite is not yet clear, but no doubt *there were reasons*. Equipped with the twelve magic passes and a blue band of office they then conducted parties which numbered anything from one to thirty persons round the hangars and told them what was what and who was who.

On the whole the great British public took an extremely intelligent interest in both the aeroplanes and the pilots. The Douglas and the Mollison's Comet attracted the most interest, while amongst the pilots Jack Wright and Colonel Roscoe Turner made a tremendous impression, principally because they were nearly always with their machines and were ready to answer the most ridiculous questions with a gentle politeness which endeared them to everybody, though their gentleness sometimes concealed an irony for the poor fish among the visitors which was only, and immensely, appreciated by the initiates.

The real rush started on Friday morning, which for rank and fashion must have run very close to Gold Cup Day at Ascot. Naturally, nearly everyone was either a very old friend of Harold Perrin's or had an introduction to him and therefore *must* see round the hangars. The hangar guides worked feverishly until their shoes felt full of feet and their voices croaked and their sense of humour died, but there were compensations. One guide received an invitation from a perfectly unknown lady to come and stay with her, and another had sixpence pressed into his hand,

At lunch time the hangars were railed off for The Prince of Wales's visit. Lovely for the hangar guides, but in the Secretary's Office the whirlpool whirled faster and faster. Members of the Diplomatic Corps with the highest credentials, influential foreigners, important members of the aircraft industry, bosom friends (mostly feminine) of the pilots, officers of the RAF, wives of Colonial Officials, stormed, cajoled, huffed, cursed or entreated the office staff to give them hangar passes and were told that if they would come back after the Prince had left of course the hangars would be open again.

While the staff was making this comforting statement for the hundredth time an air force orderly dashed in clearly bearing grave tidings, which turned out to be the startling news that their Majesties were honouring the camp with a visit and were due almost any minute.

The formula for dealing with importunate hangar-hounds had therefore to be hastily changed to telling each person in strict confidence that the King and Queen were coming, but that after their departure something would be arranged. Actually the King and Queen did not leave until about 16.00 hours, by which time the Authorities decided that the chief people concerned, namely, the pilots and their mechanics, were entitled to a little consideration and the hangars were finally closed.

By this time officials at the hangars and tarmac who had started the week polite and even humourful had completely lost their senses of humour and proportion and some also lost their heads. Even making allowances for strain and the additional difficulties of unexpected Royal visits one or two of them, in usurping the functions of the Aerodrome Control Officer, were unpardonably offensive to inoffensive members of the public. But one, at least, met his match in a highly coloured lady visitor who silenced him with a few well-chosen and even more highly coloured words.

With darkness the nightmare madness of the whole thing became intensified. At the Secretary's Office pilots came in and out having their logbooks and Customs' carnets seen to, and producing grim little revolvers to be sealed. At the telephone the Chief Constable of Suffolk was telephoning urgently for police reinforcements and finding that he was issuing orders to the Italian Embassy, who in turn were trying to get onto a totally different office. On making a second attempt he was given the delightfully farcical information that he could not get Mildenhall 5 unless the call went through London!

In a corner the Irish Swoopers were holding a council of war in undertones. In another Mr Reynolds was making abstruse calculations about the start. At a table the extremely patient but very hungry representative of HM Customs was patiently waiting for his last pilots, who were eventually discovered to be sleeping peacefully in their hotel at Thetford, having entirely forgotten to clear Customs.

At about 23.00 hours the Secretary was besieged by an infuriated mob of pressmen demanding the *truth* about *The Irish Swoop*. Then Jack Wright turned up, having lost 1,000 letters which he was taking with him, and on the sale of which he was depending for his return fare to the USA. At one time it looked as if he would have to become a naturalized Australian citizen, but eventually, after a frantic hunt, they were discovered with a Mobiloil man, and we believe that they were given to Roscoe Turner to carry in the Boeing."

The inspection by the King and Queen and the Heir to the Throne rounded off a week of action and eager anticipation. The coming of the new dawn would see a great plan unfolding as competitors left on their long journeys, headed east. The reality of the situation was close at hand; for some there would be triumph and satisfaction, and for others nothing but frustration, danger, bitter disappointment and even tragedy.

9

Take-off at Dawn

In the excitement of the moment and after all the preparation, Roscoe Turner flooded the engines of the Boeing 247 when preparing to taxy out to the starting line. There was near desperation by the hangars as 6.30am ticked closer but at 6.15am the engines fired and the aircraft travelled across the grass aerodrome at near flying speed to take up her position.
(Gerald Lambert for Flight.)

WITH NEIGHBOURLY assistance, Donald Parker, living at the Mill House in Mildenhall, and brother of Short Bros. test pilot John, managed to provide overnight accommodation for much of the boards of directors of Short Bros. and Imperial Airways. Oswald Short and others collected supplies of clean straw and billeted themselves in the stables. Next morning with the benefit of local knowledge, the party managed to get to the aerodrome by a circuitous back-route after traffic had clattered to a standstill outside their front gate.

Scott and Campbell Black together with Cathcart-Jones and Waller had been accommodated in Cambridge, 23 miles from Mildenhall. On Friday night Jim and Amy Mollison stayed with Walter Clarke and his family in the village, having booked full-board accommodation for a period of three weeks at a total cost of 7/6 each, although they rarely stayed. Jim's agent, Bill Courtenay, was required to get two untaxed cars back to London after they left.

On the eve of departure, the bogey of American aircraft and their 'substantial conformity' with the ICAN standards raised itself again. After arrival at Mildenhall, Colonel Fitzmaurice had presented American certification for *The Irish Swoop* (described by the Clerk of the Course as 'a very ambiguous certificate of compliance') which permitted operation of the aircraft at an all-up weight of a little less than 5,500lb. However, his proposed fuel uplift from Mildenhall of 120 US gallons in the rear tank and 480 US gallons in the front would have raised the Bellanca's weight to 8,350lb at which the machine had not been tested to meet the ICAN requirements in respect of landing performance. Take-off trials in the USA had been completed satisfactorily at the higher weights, a fact endorsed by the American Assistant Secretary of Commerce, Mr E Y Mitchell, who had signed the Certificate of Airworthiness on 3 October under the simple statement:

"The airplane was flight tested for this higher gross weight and conforms substantially with all international requirements except length of landing run."

The Stewards of the Race sub-committee cabled the US Department of Commerce at 6.00pm on Friday asking for confirmation of the maximum weight at which the Bellanca Monoplane 'conformed substantially' and whilst waiting for a reply decided that Fitzmaurice should be given the opportunity of flying the machine in the Race at the weights already certificated. Before he could be contacted the answer was received from America at 7.45pm. It

A crowd surrounds Black Magic *as Jim and Amy Mollison make final preparations to leave. The space between the first two Comets on the line was due to the late arrival of the Boeing 247. The West and East hangars can be clearly seen to the north. The aircraft are all facing to the south west, into the prevailing wind.*
(National Library of Australia, Canberra.)

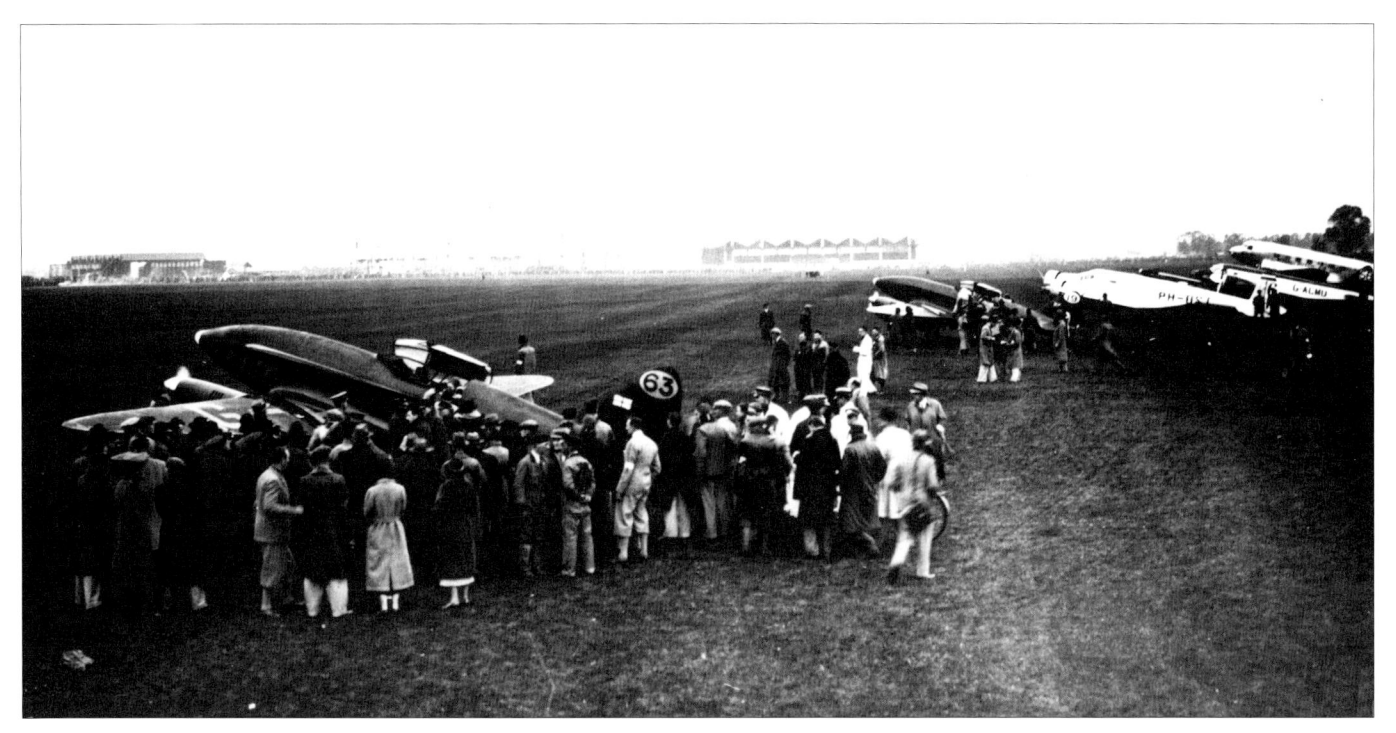

With the Boeing 247 safely in position all competitors were, as briefed, positioned in starting order with engines running by 6.30am. The only gap now was caused by the late withdrawal of The Irish Swoop, *Race No. 29. (National Library of Australia, Canberra.)*

confirmed what was already known, that the aeroplane conformed to all ICAN requirements only at the lower weight. Fitzmaurice confirmed that due to shortage of time, he had curtailed the flight trials in the USA on the grounds that the test pilot was incompetent and might damage the aircraft. Having arrived so late in England, Fitzmaurice was hoping to resolve the matter by negotiation as there had been no time to conduct landing trials before scrutiny at Mildenhall, and the Rules precluded performance flight testing once there.

Fitzmaurice was called before the Stewards under the chairmanship of Colonel Moore-Brabazon to present his appeal against probable disqualification and to argue his case for eligibility at the higher gross weight, during which he cited the efficiency of the fuel dumping system which could discharge 500 gallons of fuel in 44 seconds. No special brakes had been fitted to permit the aircraft to land in compliance with the ICAN requirements due to the efficiency of the dump valve and the belief that the aircraft would be landing at light weights at the end of long flights. It was this dump device, he insisted, that had caused the US Department of Commerce not to ask for full load landing tests, but he failed to convince the officials who rejected the appeal and insisted that he could not operate at a maximum weight greater than 5,458lb. The Stewards then convened to analyse the chain of events which had led them to this situation only ten hours before take-off. They concluded that they had acted properly and within the Rules and could make no concessions. This view was communicated to Fitzmaurice, who decided that he would operate at the lower weights but after consultation with his sponsors they decided that to commit the aircraft to 19 landings en-route was clearly unacceptable, and they had no option but to withdraw. After the enormous expense and effort of safely reaching Mildenhall when many others had failed, *The Irish Swoop* was to remain in the corner of her hangar, a monument to optimism born from the interpretation of an imperfectly drafted Regulation. As consolation the Race Stewards refunded the £5.00 protest fee.

Fitzmaurice later claimed that on arrival at Mildenhall he had given a copy of the aircraft's Certificate of Airworthiness to the Clerk of the Scales with an invitation to inspect the dump valve and had been assured that everything was in order. He fully understood why the sub-committee should have been concerned about an aircraft crashing on take-off due to overloading, views they had expressed a year previously. He also suggested that the sub-committee's cable to the Department of Commerce in the USA was worded to give the reply they required and that "they feared my competition." In addition he expressed his belief that the Race had become "a grim industrial struggle between the British aircraft industry and that of America."

Following his withdrawal, some American critics admitted that Fitzmaurice had no reason to expect to be allowed to race without restriction under the terms of his special certificate when other competitors also had had their loads reduced, namely Stack, Woods and Roscoe Turner. On behalf of the Race sub-committee a statement was issued that confirmed that no discrimination had been exercised against Fitzmaurice; the Rules of the Races were designed for safety. No other competitor had protested against the Bellanca but a number had waited anxiously for the sub-committee's decision with a view to safeguarding their own interests.

Various rumours then spread which suggested Fitzmaurice would fly the route anyway, but not as a competitor, and operating at his higher take-off weight. An alternative suggested that he might possibly try to demonstrate the aircraft in a fully loaded condition at Mildenhall for the benefit of performance measurement by independent witnesses. Neither of these rumours proved to have foundation. There was even a suggestion that the Parliament of the Irish Free State was to pass special legislation, but that would have made no difference. According to the Rules *The Irish Swoop,* dubbed *Shamrocket* by a largely unsympathetic press corps, was ineligible for the Speed Race as presented to the scrutineers and could not be permitted to take any official part in the proceedings. With a potential range of 4,000 miles and a top speed of 273mph *The Irish Swoop* was considered a hot favourite in the Speed Race and disqualification of the entry was regarded by most sporting aviators as a matter for much regret. Only a Nazi newspaper published in Germany came to the aid of a bitterly disappointed crew, suggesting that the entry had been eliminated because the British were afraid it would achieve more than their own entrants.

Some crews had left the generally spartan facilities of the aerodrome's tented quarters on Friday evening, 19 October, heading for pre-arranged and mostly secret addresses, but most of the competitors were forced to spend their last hours of relaxation cat-napping in the Royal Aero Club's imported armchairs. At the twelfth hour and only after some considerable persuasion, the Air Ministry had reluctantly agreed to unlock some of the newly completed but still empty 'permanent' facilities.

Bernard Rubin, still suffering from the illness which had prevented him from crewing his Comet, requested through de Havilland's Business Manager, Francis St.

Barbe, permission to site a caravan within the Royal Aero Club's tented city. Rubin's doctors had permitted his attendance provided he refrained from physical exercise. When the sub-committee realised the caravan was almost certainly intended as overnight accommodation the request was refused on the grounds that Air Ministry permission to use their site expressly forbade such a proposal.

C G Grey had his views of course and his 'on the spot appreciation' was published immediately after the start:

"On Friday night the competitors went to bed early. As a special concession from the Air Ministry some of them were actually allowed to sleep in Air Ministry buildings. Why on earth the whole lot could not have been accommodated in the same way instead of having to go, some to Cambridge, some to Bury St. Edmunds, and some to Ely, is one of those little things which only a cast-iron official mind can explain.

One would have thought that in a great International event of this kind the Air Ministry would have extended something like government hospitality to the competitors. Considering that most of the buildings at Beck Row bear plaques on which the figures 1931 appear, there is not even the excuse that the buildings are not fit for human habitation.

By about 11 o'clock those people who had lodgings outside the aerodrome had gone off, a good many hundreds slept more or less comfortably in their own cars in the car park. But there were still dozens drowsing in chairs and on the floor in the Aero Club Pavilion.

At or about 03.30 hrs on Saturday those of us who were sleeping in the car park were awakened by cars streaming by on the roads and coming into the park behind us. When we tumbled sleepily out, the whole sky was ablaze with the reflection of headlights in every direction. A young woman in the car next to us woke up, looked round the parked cars and at the headlights all round and remarked: 'Dawn among the refugees!'"

Harold Perrin, the unflappable and stalwart Secretary of the Royal Aero Club who had borne the brunt of the past years' preparation, slept soundly upon the table in his office, with Colonel Bristow, a Race Steward, occupying the secretary's desk. During their slumbers perhaps they recalled the words of Sir MacPherson's Eve-of-Race goodwill message:

"Of the many factors to play their part in the High-speed Race, the attempt to reach Melbourne from London in something like three days will inevitably call forth, in the highest degree, the skill, resourcefulness and endurance of pilots; the ingenuity and faithfulness of the aircraft and engine construction; the expert mechanical care of engines that must function almost non-stop half-way round the world; the sustained efficiency of refuelling services during day and night; and the close co-operation of wireless and meteorological stations along a course which crosses three continents and spans the seasons from autumn to spring.

The magnificent response by leading airmen of the world who have so generously come forward to make this event the greatest race in the history of aviation is very gratifying to the Air Race Authorities in Australia.

The full growth of aviation must be world-wide and the international support that the Race has evoked since its inception convinces me that such a contest must help to broaden the mutual basis for friendly exchanges of services and understanding between nations quite apart from quickening the air-mindedness of their people. I but express the feeling of all Australians in wishing you a safe and speedy flight to Melbourne where a warm and hearty welcome awaits you."

With the last hours of Friday slipping away, the roads for miles around Beck Row and Mildenhall Aerodrome became progressively more choked with slow moving then stationary traffic. There was little or no police control, apart from an elderly rustic uniformed with a Special Constable's armband and his gnarled walking stick, who attempted to see fair play at an intersection in the village. Mr Leonard, a resident of Folly Road, awoke to find two enterprising strangers had arranged for a number of cars to be parked on his land and were pocketing half a crown a time.

As predicted, party and theatregoers from London's West End, many still in evening dress, were noticeable amongst the throng, particularly those who had hailed London taxis. Neither the vehicles nor their passengers were in keeping with their rural surroundings, but these were no ordinary times.

With the clock ticking inexorably towards 6.30am, those immobilised in the traffic stream and within walking distance of the aerodrome abandoned their transport and took off on foot across the fields. A young shepherd, Edgar Taylor, had walked 30 miles across the countryside from his farm at Hingham near Watton, and after the start had to walk back again. One observer likened the scene to a British Army advance on the Western Front as waves of spectators trudged towards the perimeter fence. Thousands more were stranded miles away and destined to remain so until long after the last competitor was safely on his way.

Brian Allen, Manager of Henly's Aviation Department at Heston, had arranged a coach party which was scheduled to leave the Motor Show Ball at Grosvenor House at 3.00am, arrive at Mildenhall an hour before the start and enjoy breakfast in Newmarket by 8.00am. It was a sporting effort at £1.0.0. per head, but the organisers were not to know that even the prospect of a late lunch in Newmarket was in doubt.

By five o'clock on Saturday morning most competitors were ready and anxious to be off. Kitted out, nervously they offered assistance to mechanics as machines were wheeled out from floodlit hangars into total darkness. Following overnight rain it was now a dry morning and warmer than of late, with a stiffish breeze from the south west which dictated a take off run towards the gun butts, the favoured direction. Final engine adjustments were completed, the activity stirring those who were still dozing in their cars and a miscellany of dark corners they had discovered around the site.

Determined to see the start, Alex Henshaw decided not to ask for permission to land at Mildenhall, which he knew would be refused, so with his father in their Leopard Moth G-ACLO left Skegness at 05.15am on a compass course for the Suffolk aerodrome. Overhead Sutton Bridge all the airfield lights suddenly flicked on as the RAF believed that an aircraft overhead at such an hour and in the gloom could only be one of their own. Henshaw got lost over Ely but 20 minutes later found Mildenhall 'under a blanket of murk and showing no lights'. He landed in what he described as 'a discreet corner' undetected and unchallenged.

Movement in the dark around and about some of the unfinished facilities was to be avoided especially in the 'Aero Club Enclosure' which included excavations in support of steelwork being erected for two additional 'A Type' sheds. The Editor of the Swiss magazine *Interavia* was one casualty and needed attention in the Station's Sick Quarters before being airlifted back to Croydon in KLM's supporting Fokker and on to a London hospital. A lady visitor from

Sir Alfred Bower, acting on behalf of the Lord Mayor of London who was indisposed, raised the Union flag just before 6.30am and on the stroke of the half hour its scything motion declared the Races were to start and Comet G-ACSP should be on her way to Australia.
(National Library of Australia, Canberra.)

Fifth scheduled starter was the Dutch Pander who followed the aborted take-off by the green Comet. The Viceroy went off in her turn but aborted in order to collect newsfilm. Lost to history is the reason why there is clearly a different starter with a new flag: Timekeeper George Reynolds with an escort from the Royal Air Force.
(National Library of Australia, Canberra.)

the Reading Aero Club fell into a hole so narrow that her arms were pinned against her sides and only with great difficulty was she extricated by friends. Freddie Meacock, Dudley Russell and two other de Havilland apprentices from Hatfield had driven up to Mildenhall in an Austin 7 to assist with polishing the Comets. They arrived in the dark and strode out towards the lighted hangars but only three young men arrived. Without a word, Dudley Russell had disappeared down an unprotected hole in the ground. Even in daylight there was danger: two spectators seeking the advantage of height offered by banks of earth, stumbled and slipped into muddy excavations.

Nobody had realised that public interest in the start of the Race, timed for a dawn take-off, would encourage an estimated 60,000 spectators into the country lanes of Suffolk, although the author of one article published the previous August had predicted just such a turnout if the event were to be run from an aerodrome away from the immediate environs of London. Remarking that only a few dozen air-minded folk bothered to set out to watch an 8 o'clock start of a King's Cup Race from Hatfield, the perspicacious C G Grey opined that similar numbers could be expected if the October dawn was to break in Hertfordshire, but should the Race start from perhaps 100 miles from the capital, then thousands would be encouraged from towns and villages within a 50 mile radius to become part of this historic event. Grey even foresaw what he termed 'the mass production of all-night parties'. The theory proved to be uncannily accurate except that Mildenhall was only half of Grey's assumed distance from Piccadilly.

The limited security force was already considered barely adequate to cope with an anticipated crowd of 10,000. Now the huge spectator army massed around the perimeter continued its surge until 5,000 individuals were astonished to find themselves within the aerodrome boundary and close to the start line. Only immediate action by police officers, now numbering 12 on foot with others on motorcycles and on horseback, a detachment of uniformed RAF Apprentices from Halton, 80 cadets drafted in from Cranwell and 60 RAF mechanics from Duxford, helped the crowd form an orderly line behind the waiting aircraft and another parallel with the direction of take-off, vulnerable to anybody swinging left. Assisted by a fire engine, order was maintained and the good natured crowd in festive mood was rewarded by an excellent view at no charge. Less impressed was the press corps whose elite photographers and newsreel crews had been selected for exclusive duties on the start line. They were overtaken by the public's advance, and lost their hard-won privilege. Newspaper reporters found themselves isolated behind rows of 'ordinary spectators' and threaded their frustration into their early despatches.

Race officials, who had already doubled the capacity of the two public enclosures and improved their security, had also considered the vulnerability of the aerodrome's southern boundary, close to the most likely starting point, and there were thoughts of establishing an additional public enclosure protected by barbed wire fences, but it was all too late.

Alan Goodfellow, Clerk of the Course, was horrified to see the aircraft surrounded by spectators and was concerned that the start might have to be delayed. The sub-committee had already been advised that if the start should be delayed for any reason for more than two hours he would impose a 24 hour postponement.

Some spectators had been hoping for a Le Mans style start, but they were to be disappointed. The take-off positions had been decided by ballot the previous evening and then changed at short-notice to create two distinct groups, allowing the fastest away first. Aircraft were required to line up from six o'clock in two rows, A and B, each of ten aircraft, with motors ticking over.

The students from the Chelsea College of Aeronautical Engineering were on duty from four o'clock and after a briefing in the Mess were detailed-off to chaperone aircraft moving out from the hangars. Their leader, Walter Clennell, remembered it had been raining and as he prepared to cross the sodden grass he found Jackie Cochran, lightly clad and wearing thin shoes, alone and in some state of confusion. Ever the gentleman, he lifted her over his left shoulder and carried her bodily to the Granville.

Loudspeakers set up around the concrete aprons boomed out details of the latest weather reports as the hangars finally were emptied and machines prepared for the start under the direction of Christopher Clarkson and his teams. Some were manhandled out to the line, some 500 yards south of the East hangar, in an effort to preserve every drop of fuel whilst, with an improvement in the light from 6.00am, others moved under their own power. All three Comets were pushed out and lined up as directed by Eric Mitchell, de Havilland's senior engineer on site. The aircraft had been vigorously polished to a mirror finish by de Havilland Company apprentices and students from the Technical School, sent up from Hatfield to help. After the Race, advertisements for Titanine Paint declared that all the Comets had been sprayed with their 'Satin Finish' products.

Due to the continuation of eleventh hour repairs to the green Comet G-ACSR, Race officials had agreed, somewhat reluctantly, to the aircraft being refuelled inside the hangar, and thanks to their understanding she was ready on time. Last to leave the security of the hangars was the Boeing 247, scheduled to be second aircraft away. Much to the frustration of his ground engineer, Don Young, Turner had experienced difficulty in starting the engines *(Nip* on the port side and *Tuck* on the starboard) due to overpriming but they eventually fired at 6.15am to the accompaniment of streaming exhaust flames, and the subsequent passage out onto the aerodrome was described as a rehearsal of the take-off run, 'swift and unceremonious'.

The outbreaks of rain from the previous evening had passed through and on what was to be a generally dull and overcast day the post-dawn sun was just breaking through. On the starting line, under what the correspondent from *Flight* magazine described as 'being fit for the occasion: layer upon layer of jagged orange clouds climbing into starlit purple', waited a party of young RAF pilots who had volunteered their services. Among them, Flying Officer Arthur Clouston, later to achieve fame as an accomplished test pilot and who would himself fly record times in a Comet. Flying Officer John Grandy, a 21 year-old Bulldog pilot with No. 54 Squadron based at Hornchurch was there, unaware that following a career which would take him to the highest rank in the Service, Marshal of the Royal Air Force and Chief of the Air Staff, in retirement he would become inextricably linked with the DH.88 Comet. Alongside stood Jeffrey Quill on whose shoulders would be heaped the onerous responsibilities of flight testing Supermarine's new fighter aircraft, the Spitfire. Quill noted that Jim Mollison looked quite sick as he prepared to board the Comet; his face ashen and without expression. Amy Mollison appeared nervous and apprehensive, and the young officers developed a united sense of pity for her as she climbed into the back seat and made ready to be first away in *Black Magic*. The Mollisons were amongst the favourites at 12-to-1 in spite of Jim Mollison's stated intention of not flying any sector which would involve a night landing. Much popular money had been put on the winner reaching Melbourne in about 86 hours.

The intrusive Tannoy broadcast the news that weather in southern Europe was not good, and might worry competitors who were heading for the first Checking Points at Marseilles or Rome. The longer range machines flying the Great Circle track directly to the first Control Point at Baghdad, were more likely to avoid the worst of the heavy cloud and rain.

At the request of Melbourne's Lord Mayor, the Lord Mayor of London was invited to drop the Union flag after which the Race would be on. In the event, London's Lord Mayor, Sir Charles Collett, was unable to attend through illness and a previous Lord Mayor, Sir Alfred Bower, acted on his behalf.

At precisely 6.30am local time, (0630 hours GMT) when, as if by command a rainbow developed overhead, timekeeper

The last four of the aircraft to start in the Handicap Race facing into wind in the south eastern corner of the airfield. Thousands of members of the public are ranged along the line of trees in the background beyond the Fairey Fox G-ACXX with the stalled engine.
(The Aeroplane.)

Third from last to leave, not counting the re-runs for the Fairey Fox and the Viceroy, was Geoffrey Shaw in the British Klemm Eagle, Race No. 47. He was waved-off by yet another starter, this time working solo with his own time-piece and flying an Australian flag.
(National Library of Australia, Canberra.)

George Reynolds stabbed his watch. Sir Alfred Bower scythed his flag through the south westerly breeze and Jim Mollison juggled the throttles of *Black Magic*, but failed fully to release the brakes. Hubert Broad was seen to be getting animated standing alongside as the aeroplane struggled forward then finally accelerated away and easily raised her tail after Mollison realised his error. Airborne after the third attempt to lift off, the heavy aircraft dipped gently onto a southerly heading to fly a track planned deliberately to cross Turkish territory, in spite of the prohibition still placed on Jim Mollison, to assure the shortest direct route to Baghdad. After the prescribed 45 second pause following the departure of the black Comet, Roscoe Turner encouraged the full 550hp from each of his once reluctant Wasps and took up the chase.

Cathcart-Jones and Waller were next away following a gap left by the withdrawal of *The Irish Swoop*, but the aircraft swung to port at an early stage in the run followed by a swing to starboard before the commander wisely throttled down and taxied back, the port engine spitting jets of white flame. As G-ACSR was returning, the Dutch Panderjager took-off, then the Viceroy lined-up and made a fast run before easing to a stop and back-tracking to the line, causing further confusion amongst the spectators. Stack was dismayed to have been declared scratch machine in the Handicap, but allegedly had been offered a substantial amount of money to carry newsreel film of the start through to Australia and 'sportingly' had agreed to his delayed departure. Having taken his own allocated place in the recorded sequence he taxied back to wait until the last machine was safely off the ground.

Scott and Campbell Black were away next in *Grosvenor House*, a machine noteworthy for her brilliantly reflective scarlet paint scheme. The two pilots were reported as being 'hard as nails' having trained exhaustively for 12 weeks with at least three hours squash practice every day, and only an occasional beer or whisky nightcap to disturb the balance. At the time Charles Scott released his brakes, G-ACSS had accumulated a total of only 3hr 30min flight time in ten trips: 65 minutes of production testing by Hubert Broad and the rest in a series of seven training and familiarisation flights divided between Hatfield and Mildenhall. The red Comet was immediately followed by her recalcitrant green sister ship which on this occasion ran straight as an arrow having lost two minutes.

The KLM DC-2 was next to go, seemingly slower than the two previous starters, an illusion caused by their difference in size. In contrast to her polished metal structure with barest black insignia, the full span registration letters painted on the top surface of the wings were bright red, adding a flash of colour as she accelerated away. The aircraft was almost late to her place at the start as police had been called to remove a young Dutch boy, Kohannes van Lee, who was believed to be attempting to stow away. Held in custody over the remainder of the weekend he was returned to Holland by airline the following Monday.

At 20 years of age an interested spectator, Pat Fillingham, was one of the many advised that if they had passports, and presumably the fare, they could have been carried as passengers! Five years later Pat Fillingham was a de Havilland test pilot, roaming the world on behalf of his company.

Peter Masefield closely watched each take-off run with a professional eye. They were all fairly interesting, in his opinion, with some more so than others! The departure of the Granville Monoplane, the next away, provided a memorable highlight. The aircraft bounded across the aerodrome putting maximum load into its undercarriage, before staggering into the air in a semi-stalled condition then climbed slowly away behind a screeching propeller, a noise quite unlike any other heard in the competition. Later, Smith disclosed that the flaps had not been working properly and Jackie Cochran said she believed the aircraft had actually stalled.

John '*Utica Jack*' Wright was next to go in the Monocoupe, first of the second group. Just before the start, funds he expected to be confirmed by cable still had not materialised, but at 6.00am Perrin personally loaned him £10, a sum with which he was expecting to sustain himself during a flight halfway round the world. Perrin told a colleague that if the Monocoupe reached Baghdad he could claim the loan from the return of the entry fee which would become due under the Rules. Wright made no pretence of being short of cash. When musing on how much the adventure with her Gee Bee was costing her, Wright had said to Jackie Cochran that all he wanted to do was "fly this airplane while it's still mine!"

At the last minute Wright had revealed he had no maps to cover the route between Darwin and Melbourne. With great benevolence, the Assistant Clerk of the Course, the Australian Flight Lieutenant Swinbourne had given him a set, purchased at a cost of £1.16s 5d from his own pocket. Perrin later assured him the expenditure could be repaid under the terms of the Club's hospitality fund!

The little Monocoupe demonstrated its nimbleness during take-off in a manner more subdued than when it first arrived at Mildenhall, and was soon turning on course. She was followed by Davies and Hill in the Fairey IIIF G-AABY then Gilman and Baines in Fairey Fox G-ACXX but their engine stalled when full power was applied, probably too rapidly in the excitement, and the aircraft was compelled to

The Irish Swoop, *attended by engineers from Rollason Aircraft Services and Pratt and Whitney Aircraft, was withdrawn after the last attempts to satisfy the issues of certification failed. The crew was keen to fly to Australia after flight trials which had been arranged to take place at Portsmouth.*
(Gerald Lambert Collection.
By kind permission of the Suffolk Record Office, Bury St. Edmunds.)

wait until all others were on their way. Parer and Hemsworth were next to leave in their Fairey Fox G-ACXO, the students from the Chelsea College having worked all night to change the radiator. Woods and Bennett in the travel-stained Lockheed Vega should have left at 6.39am then the New Zealand Dragon Six, *Tainui,* off on its delivery flight with Hewett, Kay and Stewart on board, but Jimmy Woods, in his anxiety to be on his way, misinterpreted the preparatory raising of the starter's flag as the signal to depart and rolled 10 seconds too early. The Race Stewards were left having to make a decision whether to disqualify the Vega or merely add a penalty at Singapore. They had plenty of time to contemplate their course of action. The Danes, Hansen and Jensen, followed in their Desoutter then Harold Brook and his lady passenger in the Miles Falcon which demonstrated the best take-off climb of all.

McGregor and Walker sat patiently in their new Hawk, seeming to mimic a flying club's local sortie rather than a trip to the southern hemisphere. They had a good view of each departure which they assessed with professional's eyes until at 06.42 and 45 seconds the dropping of the starter's flag indicated their turn to apply full power. The Hawk was still within sight as Shaw in the British Klemm Eagle was winding up the undercarriage, followed in turn by the Stodart cousins in the Courier.

Final starter in the scheduled order was the young and popular Jimmy Melrose in his Puss Moth, flying with his half-eaten fruit cake. He was followed by Gilman and Baines in their Fox which had not been ready due to that reluctant engine and had accumulated handicap time even before departure. The Viceroy's crew, who had now waited 13 minutes after their 'official' start, picked up the newsreel films which were snatched from the cameras and roared off after the pack to the accompaniment of loud cheers from the crowd. The lead aircraft was already 50 miles ahead and accelerating away from the back markers.

Squadron Leader W Helmore who was due to broadcast a description of the start on the Empire Transmitter located in London had an aircraft waiting for him at the local landing strip but as the roads were choked he had to run almost two miles to reach it. He flew to Hatfield and then on to Hendon, arriving late at the studio at 9.30am. The BBC considered the report so significant that the transmitter's operating hours were extended.

In charge of the de Havilland Comet repair team, Bob Robson was so exhausted after continuous day and night duty that he fell asleep just before the start and missed the sight of his beloved charges roaring off on that historic morning. Not until the following week was he able to catch up on his lost opportunity when news film was shown at his local cinema.

Apart from *The Irish Swoop* brooding in a corner with her team of frustrated supporters and two police guards to ensure that the excess petrol was unloaded, and a few liaison aircraft which soon would be gone, Mildenhall Aerodrome was suddenly empty and strangely quiet. The tens of thousands who had brought this area of Suffolk to a standstill had witnessed a unique spectacle lasting barely 17 minutes and now were faced with the enormity of disengagement.

As officials drifted into a post-departure breakfast prepared at a still early hour by the splendidly accommodating NAAFI organisation, the press correspondents compared notes and adjudged affairs thus far to have been a success, most things considered. With assured expectation they now awaited further developments.

The starting officials agreed that apart from their decision on the Lockheed Vega, which was impose a time penalty only, no allowances were due to be telegraphed through to Singapore where time adjustments were to be made. Alan Goodfellow noted in his diary: "And so back to our respective offices and tasks after a strenuous week of work and good comradship which few of us are likely to forget."

The apprentices from the Chelsea College, exhausted after their week's continuous effort, had one final duty. They drained most of the fuel from the disqualified Bellanca and at midday left in their coach bound for Brooklands, unaware that some competitors already were in difficulty.

10

Ever Eastward

Following their departure from Mildenhall the Race contestants were quickly spread along the route to Australia. Those flying on Handicap took full advantage of the Rules and made progress at a more leisurely pace. John Polando and Jack Wright enjoyed a cup of coffee whilst their Monocoupe 'Baby Ruth' was being refuelled at Marseilles.
(National Library of Australia, Canberra.)

Lockheed Vega G-ABGK was a centre of attraction when she arrived at Marignane Aerodrome, Marseilles, after a four-hour flight from Mildenhall. Like several others taking the same route the aircraft spent much of the time cruising above cloud.
(National Library of Australia, Canberra.)

NEVILLE STACK landed the Viceroy at Abbeville on the north French coast an hour after leaving Mildenhall, forced down by a combination of appalling weather and a cockpit filled with smoke from a faulty electrical circuit. Airborne again just after lunch the aeroplane was diverted into Le Bourget Airport, Paris, where the crew elected to remain overnight. Stack was reported to be worried and exhausted due to the financial burdens he was carrying, and was becoming over-concerned at what were considered to be trivial defects on the aeroplane.

The crew of the New Zealand Hawk, 'the best possible value for £1,000', had elected to fly at 1,000ft, and as they levelled from their brief climb, the Manawatu Club instructor shouted to his Captain, "Well! On our way!" It was a cry of relief as much as the anticipation of a successful aerial voyage and a safe homecoming.

McGregor was to fly the Hawk at each landing and take-off operating from either cockpit, and to share en-route control with Walker who otherwise was responsible for navigation. Several crews had elected to follow the same split-duty option. After leaving Mildenhall, they set course for Marseilles but ran into the troublesome cloud sheet near Abbeville which forced them down almost to tree-top height. Establishing a positive fix on his position McGregor immediately climbed on instruments to 9,000ft and cruised blind for an hour before deciding that it would be hardly less cold at a higher altitude. At 10,000ft the aircraft emerged from the clammy vapour into brilliant sunshine.

The thick stratus persisted to within 90 miles of Marseilles where they landed safely four and a half hours out from Mildenhall. With log book entries verified and men and machine refreshed, the Hawk was airborne for Rome within 30 minutes. The fuel consumption had averaged 10 gallons per hour at full rpm and the pilots were happy with that knowledge. On the ground they had great difficulty in hearing anything that was said as a result of the constant blattering from the engine and propeller wash which was transmitted through their open cockpits, but once airborne, communication by Gosport tube was entirely comprehensible. The phenomenon was to remain with them both for more than a fortnight.

Cathcart-Jones and Waller wire locked their throttles into the fully open position as suggested by the de Havilland design team and ten minutes after take-off overtook the red Comet as they climbed straight to their cruising altitude of 10,000ft. Flying above the advancing blanket born of an Atlantic depression, and which now appeared to smother the whole of Northern Europe, their intention was to fly directly to Baghdad. They were forced to rely on dead-reckoning navigation as far as the Black Sea, east of Constantinople, where the cloud broke at last, permitting their first positive fix since before crossing the North Sea. One en-route sighting had been possible: the tops of the Alps had pierced the clouds at an estimated 100 miles off their starboard side.

Scott, Campbell Black and the Mollisons had also set off with the intention of flying directly to Baghdad, and under the same prevailing conditions in identical aeroplanes, their race was to develop into a private contest of piloting and navigational skills.

Woods and Bennett were flying steadily in the Vega and just before 10.30am, almost four hours after leaving Mildenhall, much of their flight above cloud and without the use of wireless communication, they touched down for their first observed stop at Marignane Aerodrome, Marseilles. They left marginally before Jack Wright and Johnny Polando in the Monocoupe who had previously landed at Lyons due to no visibility. In the same foul weather Gilman and Baines in their Fox also put down at Lyons and reached Marseilles half an hour after the Monocoupe had departed for Rome. Already through the French Checking Point were the Dutch Pander and Jimmy Melrose's Puss Moth, the Danes Hansen and Jensen in the old Desoutter and the Stodarts with their Courier which had suffered a broken cockpit window. The cousins had shipped a lot of water and lost several maps through the gaping hole, and

The Control Officer at Marignane Aerodrome, Marseilles, attracting the attention of Michael Hansen as he taxies the Desoutter towards the concrete apron where Race formalities were to be completed.
(National Library of Australia, Canberra.)

Courier G-ACJL on arrival at Marseilles showing evidence of a wet aerodrome surface through the accumulation of mud on the tyres. The front part of the sliding window on the co-pilot's side of the cockpit appears to be missing.
(National Library of Australia, Canberra.)

David Stodart with his coffee system provided by the aerodrome catering department looking relaxed after the long flight in the early morning from Mildenhall.
(National Library of Australia, Canberra.)

discovered even at this early stage that their remaining charts were probably of too small a scale.

Gilman and Baines took off from Marseilles, but returned shortly after with a snag which delayed their further departure until late the next morning when they headed for Rome. Due to weather, Harold Brook put his Falcon down at Plessis-Luzarches, two and a half hours out of Mildenhall. He was soon airborne again bound for Marseilles, but was forced to divert into Istres, a military airfield 20 miles from his destination, due to shortage of fuel, landed heavily and remained there with his passenger overnight.

It might have been at Istres that Brook realised he had left a small suitcase containing a Very pistol, iron rations and papers in the hangar at Mildenhall. The Royal Aero Club was alerted to the loss through Brook's brother in Birmingham but in spite of diligent searches that continued after occupation of the site by the Royal Air Force nothing was ever found.

The Fairey IIIF flown by Davies and Hill lost indications on all her blind flying instruments over the English Channel, refuelled quickly at Bourges and routed through to Rome, but less lucky were Ray Parer and Geoffrey Hemsworth in their military-surplus Fox. In mid-Channel the Felix engine began misfiring and Parer instinctively turned back towards England, but he changed his mind, maintained the bank, completed a full circle and nursed the machine over the water to cross the French coast near Boulogne, where he made a successful landing in a field at Samur.

An officious gendarme delayed mechanical inspection for an hour until a telephone call from Paris confirmed they were Race entrants and should be allowed to proceed. Now the new radiator was leaking as badly as the old one so it was topped up, and when the engine was started it seemed to run smoothly. The weary old biplane took off for Abbeville in search of engineering facilities but landed prematurely at Beauvais just as the mis-firing recommenced. All indications pointed to the radiator again, so the offending appendage was simply removed and sent into town for some soldering repairs to be carried out in a local garage. Prudence suggested a more thorough investigation should be conducted in Paris where the disappointed pilots arrived in good time for lunch after two more forced landings. Here they abandoned all thought of further competitive flying for the time being.

Hewett and Kay in the Dragon Six had also made a precautionary landing in

Black Magic *flew from Mildenhall directly to Baghdad where she arrived in darkness to be met by an army of officials and enthusiastic helpers from civil organisations and the military. The aircraft had travelled 2,500 miles in just over 12 hours.*
(National Library of Australia, Canberra.)

During their two hour turnround at Baghdad Amy Mollison put off meeting Iraqi Government officials until after she had enjoyed a hot bath. In this photograph she is seen holding the fur-collared flying suit she was wearing on departure from Mildenhall. Jim looks as though he is ready for another day at the office.
(National Library of Australia, Canberra.)

London was anxious for news as nothing had been seen or heard of any one of the trio since departure. But 12 hours and 40 minutes after its retarded take-off run at Mildenhall, the first of them bored into the circuit at Baghdad. In the darkness, Jim Mollison switched on the landing light in the elegant nose and flew three circuits, gradually losing height before making a perfect landing.

Black Magic had flown 2,500 miles non-stop at an average speed of 200mph. Amy and Jim Mollison appeared to be happier and in better spirits than when last seen at Mildenhall, and within a cheering assembly the Mayor of Baghdad was on hand to welcome them and offer congratulations. There were messages from King Ghazi and invitations to make a speech to the excited crowd. An Iraqi cabinet minister was anxious to meet Amy but was told she was taking a hot bath, and that had to come first.

Jim Mollison reported that the greater part of the flight from Mildenhall had been at 12,000ft due to the weather. He had been alarmed at the petrol state as the Comet approached Aleppo but he and Amy had decided to press on and they arrived overhead Baghdad with about two hours' supply remaining.

Pleasantries completed and 230 gallons of 72 octane fuel signed for, the Mollisons taxied out just over an hour after landing but were delayed in taking-off waiting for the dust cloud they had created to settle. Airborne again for Karachi, now almost two hours after their arrival, they missed by 12 minutes the entrance of *Grosvenor House*. Scott and Campbell Black had made excellent time even allowing for a precautionary landing at Kirkuk, an RAF Station 150 miles north of Baghdad which had appeared below the aircraft at a time when the crew were growing concerned at the lack of a positive fix. They were running

France, forced down by the weather at Boulogne, but they were able to press-on and reached Rome by early afternoon. Shaw in the Eagle contrived to spend Saturday night at San Feliu de Guixols, a tiny resort on the Spanish Mediterranean coast, 150 miles across the Golfe de Lion from Marseilles. He sent a telegram to London to confirm he had forced landed and that both he and the machine were undamaged. Meanwhile he was waiting for permission to leave for Marseilles.

The long range machines had been setting a cracking pace towards Baghdad, with the exception of the Granville Gee Bee which was on course to a planned landing at Bucharest. The KLM DC-2 flew directly to Rome and effected a turn-round in less than 30 minutes, passing through Athens in the afternoon ahead of the Boeing, reaching Neirab Aerodrome, Aleppo, in the late evening. The Pander Postjager was through Athens and on its way via the Persian Gulf to Baghdad.

From Mildenhall, Turner and Pangborn had elected to fly their Boeing 247 directly to Athens. There they touched down on what was alleged to have been the only paved runway at any aerodrome (apart from Bangkok) used during the Races. They arrived 45 minutes after the DC-2 had left, pleased with their accurate navigation. The crew arrived complaining of splitting headaches. The heavy cloud cover encountered after leaving Mildenhall had forced them to fly at 15,000ft and headwinds had lengthened their journey. Without an oxygen supply there had always been the dangerous probability of hypoxia. The Boeing spent an hour and a half on the ground before taking-off for Baghdad, a flight estimated at some eight hours. "Leaving Athens was like kissing goodbye to civilisation," said her captain.

Observers along the route were watching for passage of the three Comets, and

short of fuel after threading their way through a series of thunderstorms sweeping across Turkey. Over the Syrian border the clouds broke and by the light of the moon, they began searching for a landing ground when almost immediately the RAF base had appeared below. They landed and took on 20 gallons of fuel before continuing to Baghdad where they were on the ground for only 23 minutes during which time the crew enjoyed a bath and a meal before setting off in pursuit of their arch rivals.

The DC-2 left Aleppo at 11.30pm and arrived at Baghdad shortly after 2.00am when she was directed to the 87 octane refuelling enclosure. A large crowd had gathered inside the area uncontrolled by the police and refuelling operations were hampered. Parmentier signed for his fuel 'under protest' as the delivery metering system had gone off line due to a build up of gas and a choked filter. Stanavo had put huge effort into planning their refuelling operations at Baghdad, paying particular attention to the siting of the enclosures not only to ensure aircraft would have the minimum of distance to taxi post-landing and pre-take-off, but also that spectators would have the best possible view.

After 50 minutes on the ground and to save crew time, the DC-2's engines were started by a representative from the Wright Corporation and as the aircraft moved away from the enclosure a few minutes after 3.00am for the five hour run to Jask, the Dutch Pander touched down.

Cathcart-Jones and Waller in the green Comet had encountered the same violent storms over Eastern Turkey and had flown at 15,000ft then 17,000ft in an effort to escape the buffeting. Waller had complained of feeling unwell, probably due to a lack of

Grosvenor House touched down at Baghdad only a few minutes after the Mollisons had left for Karachi. The aircraft had landed at the RAF station at Kirkuk for a top-up of fuel and was on the ground at Baghdad for less than 30 minutes during which time the pilots each had a bath and a meal.
(National Library of Australia, Canberra.)

The KLM DC-2 arrived at Baghdad at 2.00am on Sunday. Although elaborate plans had been made for the reception of Race aircraft, the public invaded the secure area which, according to Captain Parmentier, delayed the refuelling of the aeroplane. PH-AJU taxied away after almost an hour on the ground and coincident with the arrival of the Pander.
(National Library of Australia, Canberra.)

oxygen, and 'Seajay' had descended to 12,000ft but they had lost visual contact with the ground and gone off track. Believing they had reached Aleppo the crew decided to change course for the run into Baghdad, but they were already well off their intended route and actually over Mosul. Mountains appeared where no peaks should have been, and the aircraft was turned again in an effort to establish a positive fix. The crew realised with a growing feeling of dread that they were lost in the darkness over some of the least hospitable territory on the whole course. Waller was on the point of suggesting that 'Seajay' should dive the aircraft straight into the ground rather than risk a lingering death after a crash landing, but the pilot elected to fly for another half an hour in the search for the flattest territory he could find.

Waller spotted a tiny pinprick of light in the great void a little distance to their east and Cathcart-Jones flew towards it, but even with the full brilliance of his landing light piercing the gloom at low level, the crew could not identify the source of the illumination nor the nature of the territory below.

Suitable terrain or not, there was little alternative but to attempt a landing. Lining up with what they both hoped was an open space Cathcart-Jones eased the now lightly loaded racer onto what appeared to be a reasonable surface where they touched down without difficulty and bumped to a stop after an uncontrollable yaw to the left. In the middle of a rocky desert, surrounded by ragged mountains, the Comet had landed in an uncluttered space beside a railway engineer's construction camp at

What could not be seen by crews in the darkness at Baghdad were the extensive preparations made by the Emergency Services of the Red Cross and Red Crescent who provided ambulances and fire engines and a substantial back-up facility for racing crews and their own personnel.
(National Library of Australia, Canberra.)

Green Comet G-ACSR landed at Baghdad just after 8.00am on Sunday, eight hours behind schedule, following a night spent in the Persian desert at Dizful, some 250 miles beyond their intended destination.
(National Library of Australia, Canberra.)

To expedite her turnround, and to offer relief to her crew, Jackie Cochran arranged for services and facilities to be established at all airfields expected to be visited by the Gee Bee. This refuelling tower was built at Baghdad but, sadly, was never used. The use other personal comforts which had been established were offered to Amy Mollison.
(National Library of Australia, Canberra.)

Dizful in Persia, 250 miles beyond Baghdad. The engineers offered hospitality to the crew, and stared in silent disbelief when they discovered the aircraft had flown directly from England.

At first light the pilots were horrified to see the true nature of their landing ground, strewn with rocks and wadis, poles and bushes, and with an open ditch running at an angle 50 yards from where the machine now stood. Cathcart-Jones realised that if the Comet had not swung on touch down the aircraft would have hit the ditch square on, losing its undercarriage at the very least. His overall impression of the site was that to have attempted a conventional arrival in daylight would have been a nightmare verging on the impossible. Luck, clearly, was on their side.

To have landed in Persia without permission could have resulted in their arrest, but while their hosts offered them cold showers and an eight course dinner, the Commandant of the local police was summoned to stamp their passports after which an armed guard was posted on the aeroplane and the pilots were shown to their bedrooms.

They both suffered a disturbed night and hardly refreshed, quietly made preparations to leave at first light when they hurriedly persuaded the machine back into the air, rejoicing at her solid construction. They were slightly anxious that in spite of their hospitality, the Persian authorities might have found some reason to detain them. The Comet landed safely at Baghdad at 8.17am, eight hours behind schedule and with two gallons of fuel remaining.

During a telephone briefing before leaving Mildenhall, the Rumanian Air Minister, Mr Radu Irimescu, had advised Wesley Smith that he should land at the military aerodrome at Pipera because the intended destination, Bucharest's Baneasa Aerodrome, was too small to handle the Gee Bee in safety. The change was not communicated to the Distributia Petrol Company who had pre-positioned three refuelling wagons and barrels of petrol thought sufficient for ten Race aircraft. First arrivals were anticipated from 3.00pm but in the absence of communications there would be no advance warning. At 4.30pm an aircraft was spotted flying towards Pipera identified as Race No. 46, and personnel and equipment were immediately transferred from Baneasa by road.

Having flown at 14,000ft in freezing conditions without oxygen and unable to reach their flasks of hot drinks, Wesley Smith and Jacqueline Cochran approached Bucharest on instruments, slowly descending over the Carpathian Mountains through a heavy overcast and switched over to their reserve fuel tank. The engine started coughing and Smith's instinctive reaction was to open both canopies in preparation for an escape by parachute. The Gee Bee was not a friendly aeroplane at the best of times and deprived of motive power she was expected to react like the proverbial brick. Smith's canopy slid back easily but the rear cockpit cover jammed and Smith remained at the controls as Miss Cochran made a desperate attempt to discover the source of the engine trouble. Only now, having to use the reserve tank for the first time, did they realise that the change-over cock had been labelled incorrectly, and instead of the tap opening the appropriate valves, the co-pilot's action had shut-off the petrol supply completely. Having established the cause and restored the flow, the engine picked up.

But their troubles were not yet over. Selecting flap for a landing approach, Smith was horrified to feel through the controls that one section of flap obviously had failed to extend and only his quick reactions at about 200ft prevented the aircraft from rolling into the ground. They managed to regain height and the pilot persuaded the extended flap back into neutral by which time the rear canopy had adopted a fully open position. A second approach had similar characteristics and the crew decided after an exchange of written notes that if they could not land at the third attempt they would climb to height and abandon the aircraft in favour of an arrival by parachute.

With no flaps extended to exercise any degree of moderating influence, the Gee Bee landed at high speed, stopping only just within the confines of the military airfield and causing damage to the undercarriage. Jackie Cochran's first confessed action was to apply her lipstick in anticipation of meeting her supporter, Radu Irimescu. With few regrets from crew or impartial observers, the machine was withdrawn from the competition. In a sporting gesture towards the only other woman pilot in the Race, Jackie Cochran cabled Baghdad to offer Amy

While the Speed Racers were already into the Middle East and Asia, those following the Handicap Rules were still pressing-on through southern Europe. With Henry Walker in the front cockpit, the New Zealand Hawk arrived in Rome three hours after leaving Marseilles and, after a 25 minute turnround, the crew elected to go on to Athens, a further 654 miles.
(National Library of Australia, Canberra.)

The New Zealand Dragon Six, ZK-ACO, was delayed in Rome by a minor engine fault which permitted the Miles Hawk to catch up and for both crews to review some of the latest male fashions being paraded.
(National Library of Australia, Canberra.)

Mollison use of all the services and supplies laid down for the expected transit of the Gee Bee.

Elaborate preparations had been made along the route estimated to have cost more than the £10,000 prize money. American ground engineers had been positioned to all planned stopping points and in addition to the special towers built for the express delivery of fuel, private baths and hot and cold showers were installed. At Darwin it was planned that Jackie Cochran would be met by a trained nurse who would provide a body massage. An American chef was positioned at Charleville with whom an order for 'chops and parsley sauce' had already been lodged.

On arrival in Bucharest Jacqueline Cochran was described as wearing a 'light blue flying suit which became her slim figure', and now was anxious to visit the local shops before they closed to buy a top coat. Her wardrobe had been sent forward to Melbourne in anticipation of celebrations at the end of the Races. The crew stayed overnight at the Athéné Palace Hotel and after buying some new dresses, Miss Cochran left Bucharest, catching the Orient Express bound for Paris while Wesley Smith's task was to get the Gee Bee airworthy and then dismantled for transport to a seaport from which she could be shipped back to the USA.

McGregor and Walker continued to make steady progress at an average speed of just over 120mph in their bargain Hawk. They reached Rome a little more than three hours after leaving Marseilles and encountered their compatriots with the Dragon Six, delayed with a minor engine snag. Having already flown more than 1,000 miles after a sleepless night, there was some debate

Donald Bennett standing outside the door of the Lockheed Vega G-ABGK on arrival in Rome whilst James Woods emerges from the cabin clutching the Race documents. By the end of the first day the Vega had flown on to Athens. (National Library of Australia, Canberra.)

about continuing, but within 25 minutes the Hawk had been refuelled and was airborne for Athens, 654 miles to the South East. Guided by the many lighthouses along the coast of southern Italy the Hawk flew mostly in darkness and arrived overhead Tatoi Aerodrome at 10.20pm. McGregor was exhausted, and inadvertently flipped off the magneto switches during the approach. The propeller windmilled until just after the aircraft touched down barely inside the boundary fence and, believing they were out of fuel, Walker gathered up the log books and ran 1,000 yards to the official checking-in position. *Manawatu* was pushed into a hangar where five gallons of petrol were discovered in the tank. There followed a frustrating 90 minutes' search for an engine fault that never existed, depriving the crew of urgently needed sleep.

By the end of the first day all machines entered for the Handicap Race could be accounted for: the Hawk and Vega were at Athens. Rome was overnight host to the Desoutter, Courier, Fairey IIIF, Puss Moth, Monocoupe and Dragon Six. The Falcon was waiting at Istres and the Eagle in Spain. Gilman's Fox was unserviceable at Marseilles; Stack's Viceroy and Parer's Fox in Paris.

In the Exhibition Hall of Australia House in London, a model of the route had been laid out. Measuring 50ft long by 5ft wide, tiny model aircraft representing each of the contestants were placed in their respective positions as each new report was received. At the end of the first day, only one aircraft had officially retired, but others were not in good health. The relative positions on the model confirmed that as expected, the high profile machines and their crews were making the headlines but, in a classic hare and tortoise analogy, the smaller fry were making ground with the minimum of fuss and great expectation.

11

In Search of a Prize

Scott and Campbell Black had flown their Comet G-ACSS to Allahabad direct from Baghdad and were unaware of the dramas behind them at Karachi. To complement his incongruous desert outfit, Scott was also wearing a pair of carpet slippers.
(National Library of Australia, Canberra.)

ON SUNDAY, 21 OCTOBER 1934, dawn broke at staggered intervals for the Race machines now spread along some 3,000 miles of the route from Mildenhall to Melbourne.

Less than half an hour after their arrival from the middle of nowhere and, following a quick breakfast with Geoffrey de Havilland's middle son, Peter, who happened to be in the area on Company business, and an old friend of Seajay's, French racing pilot Edouard Bret, Cathcart-Jones and Waller left Baghdad at 8.39am for Karachi. Accompanied by a plume of blue smoke and cockpit indications of falling oil pressure on G-ACSR's starboard engine, Waller turned back to the aerodrome, encouraged by Seajay who was convinced the aircraft was on fire. They abandoned an attempt to dump fuel when the cocks only

Woods and Bennett in the Lockheed Vega touched down heavily in a patch of loose sand on arrival at Aleppo and the previously sticking oleo may have been a contributory factor in turning the aircraft onto her back. Damage was considerable and both pilots were injured.
(National Library of Australia, Canberra.)

released a spume of petrol that narrowly avoided a hot exhaust pipe.

The aircraft returned to land with 260 gallons of petrol still on board and Waller found that at the approach speed of 135mph the Comet was more controllable than he had been led to believe. Only Hubert Broad who had tested all three machines at Hatfield had previous experience of maximum weight landings. With both pilots holding on to the controls, the throttles were closed and the ignition switches turned off immediately the Comet crossed the boundary fence and the aircraft was allowed to roll-out with gentle braking after touching down at about 100mph.

The de Havilland Company had placed servicing engineers at each of the Control Points primarily to expedite the passage of the three Comets but with a brief also to assist aircraft with de Havilland engines and any other competitor should circumstances permit. Their man in Baghdad examined G-ACSR's temperamental engine and discovered a partial seizure caused by insufficient lubrication and was soon busy changing number four piston and cylinder on the starboard engine, working in the full glare of the sun in the middle of the aerodrome with no facilities whatsoever. Cathcart-Jones later described it as "one of the finest pieces of servicing I have ever seen." With the assistance of an American ground engineer who volunteered to help, the job was completed in six hours and the green Comet left for Karachi at 3.15pm, eight flying hours away along the length of the Persian Gulf.

Woods and Bennett made good time in their Vega, transiting Rome and arriving at Athens in the dark, leaving again after a brief rest for Neirab Aerodrome, Aleppo in Syria. At 10.00am they were circling the aerodrome trying to decide which would be the best direction for landing on what was reported to be an unstable surface. They were unlucky in that their eventual choice took them through a patch of loose sand. They touched down heavily and the undercarriage dug in causing the machine to flip over onto her back.

Bennett was flung the entire length of the cabin, injuring a knee and crushing three vertebrae; Woods gashed his forehead. They were lucky, but the aircraft was badly damaged and out of the Race. Afterwards, Bennett considered that a sticking oleo may have been a contributing factor. They had released a binding oleo at Heston the day before their arrival at Mildenhall and he believed the leg may have stuck again on take-off from Athens. The landing at Aleppo was firm and the substantial gear did not fully absorb the energy, putting them into an uncontrollable situation.

Woods scribbled a message which he asked Mr Mazloumiam, Liaison Officer between Captain Schurck, French Air Force Control Officer and the Royal Aero Club, to cable to his wife: *"Regret u/c collapsed Aleppo M/C out of the Race inform MacRobertson + Horry Love Jamie."*

Continuing from Athens on Sunday morning after a six hour overnight stop, McGregor and Walker flew the Hawk to Nicosia, much of the sector being over water. The turnround was expeditious, only 15 minutes, as time spent on the ground in Cyprus counted against them. Now the leg into Aleppo was flown in driving rain and poor visibility. At first they had difficulty in locating the airfield, but their attention

The tried and trusted method of recovering an aeroplane from the inverted position on the ground is to stand her tail up by the judicious use of managed manpower manipulating ropes, and to pivot the aircraft on the propeller boss.
(National Library of Australia, Canberra.)

James Woods, with a bandaged head, surveys the badly-bent propeller whilst the displaced engine in its mountings dribbles oil into puddles on the concrete apron.
(National Museum of Australia, Canberra.)

eventually was drawn by the sight of the Vega which had crashed three hours before, and they landed warily.

Sportsmen that they were, the New Zealand crew delayed their departure for Baghdad until after Jimmy Woods had written a letter to his wife in St. Kilda, Victoria, explaining his unhappy predicament. His faith in the Hawk was well justified for a week later the letter was delivered to his home address, postmarked Melbourne.

After narrowly missing a mountain peak while cruising at 5,000ft in broken cloud after leaving Athens, Roscoe Turner and his crew arrived safely in Baghdad at 3.00am. Their first approach was too high, too fast and downwind. On the second Turner attempted a touch-down close to the lights indicating the airfield perimeter which caused concern to those watching as the aircraft flew close to structures erected in the refuelling enclosure. The Boeing crew complained that they had been unable to raise the station by wireless. Their aircraft carried sophisticated communications equipment operated throughout the Race by its designer, Reeder Nichols. During the flight across Europe Nichols had been in constant touch with marine stations and an American news agency which was contracted to buy telegraphed words at a flat rate. An unforeseen drawback was that the loop aerial mounted on the cabin roof could not be rotated; the aircraft itself was required to manoeuvre when using the on-board direction-finding facilities, a feature which was thought to have contributed most to their communications' problems. The Boeing was to go off-course frequently during the Races, most notably at a particu-

larly critical time during the latter stages in Australia. After exactly half an hour on the ground in Syria, refuelled and refreshed, the team left for Karachi where they arrived ten hours later having seen nothing below them in all that time except what they described as 'rocky desolation'.

From Baghdad the Mollison's *Black Magic* was clearly in the lead and going well and Jim had made the decision to fly to Karachi rather than risk the direct flight to Allahabad. After a seemingly uneventful trip they landed heavily at Drigh Road Aerodrome at 10.15am a little over 20 hours since leaving Mildenhall, establishing a new record-time from England to India in the process which, possibly, was one of the reasons for that routing. They were away again after an hour's rest, determined, in spite of the extra sector, to keep their current advantage through the Control Point at Allahabad, but their heavy landing had distorted the undercarriage and it refused to retract into its housing. After ten minutes airborne they decided there was no alternative but to return to Karachi. Flight with the undercarriage down would have slowed them intolerably and reduced their range. In addition, the air-cooled engines would have overheated unless the wheels were at least partially stowed. The lack of engine cooling was a problem already experienced on the ground during prolonged taxying.

The undercarriage fault was rectified by engineers organised by de Havilland's Indian Associated Company in Karachi who also had responsibility for Allahabad, Calcutta and Singapore. The Comet took-off again at 6.30pm, only for the crew to discover that they had left behind the sector maps specially prepared for them by the Aviation Department of the Automobile Association. Amy later told a reporter that they had returned because Jim believed the operation of the undercarriage was so smooth it was not working at all and he could not trust it. Whatever the reason the black machine was forced to land back after an hour's flight, this time at a heavy weight and correspondingly high landing speed. The Comet rolled to a stop in the middle of the aerodrome where it remained for some time with engines idling. An official drove out to see if all was well and what assistance might be rendered but above the musical tick-over of the engines, he could gauge the magnitude of a furious argument between husband and wife. Discreetly, he withdrew from earshot, and eventually the aeroplane taxied in and shut down.

The combined delays ensured that a night landing was now in prospect at Allahabad and as he had previously declared, Jim Mollison was not prepared to take that risk although at Baghdad he had had no choice. He promised to make a decision at about midnight. Meanwhile, a ground mist had developed which effectively cancelled any immediate plans they might have had for departure. Reluctantly, Jim and Amy were forced to catch up on some rest. Their enforced nap eventually extended to 15 hours. There was already some speculation about Jim Mollison's night vision and it was no secret that night landings were beginning to cause him problems, a condition which was to remain with him for the rest of his flying career.

Cathcart-Jones and Waller might have overtaken *Black Magic* at Karachi had it not been for a loose petrol filter on the starboard engine which caused an infuriatingly obstinate petrol leak which delayed them. During the rectification, Jim and Amy

James Woods with the bandaged head and the Control Officer at Neirab Aerodrome, Aleppo, Captain Schurch, greeting Malcolm McGregor on the arrival of the Miles Hawk. The New Zealand crew had flown in rain and poor visibility from Nicosia and managed to locate the aerodrome after seeing the overturned Vega. (National Library of Australia, Canberra.)

The Boeing 247 arrived at Karachi after a ten hour flight from Baghdad having seen practically nothing en-route to assist in navigation. The fact that the loop aerial on top of the cabin did not rotate meant that the aircraft itself was forced to manoeuvre when using the radio direction-finding apparatus.

Black Magic damaged her undercarriage in a heavy landing at Drigh Road Aerodrome, Karachi, on arrival from Baghdad. Following rectification by de Havilland engineers she left for Allahabad but returned after an hour. Jim said he had left his maps behind but Amy suggested Jim did not trust the smooth operation of the undercarriage.

Grosvenor House at Bamrauli Aerodrome, Allahabad, being refuelled from some of the extensive supply of barrel stock petrol laid down to cover all contingencies but also in the expectation of a much larger field.
(National Library of Australia, Canberra.)

Due to her greater uplift fuel was supplied to the DC-2 at Allahabad from a bowser. To expedite turnrounds and avoid congestion all aircraft at Bamrauli Aerodrome were directed to stop within clearly defined zones marked out in white paint.
(National Library of Australia, Canberra.)

Mollison left for Allahabad, taking-off in a cloud of dust after which the aerodrome was immediately shrouded in a blanket of sea mist, reducing visibility to 20 yards. The foggy conditions persisted all night but lifted suddenly as the sun rose above the horizon to signify Monday morning. Seajay and Waller had been waiting in the cockpit of G-ACSR and at the first signs of improvement they started their engines and took-off.

Unaware of the dramas at Karachi, Scott and Campbell Black, operating direct to Allahabad from Baghdad, took the lead and in spite of flying off-course for a time, they safely arrived at Bamrauli Airfield at 2.18pm, a 2,300 mile non-stop flight completed in just under 12 hours. The turnround was swift, less than 40 minutes and soon they were on their way to Singapore, seemingly untroubled and convinced in their own minds that they could not be beaten.

An encampment had been set up at Allahabad consisting of over 100 tents and marquees offering hospitality to invited guests and Race crews. Jacqueline Cochran's enclosure provided a special refuelling service, spare clothes, tools, aircraft parts and refreshments. Everything was overseen by a resident manager and Royal Leonard was on standby to take over from Wes Smith as relief pilot, conforming to the Rules and Regulations which had accepted Jackie Cochran as aircraft commander. Thousands of spectators were entertained between Race arrivals by stunting aeroplanes or local musicians who supplemented the many gramophones.

Although the Maharaja of Jodhpur, His Highness Unmaid Singh Bahadur, had made his private hotel available to Race competitors at Jodhpur at no cost, he was convinced most aircraft would actually route through Allahabad and had arranged for a compound to be organised there in which he would be in residence between 19-23 October. The Viceroy and Countess Willingdon also were present, accommodated in their own special train standing in a siding adjacent to the aerodrome boundary.

From Baghdad, KLM's DC-2 had been ahead of the Boeing by almost two hours and even allowing for a refuelling stop of 25 minutes at Jask from 9.00am, still beat their rivals into Karachi. As no buildings were available at Jask and the weather was ideal, the Control Officer, G B Gelly, hired a tent which he pitched on the aerodrome and slept in it at night in case of an unexpected arrival. Hurricane lamps were left burning as a signal. The Customs officials declared their interest by leaving all their paperwork in the care of the Control Officer for his personal supervision.

The DC-2 was the first of eight aircraft to pass through Jask, more than had been expected. Wanting to be efficient and correct, Gelly added a three minute penalty to Parmentier's time as the KLM captain seemed to be more interested in talking to people who congregated around the aircraft than checking-in. Parmentier later claimed he could not find the tent although Gelly confirmed it was clearly marked with a giant 'C' and a green flag was being flown as required by the Rules.

To reach India, the DC-2 had flown six sectors to the Boeing's three, yet had increased its lead by an additional 90 minutes. Between the twins, the three-engined Pander which had left Baghdad at 3.30am had arrived at and departed from Karachi so that at one stage all three fancied machines were in convoy on the leg to Allahabad.

First to land there was the DC-2, in at 7.10pm and away an hour later heading for Calcutta. The Boeing arrived eight hours behind her in darkness having wandered off track again and due to a stronger than anticipated tailwind overshooting the airfield by 200 miles to the southwest. Reeder Nichols had called Allahabad asking for a radio transmission from which he could take a bearing. His request was answered by a blind broadcast advising that the KLM crew had been able to see the aerodrome beacon from 70 miles and that 'Colonel Turner is lost. The Americans are overdue.' At the time of her arrival in the area the Boeing was encircled by electric storms and sporadic lightning flashes which had been misidentified as the elusive aerodrome beacon.

Repeated requests for bearings were answered in the same manner. With the aircraft now completely lost, the crew discussed the possibilities of baling out or trying to land at the first aerodrome that came into view or even ditching in a river. Turner asked Nichols to transmit an SOS and to report that only ten minutes' fuel remained on board. This message was received and although those on the ground considered this to be improbable an order was passed that all other radio traffic should be stopped until communication had been re-established with the Boeing and some degree of positive assistance rendered to it.

Captain Ede, Deputy Director of Civil Aviation in India, together with an RAF Signals Officer, estimated that from what they already knew the Boeing was probably beyond Allahabad. A bearing was transmitted and the aircraft was directed to fly North West. The Boeing crew soon identified the Soune River beneath them and then at a range of 60 miles the aerodrome beacon which was being intermittently flashed on and off.

Speaking to a journalist from his home town's newspaper after the Race, Turner said, "We coasted down to a landing and taxied up to the gas pumps in front of the hangar. As we slowed to a stop our two motors gave their last gasp." However, though the gauges showed empty, about 100 gallons of petrol were estimated still to be on board, but even so replenishment seemed to take longer than usual. Later it was discovered that fuel delivered into the tanks authorised for use by the Race officials at Mildenhall had been siphoning to another (whose external filler cap remained sealed) as the result of an interconnecting valve having been left open by mistake. The fact that fuel gauges for each of the tanks fitted to the aircraft were registering full was thought originally to have been caused by a mechanical fault.

As it was impossible to load fuel through sealed caps, nobody immediately checked on how much petrol should have been delivered, so the machine was signed up by the Inspecting Officer as suitably prepared for the next sector to Singapore.

And it was there on arrival that the Boeing crew discovered that the Panderjager, which had been running ahead of them, was delayed at Allahabad, and in the most dramatic and conclusive fashion.

It was 9.35pm and pitch dark as Gerritt Geysendorffer and his co-pilot Dick Asjes prepared to land the mailplane within the confines of the 900 yard square of baked mud that constituted Bamrauli Aerodrome, Allahabad. The field was adjacent to the main railway line to Cawnpore, which offered some navigational assistance but otherwise there were few facilities for night operations apart from the luxury of an illuminated landing T, a rotating beacon and a mobile flood-light.

The tri-motor flew a steady approach but observers on the ground were horrified to see at a late stage that one undercarriage leg was only half extended and the other not visible at all. Geysendorffer put the aeroplane firmly down and with wings level stretched the landing roll as long as he could, balancing on one wheel, but as the machine lost speed, the port wing tip brushed the ground causing the Pander to slew round in a shower of sparks and dust. Emergency services were soon on the scene, heartened at least by the sight and sound of the uninjured Dutch crew squeezing themselves out of the cockpit windows, cursing both the aircraft and their luck. When the air was already blue with choice profanities, it was no consolation to be told that only one of the Comets and the DC-2 were ahead.

The mailplane was not as badly damaged as might have been supposed after such an undistinguished arrival; the propeller blades of the port and centre engines were twisted and there was damage to the port wing and engine cowling. The casualty was recovered with the aid of an RAF detachment whose airmen dug a pit under the port wheel to allow it to be fully lowered.

Provisioning the refuelling tower at Allahabad, built for the convenience of Jackie Cochran's Gee Bee, but which was to remain unused.
(National Library of Australia, Canberra.)

The Pander Postjager following her untidy arrival at Allahabad. The bent tip of the port propeller can just be seen underneath the cowling of the centre engine whose own propeller was less badly damaged. Some of the tented encampment which was established around the perimeter of the aerodrome by spectators can be seen behind the aircraft.
(BAE Systems.)

It proved impossible to repair the undercarriage which consequently was locked in the down position. Meanwhile Captain Geysendorffer travelled by train to Calcutta, taking with him the damaged propeller blades for repair by the KLM Service Centre.

During the day after the start from Mildenhall the leading machines were pushing on across the Indian sub-continent and into South East Asia. If the present rate of progress were to be maintained, 'King' Cole's prediction of a winning time of about four days would prove to be rather pessimistic.

The racing crews discovered for themselves that communications between Control and Checking Points were very unreliable, and they had the greatest difficulty in obtaining accurate position reports for any of the other competitors. In most cases the only positive news came from a crew's own observation of another aircraft transiting or lying abandoned as the route unfolded before them.

There were amongst them those who, like the crew of the Pander mailplane, were having absolutely no luck at all. Parer and Hemsworth had fallen in with a group of sympathetic Imperial Airways' engineers at Le Bourget, but even their assistance with plug changing and engine running could not prevent the curious misfiring from reoccurring almost regularly at 30 minute intervals, and eventually a fuel expert was summoned. He declared there should be no further trouble with the Fox after he had purged the system, and that would take about an hour.

On Sunday morning following from his overnight stay at Istres, Harold Brook discovered a fairing on the Falcon's fuselage was broken; an item of little consequence apart from causing a minimal increase in drag. But the matter was viewed seriously by a veritable tide of French engineers whose collective opinion was that a fuselage longeron must have broken, and the aeroplane should be grounded. To investigate, somebody cut an inspection hole in the plywood fuselage decking and after shining a torch into all the dark corners and probing where they could reach, declared nothing to be wrong. Now it was necessary for somebody to demonstrate the art of splicing a plywood patch into the decking to repair the vandalised fuselage. On the point of withdrawing from any further participation in the Races, an infuriated and exasperated Brook was placated by the sight of Stack's Viceroy as it appeared overhead, having got away from Paris but in a less than healthy condition.

The Falcon's total delay at this 'unauthorised airfield' lasted until Wednesday 24 October, by which time the aeroplane was almost certainly out of contention. Brook was angry with the French authorities and his only incentive to proceed was fulfilment of his promise to deliver his passenger to her relatives in Melbourne. The Falcon left Istres for Pisa and then Rome which was reached the same day.

Shaw's Eagle picked up the route again at lunchtime on Sunday when the machine flew into Marseilles from its night-stop in Spain and left early next morning for Italy. The aircraft landed in Rome just before 7.00am, refuelled in 12 minutes and set course for Athens where the pilot had scheduled himself for a good sleep.

The Roman Sunday was busy with a stream of competitors taking advantage of the calmer conditions of the early morning. Hansen and Jensen in the Desoutter had rested sufficiently since their arrival from Marseilles to leave just after 1.00am, five minutes later than Davis and Hill in the Fairey IIIF, both aircraft heading for Athens. The Stodarts in their Courier were away at 7.00am followed by the Dragon Six and Jack Wright and John Polando in the little Monocoupe. Polando left behind an astonished Italian policeman who had received an affectionate embrace from the pilot to win a wager with some friends in New York. Melrose waited until mid-afternoon and followed the procession on its way across the Adriatic Sea towards Greece.

The New Zealand crew delivering their Dragon Six arrived at Tatoi Aerodrome in Athens at lunchtime, rested during the afternoon and evening and were away again before dawn on Monday, headed for Baghdad. Maintaining much the same schedule was the Courier, except that the Stodarts had elected to land at Aleppo first. Jimmy Melrose arrived at Athens at 6.34pm, four minutes behind Davies and Hill in the Fairey IIIF, having put down at Janina owing to a shortage of fuel and where he took on just over 18 gallons. Unusual for the young Australian, after a four hour rest he decided to continue through the night to Nicosia where he refuelled with 24 gallons of petrol and left after ten minutes for Aleppo. Wright and Polando touched down in the early afternoon and departed with the dawn on Monday to Nicosia, running an hour ahead of the Fairey IIIF which had chosen to follow the same route. The Falcon arrived from Marseilles via Pisa and planned to stay overnight.

Inbound to Athens a drive shaft had sheared on the Monocoupe's fuel pump, and the aircraft was persuaded with difficulty across the Adriatic to a forced landing in Southern Aetolia. Almost all the fuel had leaked away, and Wright was reduced to using sign language to try to convey his requirements to the local peasant farmers who had quickly gathered and showed great curiosity. As the result of patient negotiation it had been possible to secure just enough petrol to enable the Americans to reach Athens but with absolutely no reserves. Once safely down, Wright was heard to say that an hour earlier he would have sold the aircraft for five cents!

Gilman and Baines meanwhile had battled as far as Rome by early Sunday afternoon, but there was still only continuing frustration for Parer and Hemsworth marooned at Le Bourget. The fuel company systems expert, so optimistic about putting the Fox back into the Race within the hour, had made little progress in his search for gremlins. The aircraft did make one attempt to leave but returned soon after take-off when the same pattern of engine misfiring re-occurred. Ray Parer now withdrew as an official MacRobertson contestant, although he and Hemsworth elected to continue their flight as soon as practicable, heading for Melbourne.

Days later, still with no obvious solution, it was decided on an impulse to examine the insides of the fuel tanks themselves. All three were found to contain petrol heavily contaminated with oil and was reported to be sludgy and hideously yellow in colour. As the engine had run perfectly well in trials it was assumed that something had found its way into the independent three tank system during final refuelling at Mildenhall. How and what was never established, and why was the Fox the only aircraft so affected? Parer made an official complaint to the Shell representative in Paris who was later advised by London that all Race aircraft using that particular spirit had been refuelled from the same tanker and there had been no other complaints.

McGregor and Walker were the first of the Handicap Race crews to touch down in Baghdad where they arrived at 5.00pm after flying from Aleppo mostly at 10,000ft on account of the heat and turbulence at lower levels. The constant running of the Gipsy Major engine at full throttle had induced severe vibration in the nose cowling which subsequently developed extensive radial cracking. Supervised by the crew, local engineers effected the necessary repairs, and after a day's rest *Manawatu* left Baghdad at 1.30am on Monday bound for

The New Zealand Dragon Six attracted a deal of interest as she was refuelled at Tatoi Aerodrome, Athens, from a bowser supplied by the local Vacuum Oil Company agent.
(National Library of Australia, Canberra.)

Harold Gilman and James Baines arrived in Rome early on Sunday afternoon in their Fairey Fox, refuelled and stayed overnight before leaving for Athens at dawn on Monday.
(National Library of Australia, Canberra.)

Bushire.

Before the start of the Race, Captain H Spooner of Misr Airwork, Cairo, had announced that his company would be operating an air service between Cairo and Baghdad for those who wished to see the Race from the air. Presumably he thought that squadrons of aircraft would be jockeying for position on every sector to a pre-arranged schedule. It was a popular misconception.

In England, as Sunday was drawing to a close, many had gone late to bed with the knowledge that the Race was already about half run. Many a newspaper Editor at his midnight conference must have agonised for flexible deadlines as news continued to pour in from points ever closer to Melbourne. By the time the morning newspapers were published their lead stories were already ancient history.

KLM's DC-2 was still hopeful of catching *Grosvenor House* even though she was five hours behind out of Allahabad where she had been forced to return after taking off without one of her passengers, and had yet to transit Calcutta and Rangoon before running down the Malay Peninsula into Singapore.

Captain Parmentier, ever the diplomat and ambassador for his company, told pressmen before leaving Allahabad that so far the trip had been the smoothest in his total experience, and the flight had been running routinely and absolutely to schedule. He confirmed that as previously notified, the aeroplane was not flying at anything above normal cruising speed, but warned his rivals in a somewhat uncharacteristic demonstration of competitiveness that after Singapore, it would be "all out."

The weather encountered by the red Comet and later the DC-2 and Boeing 247 across the Bay of Bengal was as bad if not worse than forecast. Charles Scott later wrote of the experience: "The spectacle in the ebbing sunlight and rising moon inspired terror in both of us. Our preconceived anxieties were soon very real and frightening. A great storm raged right and left of us as clouds swept in battalions above, and still more ominous clouds rolled below. Steadily we flew on between them, with no chance of four hour shifts. Both of us were hard on the job with feet on the rudder bar and hands on the control column."

Grosvenor House descended through a break in the cloud to emerge into clear air north of Akyab, a designated emergency airfield in Burma, situated on the east coast of the Bay of Bengal, midway between Calcutta and Rangoon. Although not a designated Control Point, officials from the Burma Oil Company, Indian National Airways and KLM operated a watch and whilst the team sat smoking in the cool of Monday evening, they saw the Comet pass over at 8.00pm. Mr R Tomsett of Indian National Airways remembered the occasion:

"The sky was clear and wonderfully favourable for the first aircraft. Later the DC-2 passed over and the sky had changed to overcast but a good ceiling was maintained for the remaining machines but the first had the definite advantage."

The Comet flew at 1,000ft across the islands to Alor Star, just south of the Siamese border on the west coast of the Malay Peninsula. Campbell Black signalled that his decision was to go on to Singapore which was as well. The Comet's crew could not have been aware that, in spite of the best efforts of the RAF, Alor Star Aerodrome was underwater and closed.

Reaching Singapore the crew may have been aware of the light beacon provided in the harbour by all the searchlights on board HMS *Terror* whose beams were directed vertically into the night sky. At Seletar Airfield the flare path had been set up incorrectly and consequently the Comet landed downwind, bouncing heavily, lucky to escape serious damage. Observers thought they detected a smoking engine which

A wireless operator at his station in the marquee erected as shelter at Darwin. The Race authorities at all airfields despatching competitors whose next anticipated landfall was Darwin were requested to send departure messages and an ETA to be received well in advance of the arrival of the aircraft.
(Northern Territory Archive Service.)

caused the aerodrome fire service to be scrambled but there was no work for them. After shutting down Scott ordered two glasses of beer and jinked about with nervous energy, anxious to be on his way. *Grosvenor House* had touched down at 5.31am, just a little before the Dutch DC-2 left Rangoon, but when the Douglas was only half-way along the peninsula G-ACSS was already island hopping south of Borneo and aiming for Darwin.

Before he left England, C W A Scott (he rarely used his Christian name) had written a comprehensive description of the different routes that aircraft captains might elect to fly and the reasons why. Familiar with some of them in a solo capacity he was now faced with another crossing of the Timor, about which he wrote:

"The Timor Sea is a lonely stretch of water. The prevailing wind is south-east, which is unfortunate if that is the direction in which you are travelling. It makes the sea crossing so much longer. Hitherto, a terrible feeling of isolation had beset me. In this Race I shall have Tom Campbell Black as companion to share my isolation, two engines and a speed of 200mph. The sea will not seem so terrible with all these compensations."

The Australian communications company, Amalgamated Wireless Ltd., had made arrangements for news of all Race movements at Singapore and Koepang to be instantly transmitted to Darwin where direction finding apparatus had been installed. In addition to similar facilities at Charleville and Melbourne, a listening watch was maintained at La Perouse, the most elaborately equipped of all the Australian receiving stations, and lookouts were detailed at the company's coastal posts at Wyndham, Broome, Geraldton and Thursday Island.

Having shown such promise at the beginning of the Race, the Mollisons had lost all their advantage at Karachi. At 2.00am on Monday they finally had taken-off for Allahabad in the vain hope of cutting into the lead then enjoyed by *Grosvenor House*. Well beyond their ETA there was still no sign of them.

It was inconceivable that two such expert navigators could have become unsure of their position for undoubtedly, both were intuitive creatures. Amy never pretended that she was a brilliant handling pilot but neither did she boast about her navigational skills which had on more than one occasion proved to be uncannily accurate. On this day, instincts failed both husband and wife. A combination of poor visibility from 14,000ft, strong northerly winds different from those forecast, maps of too small a scale for the area and both compasses which had gone off line, resulted in another savage argument and near disaster. They were forced to circle until there was sufficient light to attempt a landing and seek help. *Black Magic* forced landed at dawn in a field near Jabalpur, almost 200 miles to the south and west of their destination; her crew completely lost and desperately short of petrol.

Grateful for assistance *Black Magic* was supplied with 27 gallons of low grade fuel sourced from the local bus station, and hopeful that something might still be retrieved from their situation, Jim Mollison took off and headed towards the Allahabad Control at full throttle and low level.

The Gipsy Six R engines fitted to the Comets were young thoroughbreds, refined in the short time available to Frank Halford and his team to operate at full throttle for long periods. But they were also creatures of temperament, liable to go lame if abused. Continuous high revolutions at low altitude burning petrol suited only to the purposes of the local transport company led inevitably to serious overheating. The warning indications came 150 miles from Allahabad when with Amy flying the aeroplane and Jim navigating, one engine began to misfire. Mollison would not throttle back, oblivious to the protests pressed ever more urgently upon him from the rear seat where his wife, a licensed aircraft engineer, was only too well aware of the seriousness of the problem. They limped into Allahabad on one engine at 5.30am, eight and a half hours after leaving Karachi, misread the signals and landed downwind. The engineers discovered that several pistons in the port engine were cracked and their cylinders badly scored. There were no spares on site with which the de Havilland party could repair the obvious damage, and it was clear that both engines were in distress. Cursing and swearing, the awful truth slowly dawned on Jim and Amy. Once favourites, record breakers to Karachi, mail carriers to the Royal family and now custodians of a crippled aeroplane, they were out of the Race.

Bamrauli Airfield, Allahabad, late choice as a confirmed Control Point, had certainly played bogey in the Race so far. The Pander had arrived in a cloud of dust, the Mollisons had struggled in and withdrawn, Turner and Pangborn very nearly missed the airfield altogether and now on Monday, the green Comet, having arrived from Karachi, was forced to return to the control desk with a propeller snag just before take-off for Singapore.

Black Magic *limped into Allahabad at 5.30am on Monday having become lost en-route from Karachi. The operation of the engines at high power, but using low grade fuel, caused damage to both which could not be rectified on site. As the de Havilland engineers expressed concern they were joined by the Maharajah of Jodhpur and some of his party.*
(National Library of Australia, Canberra.)

At some time during her refuelling stop at Allahabad, the propeller pitch change valves of the port and starboard engines of green Comet G-ACSR were transposed and a propeller altered pitch when taxying out for take-off. Ken Waller and Owen Cathcart-Jones were delayed for two hours while the engineers were engaged in the frustrations of fine tuning.
(National Library of Australia, Canberra.)

The crew realised as they were taxying that one of the propellers had altered pitch. Investigation revealed that during their turnround, somebody had inadvertently transposed the pitch change valves of the port and starboard engines, for which they were individually fettled. The de Havilland team spent a frustrating two hours finely adjusting the automatic change mechanism before they were satisfied and the aircraft could resume her quest for a prize six hours after arrival.

In a classic hare and tortoise situation, some of the slower machines had been making such steady and continuous progress that the green Comet had barely left the Indian Control Point before the Miles Hawk arrived at Jodhpur. Others were well placed in Syria, Iraq and Persia. In Mildenhall, the village shops were opening for business as usual. It was Monday morning.

The bad news came early in the day. Harold Gilman and James Baines had left Rome at dawn bound for Athens. Four hours later due to what was thought to be an engine problem the aircraft attempted to land at the emergency aerodrome at Palazzo di San Gevasis, 90 miles from Foggia near Taranto in Southern Italy. The old Fox appeared to have stalled during her landing approach, hit the ground hard and in a flat attitude about 100 yards short of the airstrip and burst into flames. A local eye-witness immediately summoned help and Colonel Bemerbi from Grottaglia Aerodrome arrived at the scene to take charge but nothing could be done for the crew who had died instantly. The news soon spread along the route creating a sense of shock and sadness, but it was to be the only fatal accident during the Races.

While Scott and Campbell Black sped over the Timor Sea towards an Australian landfall at Darwin, the *Uiver* was preparing to leave Singapore for her sector to Batavia, the last part of her flight coincident with KLM's regular service to the East Indies which the company so dearly wanted to extend through to Australia. A large proportion of the 22,000 items of mail carried on board the DC-2 from Mildenhall was destined for the Dutch colonies which KLM hoped to disembark for local delivery and permission was sought to take on an equivalent load to maintain the declared weights. Under the Rules the Race Committee found it was impossible to grant this request (although substitute crew members were allowed) so an additional quantity of mail was onloaded and carried through to Australia.

The Boeing, en-route to Singapore from Allahabad, had been forced to divert from Rangoon owing to a long detour around the weather resulting from the typhoon off the Malay Peninsula and to land at Alor Star just before 6.00pm in spite of the atrocious aerodrome surface conditions.

"Tropical rains so thick with water that it is impossible to fly through them. You must go above, or around," said Roscoe Turner, adding that the tiger infested marsh along the Bay of Bengal was "so thick with reeds that once your ship lands you cannot take it off; so deep and slimy that you cannot wade to safety."

Arriving at Singapore in the dark at 9.30pm like Scott and Campbell Black before them, the crew of the Boeing was faced with what appeared to be an ambiguous flare-path. Turner was unsure whether there were local hazards and should he land to the left or right of the single row of lights? He and Pangborn later realised they had been laid to indicate the runway centreline.

At Le Bourget Airport, Paris, Parer and Hemsworth, no longer official contestants,

Four hours after leaving Rome for Athens, Harold Gilman and James Baines were killed when their Fairey Fox G-ACXX crashed during an approach to the emergency aerodrome of Palazzo di San Gevasis. The aircraft appears to have hit the ground in a flat attitude after which a severe fire developed in the centre fuselage where the fuel tanks were located.
(National Library of Australia, Canberra.)

crossed the area at a cautious 5,000ft.

Officials at Jask had observed several Race entrants flying over at high altitude en-route to India. They were in constant receipt of telegraphic reports of aircraft movements from the west but were unable to get details of arrivals at Karachi. The Control Officer was not impressed:

"I do not think that the officials there quite realised how important these reports were had there been a forced landing in a country infested with armed tribesmen and not a single policeman for 600 miles, where water was at a premium and food unobtainable except in the larger villages sometimes 50 or more miles apart."

The Hawk arrived at Jask almost nine hours after leaving Baghdad, and following a 40 minute stop the crew intended to go on to Karachi, a further four hours flying east along the Baluchistan coast, running almost parallel with and just north of the Tropic of Cancer. The flight was uneventful and they arrived in India at teatime, to be greeted with the news that, like the Mollisons some 30 hours before, McGregor and Walker had set up a new class record from England. Not finished yet, the team flew a further sector to Jodhpur where they arrived in darkness at 9.00pm on Monday. The Maharaja's car was sent to collect them and they were treated to a bath, a meal and a long and refreshing sleep in their host's private hotel. In spite of the lavish hospitality and facilities arranged at Jodhpur, the Hawk was the only Race entrant, apart from Harold Brook and Ella Lay in the Falcon who arrived early in November, to transit and to take advantage of them, most preferring to operate through Allahabad.

Meanwhile, others were suffering problems and delays. Davies and Hill in the Fairey IIIF left Athens at dawn on Monday and following safe arrival in Cyprus at 9.00am planned to go on towards Baghdad. During the take-off run an aileron-balance control bungee had broken, and the nearest spares were located at Aboukir in Egypt. Nine days later, after a local garage me-

were still grounded. Their one-time companions in adversity, Stack and Turner in the Viceroy, arrived in Rome during the morning from an overnight stop at Marseilles where they decided that due to the continuing poor performance of their aeroplane, they too would retire from further competition and fly back to England to consult their lawyers.

Now it was the turn of Baghdad to witness a flurry of activity. The Desoutter arrived from Aleppo at lunchtime on Monday on its way to a nightstop at Bushire on the eastern side of the Gulf, whilst the Stodarts in the Courier which arrived an hour after the Danes, stayed for a scheduled period of rest. They were joined later by Jimmy Melrose in his solo effort with the Puss Moth. Some hours earlier the Dragon Six had landed, flying on one engine. After a minor adjustment the crew had a bath, a meal, cold drinks and suitably refuelled left for Bushire from where they decided that as matters were progressing so well they would go on to Jask, well behind their compatriots in the Miles Hawk who had taken-off from Baghdad during the early hours. After a four hour flight the Hawk had landed in darkness at Bushire aided only by the light from a few feeble hurricane lamps and left again within 45 minutes for Jask, a tiny British possession beyond the Straits of Hormuz at the mouth of the Gulf. The New Zealanders had been warned about the local tribesmen who had adopted a habit of shooting at passing aircraft, and although McGregor dismissed as 'unlikely' the chances of being hit, he

Not the popular image of a pilot racing halfway around the world, David Stodart takes time at Aleppo to complete paperwork for the Control Officer who experienced visits from nearly all the competitors in the Handicap Race.
(National Library of Australia, Canberra.)

Jack Wright and John Polando arrived at Aleppo from Nicosia on Monday but suffered a petrol feed problem which caused them to delay departure until the following day. They managed to fill their time on the ground by cultivating new friends and interests.
(National Library of Australia, Canberra.)

Two sick aeroplanes were left at Allahabad as the other speedsters headed for Australia. Black Magic can hide her troubles but the bent propeller on the Postjager's centre engine is evident for all to see.
(BAE Systems.)

chanic had made and fitted a replacement part, they were able to resume their journey but with little hope of qualifying within the time limit. Wright and Polando took 15 minutes to refuel in Nicosia, leaving less than an hour before the old Fairey biplane arrived. They made good time to Aleppo in northern Syria but were suffering from a petrol feed problem, possibly associated with their earlier difficulties, and were delayed until Tuesday morning before setting off for Baghdad.

But before Monday was over, two Race machines had landed on Australian soil. In London's Australia House, aircraft designer and racing pilot Edgar Percival moved the tiny aircraft models on the giant map in the Exhibition Hall to mark a very significant moment in aviation history. In the outside world, all eyes now were on Australia.

12

A Dream comes True

Charles Scott and Tom Campbell Black during their turnaround at Charleville on Tuesday 23 October looking tired and anxious. Would Charleville be their last stop before Melbourne? Would the Comet's engines run sufficiently well to see them safely across the finishing line?
National Library of Australia, Canberra.

DARWIN WAS ALWAYS likely to have been a focal point of world attention as the Australian landfall for all competitors even though, once established on the continent, there was still well over 2,000 miles to the finishing line at Flemington. The petrol companies were particularly anxious to ensure no racing aeroplane in transit lost time on their account and went to enormous trouble in an effort to cover all possible contingencies. Both Shell and Vacuum Oil sent special refuelling wagons to Darwin. It was essential to ensure they were in position well outside the rainy season lasting between November and March during which period overland transit was impossible.

The steamer *Mangola* arrived with supplies of oil, some special aviation spirit and support equipment shipped in from Melbourne, Sydney and Brisbane to cope with the rapid turnround of racing aircraft. It proved impossible to make an accurate assessment of required quantities so on best guesses Shell provided 16,500 gallons of 77 octane and Ethyl 87 octane which arrived in July, shipped in from Singapore on board the SS *Volsella*. In addition, 10,000 gallons were delivered for Newcastle Waters, travelling from Darwin by train to Birdum and from there by AEC Roadtrain.

The Shell company was contracted to supply fuel for Jackie Cochran and an American engineer, M G Clarke, erected a special rig similar to others back along the route and sadly all redundant. L E Taplin was drafted in to supervise arrangements for Bernard Rubin's green Comet and H E Gordon Nicholls was in charge of the Vacuum Oil company's operations. Shell suffered a tragic loss when their nominated representative, H E Hendrickson, en route to Darwin from his base at Brisbane, was killed along with another passenger and the pilot when their DH.50J mailplane, operating a scheduled service, struck the ground in open country during a severe dust storm 15 miles south east of Winton. Mr Hendrickson was replaced at short notice by Mr F Briggs.

Electric lighting became an integral part of the Darwin operation. Two mobile stands were imported to illuminate the runway, sited in accord with the prevailing wind. Orange lights marked the aerodrome boundary whilst red warning lights were carried on the roof of the Fanny Bay Gaol situated between the aerodrome and the sea. A rotating aerodrome beacon lit up the clouds and its beams were reflected from the walls of the prison. A detachment of the Royal Australian Engineers under Captain R R McNicoll erected a Clarke Chapman searchlight at Dudley Point which was scheduled to operate between 7.00pm and 11.00pm on Monday and Tuesday, 22 and 23 October, and which was supplemented on Tuesday and Wednesday by a second light directed horizontally from Dudley

Amongst other facilities prepared at Darwin the Shell Company positioned a Sussex bowser with a capacity of 700 gallons which could deliver fuel simultaneously from four hose locations. It was essential that all vehicles despatched to Darwin by road arrived well before the start of the rainy season.
(National Library of Australia, Canberra.)

As part of the immense contribution made by the fuel companies, 10,000 gallons of aviation petrol were shipped from Darwin to Birdum by the railway and thence to Newcastle Waters by AEC Roadtrain. The logistical exercise was planned when many more thirsty aeroplanes were expected to take part.
(National Library of Australia, Canberra.)

The first contestant to reach Australia was No. 34, Comet G-ACSS, Grosvenor House, *which touched down at Darwin just after 8.00pm on Monday. The port engine had been nursed for several hours during the flight across the Timor Sea.*
(National Library of Australia, Canberra.)

The Control Tent at Darwin was equipped not only with a wireless facility but also a table covered with charts and navigation instruments for the use of flying people in a hurry. Arriving aircraft were inspected by the health authorities in an isolated compound before release.
(Northern Territory Archive Service.)

Point straight towards the aerodrome.

HMAS *Moresby* had docked at Darwin prior to leaving for Wyndham in Western Australia to survey the harbour from where she sailed to her pre-determined position in the Timor, acting as a track-check vessel outbound from Koepang to Darwin with orders to display all her lights.

Probably unaware of just how fatigued the leading crews were likely to be, on Monday Race Control at Melbourne sent an optimistic request by cable addressed to nobody in particular at Darwin Aerodrome where it was delivered to Vacuum Oil's Assistant Representative Mr W Hannaford:

"Centenair Melbourne to Darwin. Request two leading machines to fly around Flemington Race Course at 1200hrs 23 October if practicable and crews to subsequently attend official reception at 1230hrs.

If aircraft arrive Melbourne before 1200hrs they will of course proceed to Terminal aerodrome which will be allocated Charleville.

Above arrangements suggested in order that crowds attending Flemington reception may see leading aircraft in flight. Racing aircraft will not land at Flemington.

Wire immediately if pilots in charge of leading aircraft will fall in with these arrangements."

The first joyous news of the Comet's arrival at Darwin was tempered when additional information gradually filtered through. The aircraft had flown for more than two hours over the Timor Sea with the port engine throttled down due to an indication of practically no oil pressure. *Grosvenor House* had landed safely in the dark within the marginal perimeter of Darwin Aerodrome, the thick layer of surface dust which had threatened safe operations softened by the first rain in seven months.

It was eight minutes after 8.00pm and every citizen of Darwin seemed to be there, brushing aside the temporary barricades set up by the police and flooding around the aircraft, eager to lay congratulatory hands on both it and the exhausted crew. According to one eye-witness, at the end of her landing roll the Comet remained stationary with the engines switched off. Charles Scott was found lying under a wing exercising his right leg which was causing him considerable pain due to cramp after the efforts of maintaining right rudder for two hours to compensate for the loss of the port engine. Race officials were concerned enough to insist that he paid a visit to the local hospital.

Scott and Campbell Black were more worried that what had all the signs of a partially seized engine would mean the end of their race but, while they refreshed themselves, Charles Tuckfield, a Sydney based engineer whom the de Havilland Company had sent to Darwin in anticipation of just such a situation, began his first inspection.

Frank Halford, proceeding by bus along Regent Street to his London office on the same morning that the Comet had arrived in Australia, was told by a conductor that *Grosvenor House* had retired at Darwin with engine trouble. A reporter from the Press Association was waiting for Halford when he arrived at his Golden Square premises and volunteered use of the open telephone line to Darwin which had been arranged by his agency. Within minutes engine designer, engineer and Race pilots were talking across 10,000 miles and suggesting that amongst other things, all six specially 'cast' pistons in the port engine should be changed for standard forged pistons if Tuckfield could make the necessary arrangements. It was also agreed that after the pistons had been changed the aircraft would take off on both engines but continue the flight with the port motor throttled back.

Grosvenor House arrived at Charleville, only 800 miles from the finishing line, with an exhausted crew and the port engine still exhibiting signs of trouble. A huge turnout of members of the local community, some of whom can be seen alongside the fence behind the wingwalkers, was summoned by a blast on the whistle of a steam locomotive standing in the town's marshalling yard.
(National Library of Australia, Canberra.)

Part of the reception committee that greeted the arrival of Charles Scott and Tom Campbell Black at Charleville. After the crew had disembarked the aircraft was wheeled into a hangar where a thorough investigation of the port engine was given top priority.
(National Library of Australia, Canberra.)

Charles Scott suffering badly from cramp in his right leg, is assisted to the Control Point by a local journalist and police Inspector O'Driscoll. Scott is still wearing his heavy coat and woollen scarf and a pair of carpet slippers.
(National Library of Australia, Canberra.)

gested the motor should be nursed. There was no point in losing the Race having reached Australia when there was so much to gain from judicious throttle handling.

Two hours and 26 minutes after her Australian landfall *Grosvenor House* was on her way to Charleville, the last of the compulsory Control Points before Melbourne. The take-off run appeared to be longer than usual but otherwise without drama, and seeing them safely away Tuckfield signalled ahead to Dudley Wright at Charleville and the de Havilland works at Mascot Aerodrome, Sydney, suggesting that engine spares might be made ready for the possibility of a double cylinder head change as he suspected burned valves in two heads at least. Tuckfield later considered the reason for the extended take-off run might have been caused by the port propeller set in coarse pitch. He remembered one of the crew using a bicycle pump and remarking that "it is not responding!"

From Darwin the aircraft carried just sufficient petrol plus a small margin for what the crew hoped would be the penultimate leg, so it was with alarm that Scott, deprived of sleep, realised he was repeatedly signalling to himself a view that they were off course and lost over Queensland. From his captain's position in the front seat, Scott certainly recognised nothing through his weary eyes even after so many years of flying operations in this same area. But maintaining their course, the line of the Charleville railway slipped into view and

However, Tuckfield's examination revealed that the engine was free turning, half full of oil, showed signs of low compression on two cylinders but definitely was not seized. After changing spark plugs and cleaning a filter that was completely clogged by what Richard Clarkson later described, in a swipe at Arthur Hagg's decision to reduce the area of the engine air intakes, as 'sludge of incinerated engine', he thought all would be well. Any further low oil pressure indications were likely to be due to a rogue gauge rather than anything more serious. He considered it unnecessary, therefore, to arrange a change of pistons now or down-route but with over 2,000 miles of flying still ahead, he sug-

Scott was persuaded to make a speech to the crowd which had gathered at Charleville to welcome the scarlet Comet and his words were broadcast by the local radio station. As was his manner, Tom Campbell Black remained silent and quickly slipped away to enjoy a quick wash.
(National Library of Australia, Canberra.)

Looking more relaxed but anxious to learn about the state of their engines, the crew of Grosvenor House *enjoy breakfast in a tent at Charleville Aerodrome. It was a welcoming and civilised gesture with white tablecloth, real china and a vase of flowers.*
(National Library of Australia, Canberra.)

the aeroplane was swung to starboard to correct the drift which had taken them too far east.

Charles Tuckfield's messages from Darwin had alerted the party of six engineers led by Dudley Wright, on loan from Qantas, who had organised a spares pool including a mainwheel complete with tyre and tube and all the necessary jacks and trestles in case a change was necessary, battery, cylinder head with valves, a piston and rings, a pair of magnetos and various gaskets, joints and other consumables.

The previous evening Wright had received a telephone call from Arthur Edwards in Melbourne seeking assurance that the engineers would do all they could. Wright confirmed that, of course, they would, in spite of never having seen a Comet except in pictures although he had been furnished with all the necessary drawings and briefing notes and was familiar with the Gipsy Six engine. As early as 10 August, the de Havilland Aircraft Company had circulated to all interested parties details of the location and capacities of fuel tanks, working arrangements of fuel caps, including the inside diameter of the filler necks, specifications for fuel and oil and a full description of jacking points, spreaders and heights required in order to remove a wheel with a deflated tyre and replace it with a serviceable spare.

At 6.50am a locomotive in Charleville's railway marshalling yard blew its whistle as a prearranged signal to indicate that a message had been received indicating that a competing aircraft was within an hour of the airfield but, due to the crew's disorientation, the red Comet did not arrive until 8.40am, making a perfect touchdown nine hours after leaving Darwin and with only 10 gallons of fuel remaining. The crew was anxious for news of their rivals and to liaise with the de Havilland engineering team which had immediately congregated around the port motor.

Scott was reported as looking haggard, worn and unshaven and could only speak in a whisper. He almost collapsed from the severe cramp which again afflicted his leg and was assisted by the local police chief, Inspector O'Driscoll, to the checking-in facilities. "Tired, tired, God I'm tired," he said, "stiff and sore all over!" But he had composed himself sufficiently to make a short address from a dais to the huge crowd which was transmitted live over the local radio. In reply, Grace Allen, the ten year-old daughter of the Town Mayor, welcomed the crew 'on behalf of the children of Charleville'.

Just 800 miles separated the Comet from the finishing line and, in addition to fame if not fortune, the crew relished the prospect of a relaxed meal and deeply refreshing sleep. Confirmation that the DC-2 had passed through Darwin and was already en-route to Cloncurry did little to raise their morale. Ninety minutes after landing, the

Dudley Wright manoeuvring the port propeller just sufficiently for the members of his team to check gaps and adjustments before refitting the cowlings. Wright had arranged for a wide variety of spares and equipment to be in position at Charleville with most emphasis placed on the necessity of having to replace a punctured tyre. (National Library of Australia, Canberra.)

Two members of Dudley Wright's team of engineers making final adjustments to the port engine of Comet G-ACSS at Charleville. The protective cover is probably in position to prevent oil splashing down onto the tyre in a similar manner to that displayed on the engineer's right arm and rolled-up shirtsleeve. (National Library of Australia, Canberra.)

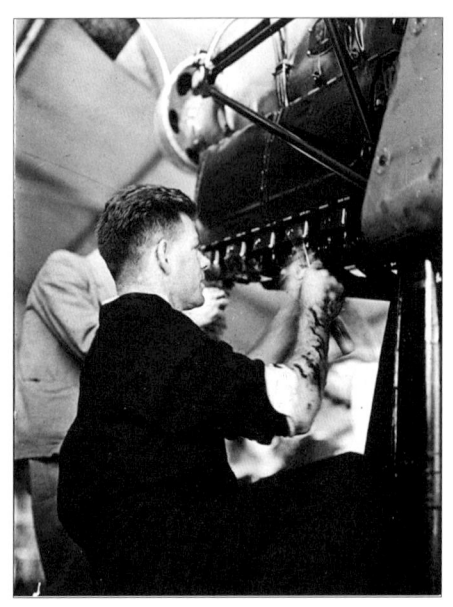

cowlings were still lying on the floor of the hangar. Once the engineers had completed their thorough external check of the oil system, tightened cylinder head nuts and adjusted valves and were satisfied, they closed up the motor which started easily and ran smoothly. Dudley Wright expressed his optimistic opinion that provided it was not over-revved, the Gipsy Six engine would continue to function satisfactorily with a reduced oil pressure.

The Comet was taken-off without any problems at 10.30am, two hours after arrival, and the crew settled onto what they hoped would be their final course, resisting the urge to fly 'all out', to quote their arch rival Koene Parmentier, but the oil pressure light began an erratic blinking and they turned back to Charleville, sick at heart. In London, Frank Halford was advised that the abort was due to the crew forgetting their maps. The engineers made further adjustments and again expressed satisfaction. At one minute to eleven o'clock Scott and Campbell Black eased G-ACSS into the air again for what they desperately hoped would be the last time.

The oil pressure warning light did not flicker and the port engine ran sweetly, although at reduced power. In order to remain fully alert during what would be a crucial final stage, the pilots each flew the aircraft in ten minute cycles as the hours ticked slowly by, conscious that on latest estimates of the DC-2's progress, *Grosvenor House* was still about nine hours ahead. Time often appeared to them both to stand still. They flew overhead Bourke, 250 miles from Charleville, but decided to continue. Would Melbourne never appear on the horizon? It was Tuesday, 23 October, 1934.

A committee based at Dubbo devised a scheme for the town of Cobar, New South Wales, situated on the Speed Route between Charleville and Flemington, to burn a series of fires set in the form of a triangle pointing south as an aid for competitors flying at night. Instead, however, an agreement was reached with the Occidental Gold Mining Company for a powerful electric light beam to be situated on the main shaft's poppet head as a better alternative. The committee provided the reflectors and a coloured globe which was installed and supplied with power by the company who maintained the beacon for a week after the first aircraft was expected. The red Comet was observed just to the east of the mine at 1.00pm on 23 October and the news was telephoned to the Race Control at Melbourne as agreed.

On Monday 22 October the KLM DC-2 had operated through Alor Star with less trouble than that experienced by some of those following behind and within three hours of leaving there had transited Singapore and was heading to Batavia. Thus far, the crew had been operating along their own scheduled commercial route, but now for the first time, the pilots would be flying over unfamiliar territory, threading their way through the islands of the East Indies and across the Timor Sea to Australia.

At 10.30pm at Rambang Aerodrome, situated adjacent to the Atlas Strait on Lombok Island, Parmentier was advised that *Grosvenor House* was in trouble. The Dutch captain who had announced that it was to be 'all out' from Singapore suggested that Scott had been driving his engines too hard, and expressed his opinion that now was the time for the DC-2 to cut into the Comet's lead.

The crossing of the Timor Sea from Koepang was completely uneventful, and KLM's 'commercial' flight arrived at Darwin at 8.57am on Tuesday, 23 October. There was time enough for all on board to repair to one of the booths which had been erected for the purpose of providing hospitality and where they took a light breakfast of cold chicken and duck. Captain Parmentier was anxious about the scarlet Comet. "Any news of Scotty?" he asked, "I like him and he deserves to win! The Comet is a fast machine in the air but on the ground also. They must take care!"

Gordon Nicholls of Mobile Oil, operating the petrol bowser which the Vacuum Oil Company had pre-positioned from Sydney five days prior to the start of the Races, delivered 267 gallons of petrol and 13 gallons of oil. The aircraft could have operated directly to Charleville but there was no point in taking risks and having been assured that fuel was available at Cloncurry they set course at 9.34am, planning to cruise at 16,000ft.

Flemington Racecourse, already famous for staging the annual Melbourne Cup, richest horse race in Australasia, was situated a little over two miles to the North West of the City of Melbourne. The site was chosen as the finishing point for the MacRobertson Races primarily for its ease of handling what was expected to be a huge welcoming crowd and the provision of other facilities appropriate to an official reception. To complete the Race and be observed by the Stewards, aircraft were briefed to fly above and along the final straight not above 200ft and across a line of white sheets pegged to the ground and neon lights sunk into the topsoil between illuminated pylons.

The prospect of Race aircraft actually landing at Flemington was never contemplated, and arrangements to accept the first competitors to arrive were made at the RAAF Bases at Laverton and Point Cook from where the weary crews were to be ferried back to Flemington in Royal Australian Air Force DH.60 Moths for the official welcoming reception.

It may have seemed insensitive in view of the general celebrations to deny spectators the opportunity of watching the Race aircraft landing, but scenes witnessed at the conclusion of similar events had been

Part of the crowd assembled at Batavia to welcome the KLM DC-2, PH-AJU, when she arrived from Singapore late in the afternoon on Monday.
(National Library of Australia, Canberra.)

No need for runway edge marking at Batavia as the KLM DC-2 touches down to great acclaim from local residents. The aircraft was on the ground for less than half an hour before leaving for Rambang.
(National Library of Australia, Canberra.)

Following a breakfast-time transit of Darwin on Tuesday, DC-2 PH-AJU flew to Cloncurry where there was an adequacy of help for refuelling and oiling before the aircraft left again for Charleville.
(Northern Territory Archive Service.)

heeded by the organisers as a warning of what could be expected if an excited crowd decided to stampede. For the safety of all, not least the arriving aeroplanes and their tired crews, the decision to receive the competitors within a controlled environment received general approval.

Just after 3.30pm a number of light aeroplanes which had been circling with intent above the racecourse, set off through a heavy rain shower to intercept a lone aircraft which had been sighted approaching from the north. *Grosvenor House* came in low and fast, Scott using full power on both engines. She crossed the line at 3.34pm leaving her optimistic escorts to trundle in well behind. Her official time was 70 hrs 54 mins and 18 secs since Sir Alfred Bower's flag had fallen at Mildenhall of which 65 hrs 24 mins 13 sec had been spent in the air.

Aware of previous occasions when racing aircraft had failed to cross the right line, or otherwise finish correctly and subsequently had been excluded, Campbell Black shouted to Scott to circle and cross the finishing line again. The aircraft was now so far in advance of the DC-2 that any errors in finishing could doubtless have been erased by a special circuit flown from Laverton.

The crowd which had been gathering in the cold greyness since 2.30am had been augmented by trainloads of spectators arriving on special services from Melbourne which commenced at 7.30am. A crowd estimated to be in excess of 50,000 voiced its satisfaction as the Comet flew her victor's circuit of the old racecourse followed by a high-speed run at low level along the straight. Hats were thrown in the air and the din was described as 'deafening'. It was a golden opportunity for the gathered press and film crews to ensure maximum coverage of the finishing dash. Meanwhile, all roads near Laverton were choked with traffic as news of the Comet's arrival was broadcast, and Victorians struggled to get close in a mostly vain attempt to witness the touchdown.

Charles Scott, who was vested with the awful responsibility of making every landing in the Comet, called upon his last remnants of physical strength and skill to put the aeroplane down safely and tidily, barely clearing the fence in the south western corner of the aerodrome. As she ran at high speed through a depression in the runway the aircraft was momentarily hidden from view by a great sheet of accumulated rainwater flung up by the undercarriage. The aircraft was pushed into the hangar where final Race formalities were to take place and where, suddenly, the excited crowds which had been allowed onto the station, broke through the barriers and swamped both aircraft and crew. Authoritative sources perpetuated the colourful invective attributed to Scott as he eased himself from the Comet's front seat: "You red xxxxxx," he is alleged to have shouted, "may I never need to step in you again." He and Campbell Black acknowledged their joint victory with a brief clasp of hands. They checked their logbook with Flight Lieutenant Henry and declared themselves entrants in the Speed event, a requirement of the organising Committee to prevent any competitor from winning more than one prize.

Members of the Royal Australian Air Force laying out the white sheets which, together with neon lights and a pair of pylons, marked the finishing line opposite the grandstand at Flemington Racecourse, Melbourne.
(National Library of Australia, Canberra.)

Final adjustments being applied to Bessoneau hangars which were erected as overflow accommodation by the RAAF at Laverton Aerodrome. It was the cost of erecting similar structures at Hatfield Aerodrome in England that caused the Royal Aero Club to reject the site in favour of Mildenhall.
(National Library of Australia, Canberra.)

News-film crews preparing their equipment in anticipation of the first Race arrivals at RAAF Base Laverton near Melbourne. All the participating aircraft which completed the route within the time that Laverton was available were housed in the permanent buildings on the site.
(National Library of Australia, Canberra.)

Having crossed the finishing line at Flemington Racecourse, Grosvenor House *landed in the rain at Laverton where she was escorted into the hangar for official scrutiny and other formalities. The members of the crew were met by their sponsor, Arthur Edwards, and two lady pilots who offered them sandwiches and bottles of beer.*
(National Library of Australia, Canberra.)

One of the two RAAF DH.60 Gipsy Moth aircraft assigned as taxis to convey the winning crew from Laverton back to Flemington Racecourse for an official public reception.

Tom Campbell Black in the front cockpit of an RAAF DH.60 Gipsy Moth at Laverton flown by Squadron Leader F R Bladin. Charles Scott was flown to Flemington in an identical aircraft under the command of Flight Lieutenant Dalton. The Race organisers may have underestimated what an ordeal this short trip was to prove at the end of a flight of already epic proportions.
(National Library of Australia, Canberra.)

First to greet them was their sponsor Arthur Edwards accompanied by Jean Batten and her radio microphone. Two Australian women pilots, Lores Bonney and Peggy Doyle, provided them with bottles of beer and sandwiches. Both pilots were shortly put into the professional care of RAAF personnel, who clipped seat-pack parachutes to their numbed bodies, and eased them into a pair of waiting Gipsy Moths, flown by Squadron Leader Bladin and Flight Lieutenant Dalton, acting as taxis for the brief trip to Flemington. It must have been an agonising experience, the final turn of the screw, having already flown half way round the world. That night the two pilots, their sponsor and Sir MacPherson and Lady Robertson were all scheduled to dine together at the Menzies Hotel.

News of the Comet's safe arrival was immediately flashed to the Royal Aero Club in London by cablegram:

"Advise Scott arrived Melbourne 053448GMT. Signed CENTENAIR."

Official confirmation was advised to the Royal Aero Club by letter from Flight Lieutenant T A Swinbourne on behalf of the Air Liaison Officer:

"I have been advised by the official timekeepers in Melbourne that Messrs C W A Scott and T Campbell Black competing in the MacRobertson International Air Races arrived at the Flemington Race Course in Melbourne, Victoria, Australia at 5hours 54

minutes 48 seconds am Greenwich Mean Time on the 23rd day of October 1934. Aircraft: de Havilland Comet."

The exultant crowds waiting at Flemington were aware that they were witnessing history. First is usually best and often unique, and there was never any suggestion that the Races might become a regular event, so they cheered, chanted and sang as the Moths from Laverton flew in to disembark their retiring passengers.

Following a short tour of the public areas in an open top car in pouring rain, Scott and Campbell Black were welcomed on a dais by the City's Lord Mayor, Sir Harold Gengoult-Smith:

"You have thrilled the world and you have earned the admiration of the whole of the British Empire and the homage of Australia by your achievement."

Sir MacPherson, the Attorney General and Chief Secretary, Sir Ian Macfarlane, representing the Premier, and the Dutch Consul, Colonel Wright, were also there. Whilst the fervour of patriotism which permeated the speeches was entirely predictable, a complete surprise was the presentation of bouquets of flowers from KLM bearing a message of homage to the winners in view of their 'navigation, technical ability and perseverance'.

Following a rendition of *'Here the conquering hero comes'* by thousands of spectators accompanied by the band of the Royal Australian Air Force, Charles Scott, never short of a word, was prevailed upon to make a short speech in reply. To anybody listening to his elegant, witty, off-the-cuff address, there was nothing to suggest he had just spent three days in a cramped cockpit. He thanked everybody for their welcome, took a newspaper out of his coat pocket and to a great cheer said: "The only reason I know I am here is because somebody gave me a paper and my name is in it!" The taciturn Campbell Black also thanked everybody but suggested he never made speeches and was not about to start now.

In Holland the victory was greeted with enthusiasm and KLM later printed posters which carried the simple headline: 'Bravo Scotty!'

Later that evening Port Melbourne was awash with light as searchlight beams from HMS *Sussex*, HMAS *Canberra*, USS *Augusta* and the New Zealand vessels HMNZS *Dunedin* and HMNZS *Diomede* were directed into the night skies as much in celebration as the navigation beacons they were intended to be.

Sir Harold Gengoult-Smith addresses the huge crowd assembled in the rain at Flemington to welcome the victors, Tom Campbell Black and Charles Scott, and to introduce them to Sir MacPherson Robertson standing hatless on the right. Scott was able to make a witty reply but Tom Campbell Black just reiterated his policy of not making speeches with the declaration that he had no intention of starting now.
(National Library of Australia, Canberra.)

The searchlights from HMS Sussex, *HMAS* Canberra, *USS* Augusta *and the New Zealand vessels HMNZS* Dunedin *and* Diomede *docked at Port Melbourne acknowledge that the Great Air Race had been won.*
(National Library of Australia, Canberra.)

13

Incident at Albury

Koene Parmentier, Captain of the KLM DC-2 PH-AJU, 'Uiver', which flew with passengers and mail along an airline route from Europe to Australia and would have arrived in Melbourne many hours earlier had there not been a little trouble near the township of Albury. Flight Lieutenant Poole, RAAF Control Officer at Charleville, about whom Captain Parmentier was uncharacteristically and publicly critical, can be seen in the background.
National Library of Australia, Canberra.

FOLLOWING A FORTY minute transit at Cloncurry, Uiver arrived overhead Charleville at 6.40pm on Tuesday having sent a wireless message an hour previously, but the aerodrome was unprepared and obeying the red Very signal which was fired circled six times in the dusk while the Control Officer arranged for runway flares to be put out. Only after he was satisfied was landing clearance approved with the

DC-2 PH-AJU during her turnround in darkness at Charleville. Always the diplomat, on this occasion Captain Parmentier expressed his concerns over what appeared to be a complete mishandling of the aeroplane's reception. Showing mistrust of the disposition of runway lights, Jan Moll decided to walk ahead of the DC-2 when she taxied out for departure, guided by the landing lights in the nose.
(National Library of Australia, Canberra.)

firing of a green. On the ground at 6.55pm, the Dutch pilots were openly critical as they considered there had been sufficient light for an approach. Whilst Parmentier was said to have been 'speechless with anger' and in the apparent absence of officials had to ask for directions to the Control and refreshment tents, he complained to the press that when the landing flares were eventually laid out they were not even on the runway. Parmentier, anxious to complete the remaining 787 miles of the final sector, was extremely disappointed when advised that Scott and Campbell Black had crossed the finishing line almost four hours previously. He was also well aware that the Boeing 247 was only eight hours behind. Nothing was assured, even now.

They were ready to leave just inside an hour, their second longest turnround since leaving Mildenhall but there were still concerns about the poor standard of aerodrome lighting and Jackie Cochran's redundant organiser, Joe Maier, offered to borrow a car to lead them out to the runway but Jan Moll decided to make a walking reconnaissance ahead of the aircraft using the landing lights for illumination.

The Aerodrome Control Officer, RAAF officer Flight Lieutenant Poole, was severely criticised for keeping the DC-2 in the circuit for 15 minutes but when approached by the press said he had no statement to make and that if a protest was raised he would submit a report through official channels.

With the exception of a build-up of thunderstorms reported 200 miles north of the finishing line, the KLM crew expected few problems and estimated to be in Melbourne at about midnight. They set course to track a little west of south, on the direct line to Flemington, but within an hour of leaving Charleville, the DC-2 ran into a series of violent electrical storms, cutting both visual and wireless contact and drifted off to the east.

The break in communications caused concern in the Radio Directing Office set up on the roof of Melbourne Town Hall, and not least in the cockpit of *Uiver* where the crew was enduring a rough passage, rocked by violent turbulence, blinded by flashes of lightning and hammered alternately by cascades of rain or sleet. In spite of preventative measures, ice built up dangerously on the leading edges of the wings and tail.

The crew, realising they were lost somewhere east of the intended track, began to circle, awkwardly situated between rising peaks of the Snowy Mountains ahead and the eye of the storm behind. They had come so far without any trouble, only to experience the worst weather of the entire Race an hour's flying time from Flemington. The machine was in fact about 200 miles north of Melbourne, orbiting within an area centred on Wagga Wagga, and extending as far as Goulburn and Albury, to east and south. At 11.25pm local time the residents of Ryan, 30 miles north of Albury, reported that a high-flying aircraft had passed over, heading south. Fifteen minutes later the machine was reported over Albury itself,

The RAAF Control Officer at Charleville, Flight Lieutenant A A Poole, testing the wind-speed for the benefit of a local journalist who subsequently wrote that this was the signal clearing an aircraft to take-off.
(National Library of Australia, Canberra.)

Captain Koene Parmentier in his tropical uniform leaving PH-AJU with the Race logbooks on arrival at Charleville, ready for a sharp word with the aerodrome authorities.
(National Library of Australia, Canberra.)

still heading south, and apparently enroute for Melbourne, but after half an hour it reappeared from the north, with landing lights switched on.

As the aircraft continued its circuitous passage, Radio Operator van Brugge re-established contact with Melbourne Control from overhead Tallangatta. He was advised of his position and directed to land at Cootamundra. A message was relayed to the airport there, and preparations were put in hand to receive the celebrity. Officials at Narromine, liaising with Melbourne Control, had requested that all available cars in the Cootamundra township be mobilised to provide emergency lighting by strategic positioning around the perimeter of the airfield. Although local drivers reacted positively the crew of the *Uiver* decided they were too far to the east and continued their search for a landmark, or particularly an airfield, in their immediate vicinity.

Arthur Newnham at Albury's local radio station, 2 CO Corowa, also broadcast an appeal for all locally owned cars immediately to converge on the racecourse, the only site adjacent to the town suitable as an emergency landing field. Seven hundred vehicles were soon heading out to the site where they were ranged around the perimeter of the track, shining their headlights towards the centre.

The idea of a town beacon had first been suggested by Flight Lieutenant Armstrong at Melbourne Control, and several other settlements adopted the same scheme. Local newspaper reporter Clifford Mott, working for Albury's *Border Morning Mail*, a publication which later was to promote him to the position of Editor, telephoned Race Headquarters in Melbourne to advise them of the situation. It was suggested he arranged for the lights to be flashed in a Morse sequence, spelling out the name of the town, rather than a sequence of haphazard flashes. Melbourne believed the lost crew might recognise the code, establish their position and then decide upon the best course of action.

The lights of the Albury War Memorial were the first to be flashed in Morse, actioned by the North Albury Battery signaller Mr R Jillard. After an hour Albury's Mayor, Alfred Waugh, the Municipal Electrical Engineer, Lyle Ferris, together with the town's Deputy Postal Inspector, Rob Turner, went to the power house in Kiewa Street, and there intermittently interrupted the whole of the municipality's electrical supply. The result was a massive light beacon, with all the street lamps and neon signs flashing on and off in the Morse code, attempting to attract the attention of racing airmen lost in the night sky with the familiar cryptographic rhythm. Aboard *Uiver*, Parmentier saw the lights but was unable to interpret their urgent message due to the constant buffeting and impaired visibility, and elected to continue flying another wide sweep.

When at last the Albury radio station established direct contact with the DC-2 and were able to advise the crew of their position and relay instructions received from Melbourne concerning the best course to fly to the finish, the pilots chose to continue on their circuit, still unsure of the relative position of the high ground to themselves and their destination. They were also becoming increasingly anxious about the state of their fuel supply and the apparently unending storms which surrounded them. After a further 30 minutes they decided to retrace their steps to the last positively identified position, the town of Albury, and seek the most suitable landing area there.

It was just after one o'clock in the morning when the DC-2 again passed over the town. The crew immediately picked up the long queue of lights displayed by the vehicles streaming out towards the racecourse, together with the pool of light established by those already in position. Hundreds of people had travelled out of the town utilising every available mode of transport, many still wearing night clothes under weatherproofs.

As the aircraft circled, to illuminate the approaches to the landing area Parmentier dropped the two 200,000 candlepower flares carried in all DC-2s in a housing behind the fuselage baggage compartment. Each flare burned for three minutes as it descended on a small parachute. At 1.30am, *Uiver* touched down, rolling to a halt only 100ft from the northern boundary fence. Fuel remaining on board was sufficient for about two hour's flight at economical cruise. In his role as Deputy Postal Inspector Rob Turner sent telegrams to the Air Race Committee in Melbourne and to Holland explaining that the DC-2 was safely down and undamaged. Now the crew was able to take stock of the position.

In The Hague, anxious crowds had gathered outside the KLM Headquarters, aware that the DC-2 had gone off schedule. They refused to leave until after Albert Plesman had received and read the Postal Inspector's telegram from Australia, and even then many would not move, intending to be on hand to hear news of the aircraft's safe arrival in Melbourne. It was to be a long vigil.

Uiver had landed on a sodden area of grass only 400 yards long. Soon after her slithering arrival there had been another deluge and any consideration of an imme-

In the darkness and aided only by a circle of lights provided by the headlights of vehicles driven out to Albury's racecourse, the DC-2 had touched down and stopped safely on the saturated turf in the middle of the track. It was a miracle aided by the skill of the crew that there was no damage.

By the cold light of day it could be clearly seen that moving the aircraft from its bogged position was going to be achieved only after considerable physical effort and a deal of ingenuity and initiative.
(National Library of Australia, Canberra.)

diate departure in the dark was quickly abandoned. The crew repaired to a local hotel for a brief rest, before relieving the aeroplane's police guard at dawn. It was evident that a take-off in her presently loaded condition would be impossible and everything that could be removed to assist in lightening the aircraft must be discarded. This included both the non-pilot crew members, the three passengers, seats, luggage, emergency kit and stores. Even the pilot's overcoats were listed for ejection. The 22,000 airmail letters from England, plus others uplifted in the East Indies, were offloaded and carried to Albury Post Office in the delivery wagon operated on behalf of the Wangaratta Bacon Curing and Freezing Works. According to the Rules, Race No. 44 would now attract substantial Handicap penalties but it was not clear whether the severity of these would jeopardise what was still the probability of second place. Now news was received that Turner and Pangborn were down at Bourke, only 200 miles to the north.

Several attempts at getting off that morning were thwarted by the glutinous nature of the field and *Uiver* sank up to her axles in the mud. Wooden planks laid on the surface in an attempt to offer some degree of stability proved ineffectual and were picked up and scattered by the wash from the propellers as high power was applied in an attempt to move forward. Ropes tied to the undercarriage legs were taken up by 300 pairs of willing hands and the aircraft was eventually dragged out of the bog, only to slither to a halt after another attempt to move under power. One last desperate attempt to move the aircraft was made by Albury's stout citizens and with great effort *Uiver* was eventually dragged to a higher part of the racecourse where the surface was drier, passing through a substantial gap which the authorities had created in a post and rail fence.

At about 8.00am Parmentier and Moll made final checks, warmed the engines, and to the accompaniment of 15,000 voices willing them on their way, finally escaped from their sticky prison. Bouwe Prins the flight engineer had been stationed at the critical point on the take-off run at which the *Uiver* had to be airborne in order safely to climb over the bordering trees. Should Prins raise his arms the run was to be aborted. Following a slow acceleration, the aircraft rose gracefully, banked steeply above the hedge, and immediately turned south. Forty minutes later she crossed the finishing line at Flemington but was recalled for a second pass by a red Very fired by an official. To the cheers of the welcoming crowd, boosted by the thousands admitted to the stands at the adjacent Royal Agricultural Showground, the aircraft completed her circuit as requested and passed low over the word *Flemington* pegged out in giant letters on the grass, this time to receive a green. A tiny Comper Swift took up formation to guide the DC-2 to Laverton where Parmentier completed a three-point landing in front of a mass of excited spectators.

Many years later and after retirement from his service with KLM, Bouwe Prins admitted that the *Uiver* had not actually been airborne at the critical point but he had resisted from raising his arms as he was sure the aeroplane was on the point of lifting. He was certain that under the circumstances if his signal had been negative, Parmentier would not have attempted a second run.

Jean Batten, who had flown her Gipsy Moth G-AARB along much of the Race route from Lympne to Sydney in May 1934, had been invited by the Gaumont British Film Company to make daily broadcasts to a network of 30 radio stations for a ten day period following the start of the Race from Mildenhall. Her first schedule was from the roof of Melbourne Town Hall on 20 October and later she maintained a flow of news and information with the aid of cables, telegrams and weather forecasts.

Surprised by the speed at which *Uiver*

had been rescued from the mud at Albury, Jean Batten was late in arriving at Laverton from where she was due to broadcast a description of the aircraft's touchdown. The DC-2 was already safely hangared and in view of the chaos surrounding the arrival of the Comet, isolated from all visitors including press when she finally drove into the aerodrome. Later receiving an invitation to meet the KLM crew her microphone was smuggled in through an open window and she was able to secure an exclusive live interview.

In contrast to the exhaustion suffered by the Comet crew who had merely catnapped during their 71 hour flight, the Dutch airmen emerged from their cabin 'as if they had just stepped from a luxury train'. Following completion of arrival formalities Parmentier and Moll were flown back to Flemington to receive the acknowledgement of the crowd many of whom had waited patiently overnight for their safe arrival. Sir MacPherson was on hand to congratulate them and emphasise that their passage had confirmed all his hopes for a speedy and regular air service from Europe. The Dutch Consul passed on Queen Wilhelmina's congratulations and news that "Her Majesty the Queen has been pleased to confer on all four crew the honour of Knight (Chevalier) in the Order of Orange Nassau. Please convey to them the Government's most sincere congratulations."

The KLM machine was operating on a commercial basis and Parmentier was anxious for news of his three fare-paying passengers and the mail which had been offloaded at Albury. Apart from that diver-

An enthusiastic volunteer force was assembled from the citizens of Albury and, having dug the aircraft out of her bog, they discovered that any attempt to move under power was non-productive. A new plan was required which would be co-ordinated by a team leader.
(National Library of Australia, Canberra.)

Using the well tried formula of physical strength applied in the appropriate position as directed by overseer Cecil Meredith, 'Uiver' was towed out of her bog to a higher, dryer part of the racecourse after everything of a non-operational nature had been removed to reduce weight.
(National Library of Australia, Canberra.)

A grim faced Jan Moll eyes-up the prospects for a safe take-off from the captain's window of DC-2 PH-AJU. Previously, the short rolling distance had been masked by darkness.
(National Library of Australia, Canberra.)

After their exhaustive efforts, the good citizens of Albury stood clear as 'Uiver' was taken up to full power against the brakes. Flight Engineer Bouwe Prins was stationed at a critical point to signal if he considered the aircraft should stop.
(National Library of Australia, Canberra.)

The mail sacks containing thousands of postal covers offloaded at Albury were carried to the post office for onward transmission by courtesy of 'the Wangaratta Bacon and Freezing Company'.
(National Library of Australia, Canberra.)

KLM's three fare-paying passengers who were offloaded as part of the contingency measures at Albury: J J Gilissen, Roelof Jan Domenie and Thea Rasche. The reduction in payload now affected the handicap allowance as though the lower weight was carried from the start.
(National Library of Australia, Canberra.)

sion, they had experienced no other trouble at any time, he assured reporters, had cruised comfortably below full speed, and believed the Race route was admirably suited as the basis of a permanent air link between Europe and Australia. KLM had been operating a regular service from Amsterdam to Batavia since 1929 he reminded everyone, and had been pressing the British Government over previous years for clearance to extend this to Australia. All requests had been refused on the grounds that the service could only be operated by Imperial Airways as part of the Empire Air Service.

A correspondent later reported that as an act of sportsmanship, the Dutch crew roused the two Comet pilots from their beds to offer their best wishes and to invite them for a celebratory drink.

Dr Plesman, forever the businessman, was already planning for the aircraft to operate a commercial service on the return journey to Holland, and in a rare expression of benevolence ordered that a gift of 1,000 Dutch Guilders be donated to the Albury District Hospital in recognition of that town's assistance in recovering *Uiver*.

The DC-2's performance had so impressed Plesman that he allocated funds for the immediate purchase of a further ten machines. Donald Douglas and his production planning staff were soon busy with enquiries and visitors as further orders for the new transport aeroplane were confirmed. The manufacturer was already considering an improved version for transcontinental overnight sleeper services under the company designation DC-3.

14
Not Disgraced

'The Young Australian' Jimmy Melrose arrived at Laverton on Tuesday 30 October, just after mid-day, the only pilot to fly solo from England. He was greeted by his sponsor, his mother Hildergarde, and enthusiastic mechanics of the Royal Australian Air Force.
National Library of Australia, Canberra.

Behind the competitors in the Speed Race a whole stream of small aeroplanes stretched back from Asia and the Far East into Europe, some taking advantage of the Handicap Rules whilst others were delayed for more practical reasons or just bad luck.
Individual competitors discovered that in many cases they were the centre of the elaborate preparations otherwise made for a racing fleet. The New Zealand Hawk, inbound from Jask, spent fewer than 50 minutes on the ground at Karachi before setting off again for Jodhpur and relaxation in the Maharaja's hotel.
(National Library of Australia, Canberra.)

MANY OF THE GREAT endeavours were still running their course when the £10,000 Speed Race prize was being claimed by the de Havilland Comet G-ACSS. While the victorious crew summoned all their strength to smile, jest and pump hands, others, less charismatically perhaps, were still setting records as they continued doggedly along the route from Mildenhall to Melbourne.

McGregor and Walker flew their Hawk into Calcutta from Allahabad, arriving an hour after the Speed Race had been won. Their previous two sectors had been flown at 10,000ft, the first in darkness above mountains and the second to avoid the blistering heat which was affecting the performance of the engine. McGregor had perfected an arrival technique attributed to Roscoe Turner, and known as the 'Turner Stunt'. The aircraft would dive from height to enter a flat spiral around the aerodrome boundary, giving the impression of great speed and a probable difficulty in landing within the perimeter. According to McGregor, the subsequent safe arrival of the aeroplane was always greeted with unsubdued relief and not a little confusion.

The Hawk left Calcutta after a hurried turnround, anxious to be at Rangoon before last light. The crew was unlucky in that due to the weather the flight across the mouth of the Ganges and part of the Bay of Bengal took longer than anticipated, and they arrived overhead Rangoon at 2,000ft in pitch darkness, where they were immediately assailed by a shower of paper balloons, each carrying a tiny lantern. They were puzzled too by what appeared to be numerous landing lights winking on the ground but which upon closer inspection were found to be bonfires scattered along the streets.

Tired and running short of petrol, a half hour circuit of the city in search of the airfield was exasperating, but while contemplating a forced landing beside the Pegu River, a red Very was fired off which identified the position of the aerodrome. Safely down the pilots were advised that their arrival had coincided with the annual 'Carnival of Lights'.

The New Zealand Dragon Six had arrived unannounced at Jask just before 7.00pm on Monday following a routine transit of 70 minutes at Baghdad. Their first attempt to land was too far into the airfield and they went round again. After circling the area the second attempt resulted in a perfect touchdown after which Hewett reported that the brakes were very poor. The crew stayed overnight in the KLM bungalow, leaving at 3.50am on Tuesday and later reporting the settlement was without water, wireless or electricity. Hewett discovered after take-off that he had left his canvas document bag in the car which had taken the crew to the aerodrome and on arrival at Karachi sent a message from the Air France office asking for it to be forwarded for his collection at Allahabad. However, the bag had been found and already sent on to Karachi by hand of Captain Gambode in command of Air France Fokker F.VIIb F-ALGS. The Dragon Six proved too fast for the Fokker which could not catch up and the bag was carried through to Bangkok where it passed into the care of another Air France service to Melbourne and was duly delivered to Hewett via the Socony Vacuum Oil Company.

In Karachi the New Zealand crew were required to spend some time adjusting the aircraft's brakes before taking a leisurely breakfast and leaving for Allahabad where

Jimmy Melrose reached Karachi late on Tuesday afternoon, a day behind the Hawk and, as was his normal routine, elected for a good night's sleep before leaving next morning for Allahabad and on to Calcutta.
(David Underwood.)

The crew of the Danish Desoutter flew a pattern very similar to that adopted by Melrose and the Stodarts, travelling by day and sleeping as the schedule unfolded at night. As far as possible the aircraft was kept under cover but this shot, taken at Karachi, shows her to be parked outside under the floodlights and amongst the puddles.
(National Library of Australia, Canberra.)

Jack Wright and Johnny Polando were still at Baghdad on Tuesday where they seemed to have lost none of their capacity to draw attention or to receive refreshment.
(National Library of Australia, Canberra.)

on their second attempt the aircraft made a safe landing at 5.00pm.

The Desoutter, Courier and Puss Moth had all transited Jask before lunch on Tuesday, each leaving after about half an hour on the ground headed for Karachi where the Stodarts found their Lynx engine was spraying oil along the length of the Courier's fuselage, and the search for the leak caused them some delay.

Wright and Polando hoped they had cured their fuel delivery problems and the Monocoupe passed through Baghdad at lunchtime on Tuesday, but they were forced down again at Mohammerah in Persia. While Jack Wright remained with the aircraft, Polando volunteered to walk to the nearest settlement in search of assistance, but far from receiving help he was arrested by two local militiamen armed with ancient rifles and held in a walled stockade overnight.

News of the Monocoupe's forced landing reached British engineers at a local oil installation, and one rode out on horseback, found Wright and took him back to their accommodation. Polando's disappearance was reported to the British Consul who liaised with the American Embassy in Tehran, and after intense diplomatic pressure, Polando was released. The Monocoupe was repaired sufficiently well to fly on to Bushire and then to Jask where they stayed overnight on Thursday, leaving for Karachi at 5.55am. By the following Tuesday they had struggled on as far as Calcutta where the valves in the Super Scarab engine finally gave up, and failing to locate replacements the team decided to retire from further official participation.

Geoffrey Shaw made good time after his excursion into Spain on the first day of the Race and the Eagle arrived in Baghdad from Aleppo on Tuesday where the engine seemed to attract a lot of attention. His next landing was at Bushire in darkness and the aircraft ran across a sunken road causing irreparable damage to the undercarriage and enforced withdrawal. (National Library of Australia, Canberra.)

The Boeing 247 arrived at Darwin at teatime on Tuesday having travelled across the Timor Sea from Koepang largely on the power of the starboard engine with the port motor throttled back. In this shot the starboard engine cowling seems to have accumulated an oily top surface. (Northern Territory Archive Service.)

The Stanavo refuelling staff at Pelambang, mid-way between Singapore and Batavia, had been on standby all week although the aerodrome was not an official stopping point. They maintained contact with the progress of racing aircraft through telegrams which were being delivered to the local newspaper office until the service was terminated on Saturday, 27 October. On that day, Stanavo heard a rumour that the Monocoupe had left Rangoon the previous day and made efforts to garner accurate information from Singapore but heard nothing. On the following Monday, having serviced not a single competitor, the company withdrew its staff leaving only five drums of aviation spirit, one case of oil and two fuel pumps in the care of a watchman.

After his problems earlier in the Race, Shaw's efforts in the Eagle were achieving some degree of respectability. On Tuesday 23 October the aircraft had flown from Athens and Aleppo via Baghdad to Bushire but, arriving in the dark the aircraft ran across a sunken road which caused irreparable damage to the undercarriage. Shaw's assessment of the situation was: "There are absolutely no facilities here for repair; not even string!" With great reluctance the aircraft was withdrawn from further competition.

After behaving impeccably, an hour beyond Singapore the Boeing's port engine lost oil pressure and was throttled back. Reeder Nichols attempted to advise Singapore by wireless but there was no reply. Instead, the propagation characteristics of HF communication were amply illustrated when a station in San Francisco reacted to his call and promised to pass on his message by cable.

Turner put down at Koepang where nothing obviously wrong could be found. It was agreed that they should take-off with normal power on both engines, after which the port motor would be throttled back, and the flight continued to Darwin and, perhaps, beyond on the maximum continuous power of the starboard engine alone.

The plan was put into effect and at 4.17pm on Tuesday the Boeing 247D landed in Australia, the crew first circling the town of Darwin which ensured a large crowd was soon on its way to greet them at the aerodrome. In typical fashion Turner exclaimed "Gee, this is a beautiful field; if a guy couldn't land in this he couldn't land anywhere!"

Due to the removal of engine cowlings and a comprehensive check of the oil systems which revealed a failure of *Nip's* oil pressure relief valve, transit time at Darwin extended to two hours before the Boeing left in a vain hope of overtaking the KLM aircraft already an hour out of Cloncurry and heading for Charleville. After achieving what they considered a safe height following a normal take-off, the port motor was throttled down and the aircraft settled on course for Charleville. However, the lakes and rivers shown on the charts were not reflecting the light of the moon as expected; it was the dry season during which the features simply did not exist. Completely baffled and seriously misled, the crew put the big aircraft into a search pattern and soon were lost.

Nichols established contact with the RAAF direction finding unit provided by Amalgamated Wireless at Charleville and whose expert services were now vital. The operators skilfully guided the casualty through the night skies until the crew was able to confirm their own position. They picked up the track to Charleville where they put down at 6.10am, four hours later than expected. Turner then created what was described as the greatest exhibition of hustling ever seen on an Australian aerodrome. He jumped down from the aircraft, jerked up the flaps of his flying helmet and strode to the Control Point tent, logbook in hand. Formalities were completed in a minute and he was back at the Boeing.

Where had the aircraft been all night? The crew explained that oil pressure had

A scene of sartorial elegance four days out from England. In spite of travelling through the tropics on arrival at Charleville each member of the Boeing crew was seen still to be wearing a collar and tie. Turner had jumped down from the aircraft, jerked up the flaps of his flying helmet and, with an escort of pressmen, had created a great exhibition of hustling.
(National Library of Australia, Canberra.)

The unscheduled arrival of the Boeing at Bourke just before 9.00am on Wednesday caused great excitement amongst the townsfolk. Inspection of the starboard engine revealed a cracked oil pipe. Permission to remove the cowlings and to carry them inside the fuselage was granted but it was soon discovered they were too big to pass through the door.
(National Library of Australia, Canberra.)

At Charleville, Clyde Pangborn, perched on staging which might have been made for the purpose, closely inspecting 'Nip', the port engine of Boeing 247D NR257Y.
(National Library of Australia, Canberra.)

dropped from 90psi to 30psi and there had been the real possibility of a precautionary landing for investigation but as that would have been in darkness making it impossible to see anything, including the engine, they had elected to cruise around until daylight.

Despite the best efforts of a posse of investigative engineers, no fault could found on the port engine apart from a general loosening of the oil pressure caps. As a precaution every cylinder head nut was check tightened.

Joe Maier, Jackie Cochran's special representative at Charleville, had been instructed to provide the Boeing crew with hot and cold food: steak, fresh vegetables, baked apples and American coffee and cream, but they were far too engrossed with their engines to stop for food and so Maier made up a package for consumption on board.

They were airborne 90 minutes after their arrival but 100 miles out Pangborn noticed traces of bluish smoke whipping back from the starboard engine, although cockpit indications appeared to be normal. The aircraft was immediately headed for Bourke, an easily identified railhead 50 miles across the border with New South Wales, where they landed at 8.58am.

Here it was discovered that oil seepage from a cracked pipe was being sprayed onto a hot manifold and vaporising. Following the realisation that they were faced with a difficult repair job the crew sought approval from the Race Stewards in Melbourne to fly with the starboard cowling removed and their request was granted as long as the structure was carried inside the fuselage. As Their Majesties had discovered at Mildenhall, the door spaces on modern aeroplanes appeared to be designed for dwarfs, and there was no question that the panel beaten artistry would ever go through the orifice except in small pieces, so it remained where it was, firmly attached to 'Tuck'.

The three crewmen held a council of war. Should they go on sedately and try to overtake the DC-2 which was now known to have diverted to a place called Albury? Should they fly flat out and risk blowing the engines and their chances, or should they try to rectify the problems at Bourke? Their decision was to go on at all possible speed and trust that the starboard engine would hold out. They left at 10.18am on Wednesday 24 October, bound for Melbourne. The engine continued to run well

Owen Cathcart-Jones in his check-tweed jacket and cricket pullover, and Ken Waller wearing a sun helmet which appears to have been shared between the pilots, scrutinising dog-eared documents at Darwin, probably a fuel uplift receipt book.
(Northern Territory Archive Service.)

The efficient refuelling service at Darwin was the product of much pre-planning and practical effort. The bowser refuelling the green Comet G-ACSR is liberally sign-written with the names of Vacuum Oil, Mobiloil, Plume and Stanavo.
(Northern Territory Archive Service.)

Following a general check on both engines which revealed no significant deficiencies, G-ACSR was ready for departure for Charleville a little over an hour after her bumpy arrival at Darwin. It was almost a day since the first prize in the Speed Race had been claimed by G-ACSS.
(Northern Territory Archive Service.)

and as they grew closer to Melbourne, the indicated oil pressure in the port motor steadily rose.

The dash into the finishing straight could now be made at speed and with renewed confidence and they crossed the line at Flemington at 1.36pm but the loss of time in Northern Queensland had been fatal for waiting to greet them at Laverton was the *Uiver*. The Dutch crew had beaten them by barely three hours. As positions stood, the Boeing was placed third in the Speed Race. American humourist Will Rogers was moved to comment that "In a race like that Melbourne hop, it's an honour to be third!"

In marked contrast to the smart appearance of the KLM pilots the Boeing arrived with three stubbly chins on board. Turner entered into humorous banter with the Press contingent in the big crowd that greeted them but left Reeder Nichols to express the sentiments of the entire crew before they were whisked away to their hotel for a bath, a shave and a meal:

"It was a long trip and very tiring. I don't think I want to make another straight away!"

Following their false start from Allahabad, Owen Cathcart-Jones and Ken Waller in the green Comet had cruised to Singapore in a twelve and a half hour flight without any further problems, and the 30 minute turn-round was as efficient as any, but shortly after take-off on the next stage to Darwin the oil pressure indication on the starboard engine began to fall, accompanied by a popping sound. After consultation with Waller, Cathcart-Jones decided it would be prudent to land at Batavia for investigation.

The landing just before dark was uneventful and inspection of the Gipsy engines by the resident KLM engineers produced little conclusive evidence of mechanical failure which tended to confirm their suspicion about the reliability of the gauges. But now the crew faced another dilemma. The aerodrome at Batavia provided a maximum distance of 900 yards from hedge to hedge with tall trees on the boundary, insufficient for a safe take-off in a Comet loaded with fuel for a direct flight to Darwin. There was no choice but to accept a reduced petrol load and plan for another landing before crossing the Timor. They elected to stop at Koepang on Timor Island, which they reached safely after five hours, refuelled and took off for Australia 25 minutes later.

The arrival of G-ACSR at Darwin at 1.13pm local on Wednesday 24 October was described as being similar to the giant frog's antics in *Alice in Wonderland* as the green aeroplane bounced across the rough surface following a firm touchdown. Both engines seemed to be behaving themselves although the crew expressed further reservations about the general efficiency of the lubrication system and the pressure gauges in particular. They left at 2.20pm heading for Charleville and the hope of a landing before nightfall. It was almost a full day since the Speed Race had been won. News of *Grosvenor House's* victory was greeted by Cathcart-Jones with an attitude that suggested he was no longer in a desperate rush to get to Melbourne. He also believed that he and Waller were no longer in contention for any other prize.

Severe dust storms were encountered in the Northern Territory along the route to Charleville and the crew decided on what they described as a precautionary landing at the silver mining settlement at Mt. Isa in Queensland, more than half way to their next Control Point. There was a long and lingering suspicion that they had mistaken

Comet G-ACSR arrived at the rain-soaked RAAF Base at Laverton on Thursday afternoon, the last of the Speed Race entrants, and the crew was disappointed and upset to be met only by about 20 spectators and a handful of RAAF officials. Ken Waller, Owen Cathcart-Jones and their sponsor, Bernard Rubin, had already decided that after a short rest and check of the aeroplane, they would fly back to England. (D R Butterworth.)

the site for Cloncurry whose aerodrome information sheet was included in the Pilots' Route Handbook, and situated 60 miles to the east. The aircraft had arrived from the north at about 8.00pm and circled low over the town for 45 minutes in the darkening sky searching for non-existent aerodrome lights. There was no beacon but plenty of other lighting giving the impression of a big town. The dusty streets filled with people and women fainted, fearing that the airmen would attempt to land behind the 'slime dam', a wet area which from the air could have been mistaken for a flat aerodrome surface. Aware of the circling aircraft, most of the population drove their cars the mile out to the airfield and positioned themselves around the perimeter, shining headlights towards the centre and making a bonfire of old tyres. It was becoming a routine procedure in which Australians could offer practical assistance to Race crews.

The locals feted the aviators after the Comet, which had landed at high speed, rolled to a stop, proffering bottles of cold beer which were both welcome and a distraction as the pilots, 'all done', just wanted to sleep before pressing on at the earliest opportunity. However, they were persuaded to take the almost obligatory tour of the local silver mines before staying the night at the home of the General Manager.

A local resident, Hector MacDonald, wrote: "At the aerodrome next morning the crew saw the hills surrounding them and had the shock of their lives as they had no idea there were such! They had been hopelessly lost and although they said they had fuel enough to go on to Charleville, they did not know where Charleville was. They did not know they were at Mt. Isa when they landed."

Having accumulated Handicap flight time at their unapproved stop, Seajay and Waller continued their interrupted journey at dawn next morning. They arrived at Charleville at 8.49am where they were welcomed by an enthusiastic crowd, one member of which described the commander as "a picture of sartorial elegance, complete with pith helmet, loud check coat, dirty sweater and baggy trousers." The outfit was all part of a carefully selected wardrobe to provide Cathcart-Jones with comfort during long hours in the air, and the sarcastic comments were summarily dismissed.

Following a quick check of the engines and an impromptu breakfast the two pilots were besieged by crowds of children requesting autographs on any slip of paper and, perhaps relieved of Race pressure, the crew willingly obliged leaving nobody disappointed. At 10.15am G-ACSR swept back into the air bequeathing only a cloud of dust to those who remained to wave farewell. She flew at maximum speed below 2,000ft, under the thunderstorms but in heavy turbulence until at 2.54pm she flashed over the line at Flemington and proceeded immediately to the Laverton Base where she touched down six minutes later. The green Comet was the fourth aircraft to arrive and in marked contrast to the earlier finishers, was received by about 20 spectators and a handful of Air Force officials. The Comet was pushed into the hangar where the weary crew disembarked and Waller told those who asked that he was too tired to talk about the flight. Cathcart-Jones particularly was disappointed at the reception and told a reporter afterwards that having flown 12,000 miles from England he thought they might have been met by more than a man and a boy exercising their dog. But the Speed Race had been won two days previously and for the moment, public interest had waned.

With the arrival of the second Comet all the high performance machines were accounted for, but no announcement about the award of Handicap prizes could be made until the competitors in the procession stretching back through Asia had been considered. And they had a further eleven days in which to complete the course at their own considered pace.

McGregor and Walker had left the festivities at Rangoon in the darkness of Wednesday morning. Alor Star, the next Checking Point, was beyond the range of the Hawk and arrangements had been made for a refuelling stop at Victoria Point, 400 miles south from Rangoon on the estuary of the Pak chan River. The aircraft arrived overhead to find the whole district shrouded in fog and, using higher ground as a guide, they descended cautiously to 50ft above the water where they spent an uncomfortable 20 minutes cruising between islands in search of the aerodrome. They located a wireless station and followed what seemed to be a major road leading inland, whereupon the airfield appeared below the leading edge of the wing, eight miles from the position marked on the maps.

Their later arrival at Alor Star caused great concern. The aerodrome was still a sea of mud in which the Royal Air Force personnel sloshed about in gumboots, and there was a danger that the liquified topsoil would jam the wheels inside the distinctive and elegant trousered fairings much favoured by the Miles Aircraft design team. They were aware that the result would be similar to landing with the brakes applied, almost certainly flipping the aircraft onto her back. But the landing was uneventful. It was the take-off after the short refuelling exercise which caused problems. The aircraft was lined up for a run which gave the longest distance across what appeared to be the least boggy part of the field. Having accelerated to about 40 mph the aircraft hit a mud hole and slewed to a stop, the undercarriage trousers distorted and as predicted, full of liquified earth. The machine was pushed back to the boundary by a group of RAF mechanics where the fairings were cleaned out and bent back into shape.

On the second attempt the aircraft left the ground after rolling for 800 yards, running alternately across good ground when they accelerated, only to be retarded by patches of mud. They cleared the boundary by a few feet, staggered over a rubber plantation, and almost immediately ran into a tropical storm of such intensity that their direct path to Singapore was completely blocked. There was no alternative but to divert via Taipang, an addition of 50 miles, but they pressed on, arriving in Singapore just before 2.00pm and in time for a relaxed lunch, leaving after 40 minutes.

All progress beyond Singapore was a great psychological boost, and the New Zealand team decided to press on immediately to Batavia. They arrived after a purely routine flight and the floodlights were switched on to aid their landing, for which service they were presented with a bill next morning for £5. 1s. 3d. It was Wednesday 24 October, five days out from England.

Behind the New Zealanders, the two Danes in the Desoutter were into Karachi eleven minutes after Melrose and his Puss Moth, all arrivals taking care to avoid the adjacent airship hangar and mooring masts. They flew on to Calcutta via Allahabad, followed by the young Australian who had established a pattern of flying by day and sleeping by night, convinced that his Handicap allowance would ensure him a place provided there were no unforeseen problems. The Stodarts in the Courier operated via Calcutta on their way to Alor Star where they landed on one of the worst portions of the airfield. The Stanavo Company immediately offered its entire coolie staff to get the aircraft to the refuelling pit and arranged for the aerodrome field car to collect the crew and deliver them to Control before making arrangements for their overnight stay.

Once again, the fuel companies had gone to extraordinary lengths to provide a service as described in a memo from the Stanavo office in Penang:

"Preliminary arrangements included the following: discussions and conferences with the various Kedah State officials concerning approval of storage of stock, our storage dump, various details as to actual refuelling of the aeroplanes, admission to the airport of our staff, admissions of lorries on the field, drainage and fire regulations at the dump, arranging to have available meteorological reports, reports on the arrival and departure of aeroplanes, and the transfer of a great deal of information to the Northern Section Air Race Committee and the sub-committees at Alor Star. It was necessary to notify and obtain permission of the Customs authorities to import stocks of gasoline, lubricating oils and miscellaneous materials; transportation had to be provided for these. Permission had to be obtained to store gasoline stocks in our 'godown' prior to transportation to the field. Drums had to be checked and reconditioned, stencilled and painted, cleaned and stacked for which extra labour was required.

In regard to the personnel we had, in addition to Mr Colyer, and Mr Moffat from our Bagan Luar installation who assisted him, we employed the company fitter, both up-country storekeepers, one of the Penang office inspectors, the supervisor of Belle Isle Service Station and his assistant to do the lubricating oil work, 15 coolies and men on the lorries hired to stay on the field. These personnel had to be equipped with uniforms which were made locally. They were trained to do the work as efficiently and quickly as possible. The fitters were particularly trained to attend to the actual fuelling, to clean and assemble filter funnels. The inspector was instructed to attend to all general supervision of the erection and maintenance of the dump. One of the up-country storekeepers was in charge of all the clerical work, the other was used for contacting Alor Star town which is seven miles from the aerodrome.

In connection with the final preparations at the aerodrome, it was necessary to obtain and put on the dump area nine loads of sand as conditions were extremely wet. Wood was purchased to build a walkway to the lorry so that drums could be loaded more quickly. Quarters had to be provided for the coolies at the airfield and arrangements to have food also. A police guard was obtained for 24 hour duty over the dump. Tents were borrowed and erected as well as tarpaulins for covering the stock from the weather. Temporary fences had to be erected; 'no smoking' signs were prominently displayed and miscellaneous materials brought in from Singapore."

In the event only six Race competitors landed at Alor Star and of these Stanavo supplied fuel only to the Hawk, Courier and Falcon. The memo continued:

"We are, of course, disappointed, after making all the preparations, that so few aeroplanes were actually refuelled by us at Alor Star. The weather conditions at the aerodrome were continually changing and advices sent to Allahabad by the aerodrome authorities at first made the field unserviceable, then dried up somewhat and a satisfactory report was sent out, later to be

The New Zealand Dragon Six ZK-ACO being refuelled at Rambang. It was a tedious and somewhat precarious operation. In addition to filling the standard wing tanks, drum stock petrol had to be pumped up to the platform for delivery to the long-range tanks situated in the fuselage.
(National Library of Australia, Canberra.)

Cyril Davies and Clifford Hill reached Baghdad on 31 October where new parts were fitted to the Fairey IIIF's aileron control system. On arrival in Calcutta on 3 November it was discovered the new parts too had broken and a further week was spent awaiting delivery of replacements.
(National Library of Australia, Canberra.)

changed again. This made our work difficult and having to attend to everything which we did in a sea of mud and sand, was most trying."

The Dragon Six left Allahabad early on Wednesday morning and flew under a bright moon direct to Akyab, calling Calcutta on the way by wireless to advise of their intentions to over-fly. From Akyab the aircraft routed directly to Bangkok and with some trepidation Frank Stewart called ahead for news of the airfield surface only to be relieved when the reply confirmed that the airfield was equipped with concrete runways.

Following what rest could be achieved in the hot and humid conditions, the flight continued in the darkness of Thursday morning direct to Singapore, over-flying the sodden aerodrome at Alor Star, and experiencing the turbulence, torrential rain and lightning associated with severe thunderstorms. On arrival at Singapore the aircraft was welcomed by an Aero Club DH.60 Moth Seaplane which guided her young cousin to the land aerodrome.

Although he was out of the Race, Ray Parer was still concerned about his aeroplane and the mounting bills for hotels and engineering assistance in Paris where he and Geoff Hemsworth learned of Scott and Campbell Black's arrival in Melbourne and acknowledged the achievement with a bottle of wine. The peculiar state of the fuel tanks was surely evidence that something unauthorised had been added, but what and why and by whom was not at all obvious. Even though some of the sediment was filtered out for detailed examination the whole matter remained a complete mystery. At last, all three petrol tanks and the fuel lines were thoroughly flushed through and refilled with an appropriate grade, after which there was no further problem and the engine ran perfectly. But it was all too late and at Ray Parer's insistence for the sheer adventure they headed their old aeroplane towards Australia.

The Fox's first landing out of Le Bourget was at Lyon, and the next day having been unable to find the aerodrome at Genoa they went in to land at a military airfield near Pisa where they were promptly arrested. Clearly, troubles for the New Guinea team were far from over, although during their enforced stay in Italy Parer managed to solve the engine overheating problems. Italian mechanics examining the radiator decided that the underslung extra fuel tank was destroying the free passage of cooling air and they proposed to reposition the tank and add deflectors. On 31 October the Fox was released to proceed to Naples, where the crew was arrested again. Diplomatic channels, opened for the Race, clearly had closed ranks already.

On the day the Fox left Pisa, Davies and Hill in the Fairey IIIF arrived in Baghdad where the temporary repair to the aileron control effected in Cyprus was replaced by a new part. They arrived at Jask at 2.40pm on Thursday, 1 November and stayed overnight, leaving at 3.45am on Friday and flew along the eastern side of the Persian Gulf and across India to reach Calcutta by 3 November. Here it was discovered that the newly replaced parts of the control system had broken again and would require repair or replacement before further flight could be authorised. It was a week later that Davies and Hill left for Rangoon but encountered a severe storm over the Bay of Bengal during which the engine began to misfire and they diverted to Chittagong. Not until 15 November did they reach Singapore where a further five days were lost while both the control system and a header tank were overhauled.

At Mildenhall, with all traces of the brief civilian occupation now removed, a Royal Air Force Heyford bomber squadron was

Miles Hawk ZK-ADJ crossed the Timor at 10,000ft, passing over the survey vessel HMAS Moresby *at mid-point. Her arrival at Darwin on Thursday, less than six days out from Mildenhall, created a number of new records. (The Aeroplane.)*

beginning to settle into its new quarters, oblivious to the dramas still being played out thousands of miles away.

There had been problems for the Hawk and her crew, who so far had enjoyed a relatively trouble free flight. From Batavia, which they left at 3.52am on Thursday 25 October, they encountered strong headwinds and put down at Surabaya where ten gallons of petrol were reluctantly provided but the uplift permitted onward passage direct to Rambang where they landed at 9.51am. During the 495 mile sector from there to Koepang across the Savu Sea, the oil pressure dropped and the machine was coaxed into a glide until pressure built slowly up again. The process was repeated continually, the last 200 miles being covered in a series of switchbacks. At Koepang the engine was found to be leaking oil at most joints and thorough external inspection was necessary. The cowling had split again but was repaired by a local Chinese using strips of aluminium salvaged from a wrecked Imperial Airways' aircraft. Following a test flight and minor altercation with what Walker described as 'a mutton headed official' the Hawk took off at 12.15am, eight hours behind the anticipated schedule, heading for Australia.

The crossing of the Timor Sea was made at 10,000ft to avoid the worst of the prevailing electrical storms and they passed directly over the beacon ship HMAS *Moresby* on station at a mid-ocean position. In reaching Darwin at 5.55am on Thursday 25 October, they created history. Not only was *Manawatu* the leading single engined aeroplane in the Race, an open cockpit machine with no wireless communication, but she had created a new record of 5 days, 15 hrs and 30 min from England, starting from an aerodrome 100 miles inland.

The unshaven crew stopped sufficiently long enough to enjoy a wash and a good Australian breakfast at the Darwin Hotel but publicly expressed no feelings about a Customs inspection charge of 8 shillings and 11 pence levied upon them. On Friday at 7.10am they were on their way to Newcastle Waters.

At Allahabad, following her landing accident, the Pander Postjager was considered serviceable for a test flight to be contemplated on 26 October, when the prospect of completing the Race during the specified time limit was still a distinct possibility. But luck was not with the Pander. When starting their engines, the Dutch pilots realised that the mobile floodlight had been set up in the wrong position, and asked for it either to be moved or extinguished. The light was situated near the eastern boundary and the aircraft would pass close to it on takeoff. At 10.40pm the light was extinguished and the Pander began her take-off run but she had just left the ground when she collided with the apparatus. Due to a misunderstanding the floodlight, a structure standing high and with its generator mounted on a trailer, was still in process of being towed from its previous position by the airfield ambulance doubling as a tractor.

Through the restricted view offered by the design of the Pander's windscreen and at high speed, Geysendorffer suddenly saw the lights of the ambulance almost immediately in front of him. A collision was inevitable and the starboard wing of the aircraft smashed into the floodlight causing the fuel tanks, carrying almost 450 gallons of petrol, immediately to erupt into flames. Burning fiercely, the Pander remained airborne for a short time until Geysendorffer put her back onto the ground. The crew made a miraculous escape through the narrow door on the port side behind the cockpit. The port engine was still running and slipstream from the propeller temporarily cleared smoke and flames away from the exit, but Dick Asjes burned his hands whilst rescuing the floodlight operator and a labourer. both of whom were badly injured, from the conflagration. The driver of the ambulance was deeply shocked but otherwise only slightly hurt. The crew could do nothing but watch helplessly as all chance of a gold medallion disappeared on the gallant Racer's funeral pyre. In Holland the news was received with dismay and the aeroplane immediately was nicknamed the 'Pechjager', the 'Unfortunate Speedster'.

Darwin's Customs officials had some hours to wait before their next fee arrived. The New Zealand Dragon Six flew into the circuit on Saturday morning and caused a flutter amongst the onlookers by overshooting and going round again, a routine which had become something of an unnerving habit. The aircraft had flown on to Batavia following a leisurely turnround at Singapore during which Stewart recorded that the tailwheel needed repairs. On Friday morning they flew in poor conditions to Rambang and left there for Darwin, operating at 10,000ft but encountered strong headwinds and diverted into Koepang to refuel. The weather forecast for the Timor crossing was poor and departure for Darwin was delayed until 3.00am on Saturday. After landing in Australia, Kay said their sector from Koepang had been one of the most pleasant on the entire trip and emphasised the point (possibly for the benefit of prospective customers,) that as the aeroplane was being delivered new, "straight from the de Havilland factory, for a potential airmail service in New Zealand," they were treating her gently. A cable was sent to Auckland advising of their safe arrival.

The Courier was next to land, a further 24 hours behind. After extricating themselves from the mud at Alor Star, the Stodarts had operated via Rangoon, Singapore and Rambang and had encountered the omnipresent storms over Timor. The aircraft had suffered a severe battering flying blind for 250 miles at 10,000ft in heavy rain, causing another broken window which the engineers at Darwin mended with a doped fabric patch. Only the searchlights being displayed by HMAS *Moresby* had permitted a track check. Although the crew also experienced some Customs problems then David Stodart was described by Darwin's aerodrome controller, Major H H Mann, as 'starchy', the two pilots expressed relief at being on the ground.

On arrival at Darwin all aircraft were obliged to park in a 'quarantine' area before release to the space set aside for refuelling and maintenance. The engine of the Courier kicked whilst the propeller was being swung prior to her repositioning and David Stodart received a blow on the arm which inflicted a nasty gash near his elbow. The wound was bound by Dr Cook but Stodart refused any further assistance assuring all concerned that it was quite comfortable and he would probably get it stitched at Cloncurry. The Courier left at 10.19am on Sunday, bound for Newcastle Waters.

The Hawk which had left Darwin on Friday morning arrived at the two sheds, bush hotel and pumping house collectively known as Newcastle Waters, three and a half hours later. The route was almost entirely featureless, the sole aid to navigation being the railway line to Birdum some 250 miles south east of Darwin, and the telegraph line from there across desolate bush punctuated with regularly spaced and mostly anonymous emergency landing grounds. As expected, a forecast dust storm was encountered 120 miles from the Wa-

The Stodarts in their Airspeed Courier were warmly welcomed by an enthusiastic crowd after the aircraft was moved out of the quarantine park at Darwin.
(National Library of Australia, Canberra.)

David Stodart, still wearing the rig of a city office worker, turning-over the Courier's propeller. The engine kicked back and the propeller blade struck Stodart on the arm but he refused any major treatment at least until the aircraft reached Cloncurry for a planned night-stop.
(National Library of Australia, Canberra.)

Kenneth Stodart preparing to swing the propeller of the Courier at Charleville early on the morning of Tuesday 30 October. Five hours later the aircraft crossed the finishing line at Flemington.
(National Library of Australia, Canberra.)

The steady progress of the Miles Hawk and the tenacity shown by her crew were examples to all and it was a cruel twist of fate that deprived them of a prize. The aircraft arrived at Laverton during the morning of Sunday 28 October having left Narromine before daylight, and her journey cut the existing record from England by two days.
(National Library of Australia, Canberra.)

Jimmy Melrose left Darwin early on Monday morning, operating via Newcastle Waters and Cloncurry for Charleville. Refuelling the wing tanks at Cloncurry was achieved by a team which insisted on leaning a long ladder against the leading edge. The Puss Moth arrived overhead Flemington a little after mid-day on Tuesday.
(Queensland Museum.)

West of Charleville on the Middle Creek Road, the old truck driven by local mailman Bert Sandes attracted the attention of Jimmy Melrose who was lost. The Puss Moth was landed on a patch of gidgee scrub next to the road and directions sought from the driver.
(National Library of Australia, Canberra.)

There was no alternative but to land and wait. They put down at Brunette Downs at 2.00pm where all hopes of a prize cruelly deserted them.

The non-arrival of the Hawk at Cloncurry and no word received from any observers resulted in Race Control at Melbourne placing RAAF search aircraft on standby until news reached them that the Hawk was safely accounted for.

The Hawk was grounded for 15 hours at an unofficial stopping place. Had they but realised, the Rules permitted a return to their last official Checking Point at which their time on the ground would not have been recorded. All the time spent sheltering at Brunette Downs was counted against them. By next morning the dust-storm had blown itself out and they were able to make good time to Cloncurry then on to Charleville. Although tantalizingly close at 787 miles, Melbourne was well beyond their range, and a landing at Narromine was planned, the last official Checking Point before the finish.

ters, and the last part of the sector was flown at less than 100ft when visual contact with the ground was intermittent.

The remaining 500 miles to Cloncurry promised much the same, most of which terrain was suffering from the scouring effects of the red dust storm. Not anxious to press on, McGregor and Walker were as-sured by experienced hands that the worst of the cloud had passed through and conditions would improve with their southerly passage. It was not so. After a 90 minute turnround at Newcastle Waters they battled through the gritty air at low level for 180 miles. It was their worst nightmare; the vital ground tracks had been obliterated.

An hour from Charleville, overhead Cunnamulla at the southern end of their guiding railway line, McGregor in the front cockpit suddenly realised he was being soaked in warm oil which was seeping through the bulkheads and immediately put down to investigate. The front seal around the crankshaft was leaking, a com-

The Danish Desoutter at Batavia ready for departure to Rambang. The sticky conditions at Alor Star deprived them of a better time and in order to reduce weight the crew was forced to forward their baggage to Singapore by train. (The Aeroplane.)

mon enough fault on a hard worked Gipsy Major engine, but there was nothing that could be done now. They continued on their journey, were reported over Bourke at 7.30pm and arrived overhead Narromine at 8.53pm in the dark, not only the first Race machine to use the aerodrome but the first ever to attempt a night landing there. As part of their preparations for the Race the local authorities had laid on electrical power to illuminate the wind indicator and Control Office and hoisted a red light to mark a line of gum trees which they believed might be a danger, but after circling twice the Hawk touched down without incident, aided by car headlights, landing over the Frangie Road following a line of flares. The crew was welcomed by the Mayor before being whisked away to the local hotel.

On Sunday at 4.10am and before daylight McGregor and Walker took-off without lighting assistance to face their final sector, the 425 miles to Flemington Racecourse. At low level and moderate speed they covered the distance in under four hours. Their total flight time since leaving Mildenhall until flying between the pylons was 118 hrs, two whole days less than the existing record. Having travelled so far, they had been robbed of a Handicap prize by a detail in the Regulations which had not been fully explained during the many months of public discussion and debate.

At Darwin on Saturday 27 October, Stan White arrived from England in his DH.60 Moth G-AAJO. He had taken off from Heston on 18 September, headed ultimately for New Zealand, and unwittingly appeared amidst the frenetic arrivals and departures of MacRobertson contestants for which reason his solo flight attracted little or no coverage or recognition. White was one of the increasing number of aviators passing through Darwin, a fact recognised by the Federal Government in 1934 who appointed Walter Dwyer to commission a meteorological office at the airport in advance of the Race. In so doing Dwyer became the first meteorologist to provide a weather service dedicated to the specific needs of aviation in Australia.

On Sunday, 28 October Jimmy Melrose was eighth to arrive on his native soil, touching down at 5.20pm. He had left Koepang at 9.00am after his usual nights' sleep with a reduced fuel load due to weight limitations but still sufficient for about 800 miles. Over the Timor gale force winds blew him south of track and, due to a haze, he had not sighted HMAS *Moresby*. Landfall was made beyond the edge of his maps and, considering that his navigation generally had been good, decided he must be north of his destination. He turned to fly south west but eventually recognised Pearce Point from previous experience over that same coastline and immediately banked the Puss Moth round onto a reciprocal heading.

Five minutes beyond what was thought to be the absolute limit of his fuel endurance, a search aircraft was about to set-off from Darwin in anticipation that Melrose might have put down at Bathurst Island, 50 miles north and hopefully not in the sea. At that precise moment, the Puss Moth was sighted high in the western sky. From Pearce Point Melrose had increased his altitude for maximum fuel economy to cover the 200 miles back to Darwin but within sight of his destination the petrol supply was down to the last few drops. Melrose had barely enough to give the engine a final warming burst before gliding down to land as the sun was setting at 5.30pm.

After taking his usual overnight rest, Melrose took off at 3.55am on Monday, 29 October, bound for Newcastle Waters where he landed at 8.00am, deciding to wait on the ground to allow the strong headwind to abate before leaving for Cloncurry and Charleville. It was on the Middle Creek road, about 17 miles west of Charleville, that Bert Sandes, the local mail contractor, was astonished to see a small aircraft circling at low level. It was Melrose who had been blown off course and was trying to establish a position. Having spotted the mail vehicle the Puss Moth was landed in a clear patch of gidgee scrub 100 yards ahead where she waited for the postman to come up.

"This young fellow in shorts and a sun helmet got out and asked his way to Charleville. He thought he was on the Warrego River but he was on the Ward! After I gave him full instructions he made a perfect take-off."

The minimal separation between Puss Moth and Desoutter at Karachi had lengthened to 24 hours by the time Hansen and Jensen touched down at Darwin during the evening of Monday 29 October. They arrived at sunset, flying low over the hangars. Jensen, who did not speak English, was presented with a Danish flag and assisted with translation by Danish speaking local inhabitants. Their progress had been interrupted in part by the mud at Alor Star where, over two days, they had tried ten times to get off, and failed. In order to lighten the aircraft all that could be dispensed with was off-loaded and sent by train to Singapore where the crew was reunited with their baggage. Operating through the island chain as far as Koepang with no difficulties the Desoutter then avoided the Timor storms by flying at low level, spotting for sharks, although the crew feigned disappointment when reporting no sightings.

First of the little gaggle of Handicap Racers traversing Australia to arrive in Melbourne were the Stodarts who had encountered some of the worst turbulence on the sector into Cloncurry that either cousin could remember and which David Stodart described as the worst stretch of the entire route with mile upon mile of featureless country. The aircraft was shaken to the core and a precautionary check of the structure revealed that some main assembly bolts and stays had been loosened in the pounding. The Courier flew over the finishing line at Flemington just before 11.00am on Tuesday, five hours after leaving Charleville, a little over 100 flying hours from Mildenhall. David Stodart was the first native-born Australian to finish and he and his cousin were greeted at Laverton by one of the largest assemblies of relations to meet any competitor.

Stodart still bore the scars of his encounter with the propeller at Darwin but said the wound had given him no trouble. He was much more concerned for the Lynx engine which he said was beginning to run hot and a few more miles might have caused serious problems.

Jimmy Melrose crossed the finishing line at Flemington at 12.08pm on 30 October, an hour after the Courier, the only pilot to fly solo from England. He touched-down at Laverton eight minutes later to a rousing welcome and was greeted by his sponsor, his mother. He said that he was sorry it was all over, an observation in marked contrast to one of many attributed to Charles William Anderson Scott who pronounced, "It was lousy, and that's praising it!"

Amongst the congratulatory messages was a telegram from the Acting Prime Minister of South Australia, Mr Ritchie: "South Australia is proud and honoured by your Centenary flight from England. A great welcome awaits you in Adelaide. Congratulations."

The crew of the Danish Desoutter decided to over-fly Newcastle Waters and go straight to Cloncurry, but it was a grave error. Due to the headwinds they ran out of fuel and landed at Malbon, about 30 miles short of their destination. The delay necessitated a landing at Cloncurry in darkness and no lights were showing when they arrived overhead but the touchdown 'went well' according to the pilot. Out of Cloncurry on Tuesday, Hansen and Jensen became lost after experiencing 27 degrees of drift and put down near the wool shed that was Evesham Station, midway between Longreach and Winton in Queensland. The station overseer, Mr A Henry, was interested in aeroplanes. His brother, an RAAF pilot, had been posted to Laverton

Airspeed Courier G-ACJL taxying-in with the aid of RAAF wingwalkers at Laverton at the end of her Race on the morning of Tuesday 30 October where her pilot reported the engine to be starting to run hot.
(D R Butterworth.)

Cousins Kenneth and David Stodart with a basketful of paperwork pose for the cameras following their safe arrival at Laverton. David Stodart was the first native-born Australian to complete the Race.
(National Library of Australia, Canberra.)

In the hangar at Laverton with the Miles Hawk and Boeing 247 parked in the background, the Stodart cousins were greeted by the greatest gathering of family members who welcomed any of the competitors.
(National Library of Australia, Canberra.)

On 29 October, when taxying out in pre-dawn darkness at Cloncurry for what was hoped to be the last-but-one sector into Melbourne, the pilot of the New Zealand Dragon Six lost sight of the guiding vehicle and failed to stop before colliding with a barbed wire fence.

In addition to a bent propeller and damage to a main spar, the tailwheel was displaced and the bottom of the rudder torn away. Although spares were ordered immediately repairs were not completed until 2 November and ZK-ACO crossed the finishing line at Flemington shortly before noon the following day.
(National Library of Australia, Canberra.)

Base as an official scrutineer of Race finishers. The Station manager directed his unexpected visitors towards the local railway line then scribbled a quick note to his brother which was received by an incredulous Flight Lieutenant Henry two days later.

Mr Henry's directions to Charleville were of great assistance. The Desoutter arrived there at mid-day and after refuelling left immediately for Narromine but ran into severe turbulence that caused the cabin door to fly open when Hansen's pocketwatch fell out. The crew stayed at the Court House Hotel until 4.30am and took-off at 5.00am, arriving overhead Flemington at 9.33am, touching-down at Laverton nine minutes later on Thursday 1 November, 130 flying hours from the start. It was noted that the tandem seat aircraft was not fitted with dual controls; Michael Hansen had flown the machine the whole time under the navigational guidance of Daniel Jensen for whom there was hardly space in the back seat. After the crew had extracted themselves with some difficulty Michael Hansen said, "There will be no more long trips for me; not today anyway, but we do intend to fly home again, probably in about ten days!"

By rights the Dragon Six, nursed by her delivery crew, should have been in Melbourne first. After leaving Darwin at 10.18am on Saturday, an hour beyond Daly Waters the aircraft followed the wrong track in the otherwise featureless terrain and became lost. They came across a remote farm building and circled overhead, dropping a message in a sweet tin to a figure which hurried outside at the sound of their engines, asking for directions to Newcastle Waters. The farmer acknowledged by drawing an outsize arrow in the dusty red soil aided by which James Hewett found the aerodrome and executed a perfect landing. They left at 6.00am next day, made slow progress to Camooweal where they landed for fuel and shortly found Cloncurry where local wind storms threatened the aircraft's security on the ground. The crew decided to stop for the night and make one long mostly daylight trip to Melbourne, starting early next morning, 29 October.

As the Dragon Six was taxying out in pre-dawn darkness to the lit runway, Cyril Kay realised that the escort vehicle had stopped, but it was too late. The aircraft continued forward, running through a wire fence which bent the port propeller, causing damage to a main spar and all but removed the tailwheel. The crew checked in at Control again to safeguard their handicap time. Spare parts were ordered immediately and repairs were effected for the aircraft to be test flown before leaving Cloncurry on Friday, 2 November, four days after her arrival. It was still necessary to transit Charleville, the last Control Point before Melbourne, where the aircraft night-stopped, arriving overhead Flemington at 11.33am on Saturday, 3 November, the last of the nine Race entrants to reach Melbourne within the limits set for qualification although others would still complete the course, eventually.

The Dragon Six touched-down at Laverton at 11.45am bearing the scars of her collision at Cloncurry and was welcomed by a small crowd of mostly New Zealanders to whom James Hewett announced his intention of delivering his charge across the Tasman within a few days.

Meanwhile, the scrutineers had collected in all the Race log books and were checking recorded times against those estimated by the Royal Aero Club handicappers.

A Melbourne-based charity discovered that, incredibly, 16 ticket holders had tied in its competition to guess the winner's flight time, and the £840 prize fund had to be divided into equal shares of £52.10s 0d, each representing a good return against a sixpenny investment. The prize cheques were accompanied by a note asking, please, that they should not be cashed until advised.

15

A Victory for Melbourne

Charles William Anderson Scott and Sir MacPherson Robertson in Melbourne after the Race and following Scott's visit with Tom Campbell Black to a city department store. Charles Scott carried the awesome responsibility for every take-off and every landing in the winning Comet having admitted at Mildernhall that before handling the Comet he had never previously flown a twin-engined aircraft.
(National Library of Australia, Canberra.)

NOT UNTIL AFTER the arrival of the Dragon Six on 3 November could officials in Melbourne confirm with certainty who had qualified for the Handicap prizes.

Grosvenor House, it transpired, had won the Handicap section in addition to the Speed Race, her actual flying time being almost two hours less than her allowance. Next was the DC-2 with a difference of plus ten hours, a figure which included imposition of the penalty incurred through the emergency off-loading at Albury, now translated into a time factor of two hours. It appeared to be a light sentence in view of the Regulations, but placed the DC-2 second to the red Comet and ahead of Melrose in his Puss Moth who had taken half a day longer than the handicapper's estimates.

Governed by the Rules the scarlet Comet G-ACSS was ineligible for more than one prize and having taken the major spoils, a cheque for £10,000 (Australian) and the gold trophy, the £2,000 Handicap Prize was presented to the KLM crew. It was later reported that Albert Plesman had donated the cash to Melbourne hospitals. Young Jimmy Melrose moved up to take second place on handicap and a cheque for £1,000. At the end of the day he was almost as financially advantaged as the winner for his uncle, Sir John Melrose, General Manager of the English, Scottish and Australian Bank, had pledged sums totalling £900 as consolation money if the Puss Moth was not first.

With only a few minutes separating their corrected times David Stodart appealed against the elevation of the Puss Moth on the grounds that the Courier had been unavoidably delayed at Alor Star due to the appalling state of the aerodrome surface. The three man Appeals Committee set up for the purpose subsequently received a tentative protest from Melrose, who would have lost his £1,000 prize had the Courier been promoted instead, claiming that some of the times of arrival and departure entered in his Race logbook did not actually tally with his own carefully recorded flight details. He had not intended to protest, he advised his friends, but if the Stodarts' claim was upheld he would have been obliged to lodge a complaint to protect the interests of his nominator. In the event the Courier crew's protest was over-ruled but the difference in flight time between them was adjusted to 14 minutes and 40 seconds.

With promotion of the DC-2 to the head of the Handicap section, *Uiver* relinquished her Speed Race position and £1,500 prize to the Boeing, while Owen Cathcart-Jones and Ken Waller moved into third place and qualified for £500. The only other Speed entrant to finish was the Dragon Six, out of contention long before she bit her own tail within sight of home.

Arthur Edwards was overwhelmed by the reception of the enthusiastic crowds and immediately offered an expression of thanks in the form of a 100 guinea donation to the City's children's hospital. He accepted the winner's cheque and according to some contemporary reports gave the gold trophy to the two pilots to do with what they pleased. Some years later, during an interview, Tom Campbell Black's widow, the former Florence Desmond, denied this and said that apart from a cheque received from Lord Wakefield for publicising Castrol Oil and other fees from advertising, personal appearances and articles commissioned by newspapers and magazines, neither pilot received any share of the MacRobertson prizes.

Messages of congratulation poured into Melbourne's Race Headquarters from every corner of the world. Emphasis was placed on the fact that it had been a British win, achieved by an aeroplane and engines designed and built by the British industry and flown by a British crew. Britain's first socialist Prime Minister, Ramsey MacDonald, conveyed his congratulations to the architects of a victory achieved by private enterprise, and companies in any way associated were quick to add their names to those of Dunlop and Wakefield who were the first to seize upon the opportunity of linking the victory with their products. Perhaps the most human message sent to Scott and Campbell Black was from His Majesty King George V:

"The Queen and I warmly congratulate you both on your wonderful feat. We are very glad we saw you at Mildenhall before setting out on your great adventure, and trust you are not unduly tired after the strain of the past three days."

It was widely anticipated that both win-

Speed Race Results

1. No. 34	DH88 Comet G-ACSS	70 hr 54 min 18 sec	(159 mph)
2. No. 5	Boeing 247D NR257Y	92 hr 55 min 38 sec	(125 mph)
3. No. 19	DH88 Comet G-ACSR	108 hr 13 min 45 sec	(121 mph)

Handicap Race Results

1. No. 44	Douglas DC-2 PH-AJU	09 hr 52 min 36 sec	(173 mph)
2. No. 16	Puss Moth VH-UQO	12 hr 32 min 14 sec	(103 mph)
3. No. 14	AS.5A Courier G-ACJL	12 hr 46 min 54 sec	(123 mph)
4. No. 2	Hawk Major ZK-ADJ	15 hr 57 min 58 sec	(105 mph)
5. No. 60	Dragon Six ZK-ACO	18 hr 56 min 52 sec	(116 mph)
6. No. 7	Desoutter Mk II OY-DOD	20 hr 59 min 45 sec	(95 mph)

One of the immediate post-Race social occasions was a private dinner hosted by Sir MacPherson Robertson. Included in this photograph are Tom Campbell Black and Charles Scott, Jean Batten, Sir MacPherson and Lady Robertson, Mr and Mrs N. Robertson, Mr E. Robertson, Miss Stevens, Gordon Taylor and Arthur Edwards. (National Library of Australia, Canberra.)

ning pilots would immediately receive knighthoods but they received no public honours at all, unlike the four members of the KLM DC-2 who were each decorated by Queen Wilhelmina, and Danish officers Hansen and Jensen who were recognised by their own monarch, King Christian X, in a ceremony at the Royal Palace in Copenhagen on 15 December.

It was particularly noted that even after the results of the Handicap Race had been confirmed and Turner and Pangborn's Boeing had moved to second place in the Speed Race, the American crew received no congratulatory messages from their government and the American press heaped all its praise on the success of the DC-2.

George Woods Humphery, General Manager of Imperial Airways, sent two telegrams, one to Scott and Campbell Black: "Your magnificent success will mark an important milestone in the history of British civil aviation," and another addressed to Albert Plesman in KLM's boardroom in The Hague: "Warmest congratulations on so successfully demonstrating the safety and efficiency of air transport."

Leaders of the British aircraft industry were generous in their praise, and letters and telegrams cascaded onto Geoffrey de Havilland's desk at Hatfield. Robert Blackburn cabled:

"I should like to offer you my congratulations in producing a first class design which has proved itself in a gruelling contest." Richard Fairey, an observer at Mildenhall, wrote, "Your enterprise and the outstanding design that you produced turned the possibility of an American triumph into an overwhelming British one. I saw and admired the Comets greatly."

With an appreciation of all the consequences, H P Folland of Gloster Aircraft wrote that from his own experience of recent air races he understood what de Havilland must have gone through but that the result must have given great satisfaction. He added, "Although built specially for the Race one feels that the machine can be used for the purpose of a fast mail carrying aeroplane with excellent results." Tommy Sopwith, writing from the headquarters of Hawker Aircraft at Kingston, expressed his opinion that: "Few records are safe for long but I venture to suggest that it will be a long time before this record is beaten. It is a wonderful example of thoroughness from first to last."

Sir Frederick Handley Page, in his own hand, wrote to Captain de Havilland: "The result of the Melbourne Race is magnificent. Without your effort it would have been a wash out for British aviation, and a walkover for the Americans. The whole effort, the rushed design and manufacture, the successful trial flights and little alteration and finally the skill and stamina of the pilots are one of the best combinations with a wonderful result that has been known in aviation. You and everyone with you deserve a thousand and one congratulations."

Captain Geoffrey de Havilland appended a hand written post-script to his formal reply: "And now we have got to try and catch up with Douglas and Boeing. You could do it but it will take time."

Under separate cover to C C Walker, Sir Frederick wrote: "If it had not been that your company made the machine and Scott flew it and thus both deserve the credit I should say in the words of Scripture 'It is the Lord's doing and marvellous in our eyes.' The result, after such a continued effort in your works and Design Office and by Scott on the flight, has shined even the most sober old stager in aviation. I do congratulate you most heartily. Give Hearle a pat on the back from me. Two or more if you think he deserves them."

The winning Comet had carried with it two copies of the current edition of *The Aeroplane,* a weekly aviation magazine registered with the General Post Office as a 'newspaper'. By reaching Australia within four calendar days, the magazine itself contributed to history, for it was the first British 'newspaper' to be delivered to Australia whilst the same issue was still current and on sale in England. The Editor, C G Grey, promoted the fact with a full spread on the front page of his next edition, in which he foresaw the day when fleets of aeroplanes developed in the style of the Comet would fly the European mails to Australia in a series of relays. "The Post Offices of the British Empire should pay for such transport in the interests of Imperial unity," he thundered. But they never did. Imperial Airways continued with their system which took two weeks, bettering surface traffic by a full month although the DH.91 Albatross, built from wood and using 'Comet technology', was conceived as a transatlantic mail carrier in 1936.

Further publishing history was created when a Bernard Partridge cartoon of 'Melbourne' welcoming a number of aeroplanes racing across the city skyline to Flemington, and carried on board *Grosvenor House,* was transmitted via the Australian Post Office's 'Picturegram Service' to Sydney for publication in *The Sydney Morning Herald* of Wednesday 24 October. The same cartoon was published simultaneously by the topical magazine *Punch* in London, the first time in history that such an occurrence had been recorded.

Part of the procession of seven cars carrying Race crews to the Official State Luncheon at the Parliament House in Melbourne. Tom Campbell Black and Charles Scott are in the first car followed by the four members of the crew of the KLM DC-2. The car carrying James Melrose was continually swamped by members of the crowd, mostly young women.
(National Library of Australia, Canberra.)

The scene outside the Parliament House following the arrival of the Race crews for the Official State Luncheon on Tuesday 29 October.
(National Library of Australia, Canberra.)

A short wave wireless link between Australia and England was operational just in time to transmit photo facsimiles from Melbourne to London. A movie film of the Comet finishing at Flemington was photographed frame by frame and the pictures relayed by the new system. In London, as the images were received each was filmed again and the resulting clip lasting less than 20 secsonds was shown in cinemas as part of the newsreels. The wireless transmission time for the project occupied the circuit continuously for almost three days.

Shortly after their arrival in Melbourne, McGregor and Walker were introduced to Sir MacPherson and Lady Robertson who had taken up temporary residence in the Menzies Hotel. The Hawk crew had abandoned practically all their kit in England due to weight limitations and now the two pilots were invited to visit a Melbourne department store and to chose new outfits in preparation for the impending whirl of social engagements which was about to engulf them and their fellow racers. In addition to wireless and press interviews and public appearances, there was mail to consider; McGregor alone accepted delivery of 1,000 cablegrams!

The greatest ovation ever seen in Melbourne's streets greeted the seven car cavalcade heading from the Menzies Hotel to Parliament House on Tuesday 29 October.

"A spontaneous burst of tumultuous enthusiasm, demonstrating the depth of feeling the world's most spectacular Air Race has aroused," was how one commentator described the public reception. The police, apparently unprepared to handle such a huge gathering, were incapable of preventing the cars being brought to a complete standstill time and again as spectators swamped the narrow traffic lanes.

Although the cheers rang out for the victors, they were equally loud for Jimmy Melrose, the youngest of all the competitors who arrived barely in time to join the celebration. He and his mother who shared the back seat of the limousine were soon under guard by six white-helmeted police officers mounted on the running-boards, posted to fend off swarms of admiring young women whose natural inclination was to embrace the young hero, and whose open car was already festooned with their favours.

"In a riot of flowers, cheers, streamers and milling thousands, 14 airmen passed by, bewildered, nervous, but very proud," it was reported from Melbourne. The convoy travelled to the City Centre where the crews were joined inside the Parliament House by Sir MacPherson Robertson, members of the State Legislature, and the Centenary Celebrations Council led by the inspirational Lord Mayor. Following a State Luncheon, the Prime Minister of Victoria, Sir Stanley Argyle, declared that their guests were some of the most remarkable men the world had seen, and that the State was proud to have been involved in such a historic occasion. The following day, the crews of the winning Comet, DC-2 and Boeing were entertained to lunch by the Federal Government in Canberra.

Three crews missed the victory parade: the Desoutter was still two days out and the Dragon Six crew were kicking their heels in Cloncurry. Cathcart-Jones and Waller were not in the procession either.

Following discussions with Bernard Rubin's representative in Australia, Ken Waller and Owen Cathcart-Jones decided on an immediate return flight to England during which Paramount news-film covering the finish of the Races was to be carried on board. At Laverton the green Comet was prepared under the supervision of Alan Murray Jones, Manager of the Australian Associated Company of the de Havilland Enterprise, special attention being paid to both engines. Early on the morning of Sun-

A VICTORY FOR MELBOURNE

day, 28 October, Cathcart-Jones and Waller were greeted at the airfield by Sir MacPherson and Lady Robertson who had travelled out from Melbourne to present the two pilots with their gold Race medallions. Without further fuss and after a brief round of handshakes, the engines were started and G-ACSR took-off at 7.00am, retracing her route and passing other competitors still on their way to the finish.

Owen Cathcart-Jones later wrote:

"This was no stunt flight or attempt at racing back to England but a demonstration of how air communications could be speeded up. We flew only from sunrise to sunset, approximately ten hours a day, and covered 2,200 miles a day, slept every night and had time for meals, baths and changes. Under such conditions, any machine with the range and speed of the Comet could do the journey with safety."

Most crews, it seems, had not arrived at Melbourne without some commemorative token tucked away. In addition to Alan Goodfellow's special postal covers, slipped into all the Race log books at Mildenhall, many other crews had carried some form of philatelic souvenirs which were light and took little space while being of some potential in raising valuable funds. Roscoe Turner gave a small number of covers which had arrived in Australia on the Boeing 247 to Cathcart-Jones and Waller to be carried on their return flight to England, two of which were added to the collection of another keen aerophilatelist, the chairman of the London Race sub-committee, Lindsay Everard.

Turner had carried covers postmarked New York on 26 September 1934, Mildenhall on 20 October and Melbourne nine days later. These were shipped from

Ken Waller and Owen Cathcart-Jones with Sir MacPherson Robertson and members of the RAAF at Laverton early on the morning of Sunday 28 October. Following presentation of their gold medallions the crew boarded the green Comet and at 7.00am took off for England.
(National Library of Australia, Canberra.)

HRH Prince Henry, Duke of Gloucester, presenting the MacRobertson Trophy to Charles Scott and Tom Campbell Black at the Awards Ceremony held during the public air display at RAAF Base Laverton on 10 November 1934. The RAAF Officer in the centre of the group is Wing Commander Adrian Cole, until recently Base Commander at Laverton, who was invited to become vice-chairman of the Air Race Committee.
(National Library of Australia, Canberra.)

Australia to the USA where they were postmarked again in New York on 4 December, a journey around the world in ten weeks.

Jimmy Melrose accepted several covers from friends for dropping off at various points en-route. On arrival in Charleville he decided to carry only one unique cover on to Melbourne but had forgotten Alan Goodfellow's two envelopes tucked into his Race logbook so completed the Race with a total of three.

On Saturday, 10 November at Laverton Aerodrome, before 100,000 cheering spectators, HRH The Duke of Gloucester made the official presentation of prizes and medallions, assisted by Lieutenant Colonel

D D Paine, General Secretary of the Centenary Celebrations Council and who was acting as temporary custodian of Sir MacPherson's prize fund cheques. Prince Henry had arrived in Melbourne on board the Royal Navy cruiser HMS *Sussex*. With his two brothers, Princes Edward and George, he briefly had been co-owner of a DH.60 Moth in England. Now, in Australia, it was an open secret that he was to assume the position of Governor General.

For the crews of the Courier and the Hawk, the new Dragon Six and the old Desoutter there was no prize, save a gold medallion for each crew member and an assured place in aviation history. Later, all participants received a scroll signed by

Gengoult-Smith and Sir Stanley Argyle, representing the Centenary Celebrations Council and Victoria State Government respectively. The scrolls certified each recipient as a competitor in the first 'across the world aerial contest' as it was described, during which everyone 'demonstrated his prowess as an aviator'. Nobody was to go away completely empty handed.

Bernard Rubin, sponsor of the third-placed Comet was in England, recovering from the illness which had robbed him of a place on board. He had been fit enough to travel to Lympne Aerodrome on the south coast of England to welcome home Cathcart-Jones and Waller when they returned on 2 November, 13 days after their departure from Mildenhall and five since leaving Australia. At Laverton it was Rubin's cousin, Mr G de Vahl Davis, who accepted Sir MacPherson's £500 cheque from the Duke, an event witnessed by his three year-old son, perched high on the elderly but broad shoulders of the Race sponsor himself. The confectionery magnate had told his mother that he must have a good view; it was an important day, and one that he would remember 50 years later.

The rain and storms which had greeted the winning Comet at Flemington and caused such concern over Albury continued throughout the Australian spring, causing many other outdoor Centenary celebrations to be abandoned. In 1934 Melbourne suffered some of the worst flooding in the city's history.

16
And After

At a time of financial stringency it was a mark of some generosity, driven by the immense pride in the Company's achievement, that Captain de Havilland and Frank Hearle granted the workforce a half-day holiday when news of the Comet's victory was transmitted to London. The still considerable numbers employed at Stag Lane, now mostly on Gipsy engine production, streamed out of the factory gates early on Tuesday afternoon.
(de Havilland Aircraft Co.)

NEWS OF THE COMET'S victory was greeted with particular enthusiasm by the workforce at the de Havilland factory at Stag Lane. The aeroplane was largely of their making, and had been shifted by lorry to the company's new aerodrome at Hatfield for final assembly and flight testing. It had been a close thing.

In an expression of some generosity, for the Company's strict commercial ethos was maintained by a strong board of directors, Geoffrey de Havilland granted the workforce a half day's holiday. It was just like being at school when the headmaster offered a snap privilege following some spectacular achievement by a member of his establishment. The de Havilland boys and girls were every bit as frisky as they streamed out of the factory at lunch time on Tuesday. That morning, shares in the de Havilland Aircraft Company Ltd. had risen to 60s 6d, an increase of 2s 9d over the weekend price.

In a joint letter to their workforce dated Thursday, 25 October, Captain Geoffrey de Havilland and Frank Hearle expressed their thanks:

"All of us must feel proud that a Comet has won the greatest race in the history of aviation. The wonderful success of Messrs. Scott and Black is due, not only to the untiring efforts of those who directly worked on the Comets, but also to the spirit of loyalty, enthusiasm and co-operation which every member of the Company displays. We sincerely thank you for your efforts in the past, and know that this spirit will carry us on to even greater successes in the future."

Not wishing to miss the opportunity de Havilland's publicity agent ran whole page advertisements in the aviation press carrying the simple message: 'Gipsy engines won the world's greatest air race'. The stylised view of the front of a cowled engine running at power gave just a slight hint that the fixed pitch, metal bladed propeller, was rotating in the wrong direction.

Many years afterwards, de Havilland's Chief Aerodynamicist, Richard M Clarkson, revealed the following in a privately circulated memoir:

"About a month before the first flight, Arthur Hagg, in order to get a mile or two more speed, reduced the cooling air intake to the engines to less than what was needed for a Gipsy four cylinder engine in a Moth, despite the fact that the Comets were to land, taxi, take-off and climb out of tropical aerodromes. Major Halford and his team were horrified. This could have, and very nearly did, lose the Comets the Race. All three of them overheated, and the winner limped home on one and a half engines. It was not really necessary; no other competitor could fly between the Control Points non-stop and none of them had a speed in excess of 180mph. We therefore only had to deliver the range, 200mph, safe handling, and to keep going to win the Race. After the event all Comets were immediately modified with about 100% larger cooling intakes to the engine."

Within days of the victory, British companies associated in any capacity with the construction of the Comet let the facts be known and editorials were published much like the following which appeared in *The Aeroplane*:

"The winning de Havilland Comet with which Messrs. Scott and Black won the MacRobertson Trophy had two special Gipsy Sixes with Ratier two-position automatic variable-pitch airscrews. The motors had B.T.H. magnetos, K.L.G. plugs, Hobson carburettors and crankshafts made by the English Steel Corporation Ltd. Light-alloy components in various forms were supplied by High Duty Alloys Ltd. and Sterling Metals Ltd. and William Mills Ltd.

Light alloy sheeting for fairings and the like was supplied by James Booth and Company (1915) Ltd.

Other important accessories were Amal petrol pumps and Auto-KIeen strainers. The machine was doped with Titanine, and had Dunlop tyres and Bendix brakes and Moseley air-cushions.

Specially selected wood was supplied by Louis Bamberger and Sons, and steel tubing by the Reynolds Tube Co. Ltd.

The Comet had Rotax navigating and signalling lamps, fusebox and switches, and Eclipse vacuum-pumps for the instruments.

The machine carried a standard Dagenite 6.C.A.7.Y. aircraft 12-volt accumulator, which has a capacity of 25.5 ampère-hrs.

The cockpits were equipped with a range of Smith's Aircraft Instruments and Husun P.4 compasses. The machine had a Smith Harley landing light.

Invaluable aids to navigation were the new Reid and Sigrist turn and bank indicator, and Sperry artificial horizon and directional gyroscope.

The seat-squabs, head-rests and sun-blinds were supplied by L A Rumbold and Co."

The Royal Aeronautical Society was moved immediately to recognise the achievement and awarded Scott and Campbell Black the Society's 'British Silver Medal for Aeronautics' in that their splendid feat had led 'to the advancement of Aeronautics'. Their gift upset officials at the Royal Aero Club who for a quarter of a century had understood that the Aeronautical Society was paramount in scientific matters; the Air League led the movement for patriotism and public education, while the Aero Club devoted itself to sport and the development of the practical art of aeronautics.

The Royal Aero Club Committee, considering that its privilege and prerogatives had been encroached upon, discussed the possibility of sending a sharp note to the Society but was dissuaded by Colonel Moore Brabazon's suggestion that the Society was acting in the spirit of the Amulree Committee, a permanent body of eight members set up a year earlier with the express intention of nominating awards for exceptional achievements leading to advancement in aeronautical science. The Club Chairman

Charles Scott and Tom Campbell Black returned to England on 14 December having travelled by sea from Australia to Naples and from there by boat-train to Victoria Station, London, where they were greeted by enthusiastic crowds of well-wishers. Their car took them directly to Grosvenor House for a reunion with many of their friends who had been involved with the Comet and the MacRobertson Races.
(National Library of Australia, Canberra.)

was a member. No doubt the Club felt cheated, having carried so much of the organisation, to see the British winners of what they may have considered to be their event, receiving the first accolade from a learned body. Early in 1935, it was announced that Scott and Campbell Black had been jointly awarded the Britannia Trophy and each was to receive a Royal Aero Club Gold Medal, with Silver presentations to Cathcart-Jones and Waller, so honour was suitably restored. The presentations were made at a dinner at the Royal Aero Club in Piccadilly on Wednesday, 3 April 1935.

One of the first of the many celebratory events that dovetailed nicely into the festive season occurred on 14 November when a dinner was held in the House of Commons in London for 'the organisers of the Race'. On the same evening a Centenary Air Race Aero Club Ball was organised in the Melbourne district of St. Kilda. Two days later the Grand Hotel, Dover, was the venue for a dinner and dance organised by the Cinque Ports Flying Club to honour their instructor Ken Waller. Unfortunately, Owen Cathcart-Jones was unable to attend and sent his apologies.

Charles Scott had employed a business manager before the Race, John Leggitt of Leggitt and Tuckett, with offices in Grand Buildings, Trafalgar Square and whose duty was to promote 'Scott (Melbourne) Flight' wherever possible. It was Leggitt who advised Harold Perrin that the two pilots were due to arrive in London on Friday, 14 December and was curious to know who would be in the welcoming party which he was sure would be arranged. The winning crew had travelled by sea from Adelaide as far as Naples from where they had continued by train to Paris and on to Calais to connect with the Golden Arrow boat-train. They were greeted at Dover by a team from the de Havilland Company, Harold Perrin himself and Major Blake representing Arthur Edwards and Grosvenor House (Park Lane) Ltd. Large crowds were at London's Victoria Station to greet them before they were driven to Grosvenor House for a reception and reunion with the many friends who had assembled there.

A month later, on 19 December, it was Colonel Moore Brabazon in his capacity as President of the Royal Aeronautical Society, who proposed the toast to 'The Chairman' at a London dinner organised jointly by the Club, Royal Aeronautical Society and Air League, together with the Society of British Aircraft Constructors. Grosvenor House was the appropriate venue for the post-Race banquet, attended by more than 1,000 invited guests, twice the number who had gathered on the eve of departure. Included in the guest list were eight of the Chelsea College students who had worked so tirelessly at Mildenhall.

"It appears that our efforts to help were appreciated, and it was very pleasant to know that we were amongst the greatest in the aviation world on that memorable night," wrote Harold Llewellyn.

Guests were invited by informal letters received from Harold Perrin. That addressed to Jim Mollison ended:

"By the way we hope on this occasion you will make a point of turning up on time - 8 o'clock!"

The principal guests were Charles Scott and Tom Campbell Black. Scott composed a humorous and fluent speech but Campbell Black, quiet and shy, rose merely to say "thank you." In turn they listened to Albert Plesman of KLM as he declared that as a result of the Races, the public would get 'better aviation'. It was hoped so. The flight of KLM's DC-2 had already drawn attention to and critical scrutiny of the timetable published by Imperial Airways for the Australia Air Mail Service.

Unable to attend due to doctor's orders was de Havilland Company Chairman Alan S Butler who wrote separately to the principal guests at Grosvenor House:

"I think it is true to say that I have never been so intensely excited for a period of three whole days in my life. We all in the Company felt the same way about it; wives and families, one and all showed the same enthusiasm.

May I offer my sincerest admiration for the great courage, skill and endurance you showed in carrying that flight to such a happy and successful conclusion for us all."

Arthur Edwards arrived back in London early in 1935 and was a guest at a cocktail party in the International Sportsmen's Club in Upper Grosvenor Street on 10 January, organised by Lord Tweedmouth, 'To meet Mr A O Edwards on his return from Australia after winning the England to Australia Air Race'.

At that year's AGM of Grosvenor House, Edwards said: "Some of you probably read in the newspapers last autumn that I personally purchased an aeroplane and named it *Grosvenor House*. I entered that aeroplane in the historic air race from England to Australia. I need not mention here what gratification it gave me to win the Race, but I am very happy to say that I was very pleased to be of service to Grosvenor House by helping to extend and consolidate its name in every corner of the world."

Sir MacPherson Robertson and his wife travelled to London to attend the wedding of Tom Campbell Black and Florence

Grosvenor House in London's Park Lane, viewed from the south west on a sunny winter's morning. The high capacity restaurant and hotel became the focus for nearly all the social events held in London in connection with the Races, both before the start and after the finish. (Nick Redman. Grosvenor House.)

Desmond which took place at St. James' Church, Piccadilly, on Saturday, 30 March 1935, followed by a reception at the Grosvenor House where Sir MacPherson toasted the happy couple with the gold cup presented to the winning crew.

Campbell Black and his new wife shortly left for their honeymoon in Morocco, the bridegroom piloting them in Lord Furness' DH.80A Puss Moth G-ABYW, loaned for the occasion but soon to become their own property.

Since the end of the Races, and to some extent while they were still in progress, aviation commentators had been busy analysing the many diverse aspects. It was estimated that the cost of buying all 19 competing machines and racing them half way round the world, amounted to £500,000, and that figure did not include the expenses of pre-race preparation and positioning to Mildenhall, nor the additional costs of getting home again from Australia.

When Harold Perrin was attending to some of his essential post-Race administrative duties he received invoices for landing, parking and take-off fees at Athens for competitors who had routed through there, the only aerodrome to make such charges: £2.12s.0d for the Pander; £2.0s 0d for the Boeing, £1.15s.0d for the Dragon Rapide; £2. 5s.0d for the Eagle and five shillings for the Desoutter!

The commentaries recognised that the achievement of the Comet was remarkable in that it had been conceived, designed against the Handicap Formula, built and flown within ten months and, with practically no test flying on either engines or airframe, had successfully raced all-out for 12,000 miles. The only other European aircraft designed specially for the Race, or in truth highly modified from its parent design, the Viceroy, had proved a disappointing failure and the subject of legal action between entrants and manufacturers. The Irish-registered Bellanca had not raced, and the Granville had expired on her first touchdown.

With political tensions just stirring in Europe, perhaps it was not entirely unexpected that somebody, somewhere, should point a finger. Supported by a German newspaper, the Italian *Messaggero*, published in Rome, printed the following story, translated by the Rome correspondent of the *Daily Telegraph*:

"The time allowed to make ready for the Race was so short that none of the competitors except the English had time to build a machine which conformed to the restrictive rules devised by the organisers.

This gives the English an advantage so enormous that only very exceptional bad luck can prevent the three de Havilland machines from winning all the available prizes. The English and English alone, doubtless warned in due time as to what was being organised, built three de Havilland Comets for the occasion, which are able to fly 2,700 miles at a speed of 240 mph without loading up.

The organising committee, as if it wanted still further to assure victory for the three favourite machines has ordained that the other competitors may not carry a load of petrol equal to that contained in the de Havilland tanks."

A more focussed opinion was expressed immediately after the Race. "As it now flies, the Comet is just a racer, but it is anything but a racing freak. With suitable modifications it seems to presage a very fine mail carrier." There were military implications too. One Editor suggested that the Comet might be converted into "a useful two seat fighter, assuming it could stand up to the strain of being thrown about, and providing the Air Ministry did not kill its performance with gadgets." Even before the start of the Races an ex-RAF officer joked to an Editor that if the DC-2 were a bomber, the RAF possessed only one type of aircraft with any chance of catching it. Equally, if the Comet's range were reduced to 500 miles it could carry at least as much bomb load as the Service's current single seat day bombers and be 50mph faster. It may have been cynical but others in authority had come to the same conclusions.

More than anything the Races had brought to public attention realisation that modern inventions such as flaps, retractable undercarriage and variable pitch propellers were practical and reliable, and offered immense opportunities for increased efficiency and even greater advancement. The wooden racer fitted with a propeller system serviced with a bicycle pump had dashed ahead of the big new metal airliner which had methodically devoured the route mileage with a commercial load on board: the classic analogy of hare and tortoise. Standard private owner aircraft had demonstrated their ruggedness, especially when handled by a determined crew, and the advantages and limitations of wireless communication at a period of continuing development were graphically illustrated. Those carrying no such equipment were obliged to navigate entirely by the traditional methods, and although there were problems all roads eventually led to Melbourne.

There was a suggestion on arrival in Australia that some of the Eastern States were considering asking MacRobertson competitors to race between cities for big cash prizes, but if the idea was ever promoted in the right circles, it received little or no support. As early as November 1934 it was believed that a race organised on the lines of the MacRobertson should be flown between England and Cape Town in celebration both of the Silver Jubilee of the King's accession and the 25th anniversary of The Union the following year. The Mayor of Cape Town even went as far as suggesting that the Race might be sponsored by the South African diamond millionaire Sir Abe Bailey, husband of aviatrix Lady Mary Bailey, but the challenge was not taken up.

The success of the Hawk was further proof of the genius of the Miles organisation. *Manawatu* with her Gipsy Major engine was a powerful force in a class usually dominated by de Havilland machines which were noticably absent from the event. Designs based on the Hawk specification continued to be built in small batches, each being a development of the last, until the Hawk Trainer, built for the Royal Air Force as the Magister, production of which ran to almost 1,300 machines.

The press' view of the flight of the Hawk was fairly unanimous and readers were reminded that the Hawk was the first of the 'Handicap only' competitors to reach Australia. "When you consider that two good hefty men in a light single engined aeroplane with about 130hp had flown close on 2,000 miles a day for five days they do deserve to be regarded as the heroes of the Race", was the opinion of C G Grey.

The Handicap Formula received as much criticism after the Race as it had endured before, although the Mayo sub-committee formed specifically to refine and improve it had been unable to do so. No formula was ever going to satisfy everybody and would always appear to treat some better than others. As scratch machine on Handicap, the Airspeed Viceroy was in an impossible position from the first mock take-off run, and the crew knew they had little chance of winning anything except publicity and perhaps a gold medallion and a short holiday in Australia. That they achieved only critical press coverage was unfortunate.

The Handicap Formula had assumed the Viceroy to be faster than any of the Comets as it was both heavier and carried more powerful engines. The 'wing-power' element in the Formula (horsepower per square foot of wing area) made no concession to the fact that Airspeed's aeroplane was a conventional twin-engined, commercial transport competing against speedsters designed for the purpose. Only the Lockheed Vega, a well-proven racing type, separated the Comets from the DC-2, while the Boeing 247D was on entrant in the Speed Race only so no calculations were made to determine her position in the Handicap league.

The DC-2 more than the Boeing had staggered the pillars of the British aircraft industry, designers and engineers, correspondents and informed observers, when she was inspected in detail on arrival at Mildenhall. The contemporary de Havilland airliner, the DH.86, designed against an Australian specification and first flown by Hubert Broad as recently as January 1934, was a four-engined, fabric-covered wooden biplane with a range of less than 800 miles, carrying ten passengers at a cruising speed of 145mph. The last of the 62 aircraft completed was flown in December 1937. The British industry was envious that a healthy market existed in the United States for production of the DC-2 type of machine, yet there was no finance for such a venture within the Empire. The mere fact that KLM had taken delivery of their first DC-2 for regular services from Europe to Batavia (and after the Race ordered additional aircraft) should perhaps have caused more concern, but the challenge which was certainly recognised was not accepted.

The situation was neatly summed up in London's *Morning Post* on 24 October:

"The results of the England-Australia Air Race have fallen like a bomb in the midst of British everyday commercial and military aviation. Pre-conceived ideas of the maximum speed limitations of standard commercial aeroplanes have been blown sky-high. British standard aeroplane development, both commercial and military, has been standing still. America now has standard commercial aeroplanes with a higher top speed than the fastest aeroplane in regular service in any squadron in the whole of the Royal Air Force."

The de Havilland Company needed all possible commercial advantage gained by their win, and on 8 November 1934, Captain de Havilland wrote to the Director of Civil Aviation at his office in Whitehall:

"Although we, as the designers and manufacturers of the Comet which won the recent England-Australia Air Race, have received many hundreds of congratulatory messages and much favourable comment in the world's press, it is unquestionable that the fine performance of the American Douglas and Boeing machines, standard models as used on the air mail lines of the United States, demonstrated how far this country lags behind in the use of high speed commercial aircraft.

The Comets, all of which established records during or after the contest, although produced specially to conform to the conditions of the Race, have practically no future as a production or saleable type. In designing the Comet, we had to have prominently in mind the speeds achieved by American commercial aircraft, and it was evident that our best chance of success against them lay in providing a machine of even higher speed combined with greater range. These attributes, however, in the form of the Comet, do not at present meet any demand of existing Empire Air Routes.

We believe that this Company has pursued the right policy up to the present time in producing types of commercial aircraft which can be operated economically with the minimum of artificial financial aid, and it will, in the ordinary course of events, follow on the same lines in the future.

It is, however, strikingly apparent to us that immediate steps must be taken in this country to produce aircraft of better performance and of greater commercial merit than the successful foreign machines in the MacRobertson Race, and their inevitable developments.

We are confident that, as a result of our experience in the development of the Comet combined with the wide knowledge we have acquired in the production of commercial types during the past fifteen years, we could produce a commercial aeroplane which would acquit itself with the products of any nation."

Captain de Havilland might have mentioned that of the 20 aircraft which started in the Races, nine finished within the prescribed time of which four were of de Havilland design and manufacture: two DH.88 Comets, a DH.80 Puss Moth and a DH.89 Dragon Six. One de Havilland aircraft only, the Mollison's Comet, G-ACSP, did not complete the course. Four other competing aircraft were fitted with de Havilland engines, three of which finished, the fourth suffering a landing accident enroute.

One particular paragraph in the letter is of great significance when viewed against pre-Race orders for the Comet and the apparent lack of spare propellers:

"It will not be out of place to mention

A rare occasion when both the DC-2 and the Boeing 247 were captured together together in the same frame. More than the Boeing, it was the new airliner from the Douglas Aircraft Company that staggered the British aviation establishment when she was available for close inspection at Mildenhall. Amy Mollison had seen the aircraft in production in California and had already sent her verdict back to her husband: "The only chance any non-American competitor would have in winning the Race would be if all theothers lost their way!"
(The Aeroplane.)

The DH.86 was designed against an Australian specification and first flew in January 1934. The fabric covered airframe was built of wood and powered by four Gipsy Six engines. The aircraft could carry ten passengers at a cruising speed of 145mph for less than 800 miles.

the heavy cost shouldered by the Company over the production of the Comets and the necessary Race organisation involved. Three machines were built, together with a reserve for spare parts, likewise ten special engines and spares, also twelve experimental variable pitch airscrews. Suffice it to say that the purely nominal prices paid by the three entrants have recovered little more than 25% of the total direct cost in which we have been involved. You will understand, therefore, how immensely important it is to us to derive some benefit from the Race."

The belief that major components of a fourth Comet were constructed before the Race against allocated build number 1999, and that two Gipsy Six R engines from the first batch were allocated to it, is endorsed. Indeed, the fourth aircraft, which might have been what was offered to Sir Charles Kingsford Smith, was completed in the summer of 1935 against a revised builder's reference, 2260, and a fifth aircraft, 2261, built at the same time, carried engines of contemporary construction, but fitted with de Havilland manufactured Hamilton Standard propellers.

One query remains over the matter of propellers. If spare propellers were held in reserve, why were these not immediately substituted when the green Comet suffered her landing accident at Mildenhall almost on the eve of departure? The answer, perhaps, can only be that they had not been delivered from France. There are conflicting reports on how many spare propellers were in store at Hatfield at the time but reason leads to a belief that there could have been only one.

After the Race was won, the de Havilland Aircraft Company's Chief Engineer, Charles Walker, expressed himself in the company's widely circulated magazine, the *de Havilland Gazette*:

"The production of a machine suitable for the Australia Race naturally occasioned much thought and discussion. The requirements were quite different from any other speed event, because compliance with airworthiness regulations was required, and, still more important, as long a range as possible is obviously desirable in a long race. Since a substantial saving in distance would ensue if a direct flight to Baghdad could be made, it was necessary to see if it was possible to lift over the airworthiness screen a sufficient weight of petrol to fly to Baghdad, a distance of between 2,500 and 2,600 miles, without landing. This meant, including a minimum safe reserve, 2,900 miles' range.

Until the time this design was started, a range under Certificate of Airworthiness conditions exceeding 2,400 miles had not been achieved, and the machines coming closest to this range were slow and therefore unsuitable. It had also to be decided whether, if 2,900 miles' range could be secured, the policy of going for the range and carrying no wireless and other equipment, on the one hand, or the policy of going for higher speed with supercharged engines and shorter range, on the other hand, was best. The second of these alternatives would permit the use of wireless, etc. It was apparent that if the long range could be obtained it was the thing to go for.

The senior players behind the DH.88 Comet project who agreed immediately after the first announcement of the MacRobertson Races that the formula for a potentially winning aeroplane should be conceived and refined: Arthur Hagg, Captain Geoffrey de Havilland, Major Frank Halford, Frank Hearle and C C Walker. At a time of financial stringency it was then a bold decision to press urgently ahead with design and construction and to sell each aircraft with the certain knowledge that the Company would incur a heavy financial loss.
(BAE Systems.)

Fuel Economy

Since economy in petrol consumption was so important, the choice of engines which would deliver the largest possible fraction of their indicated power to the air frame as thrust horse-power was necessary. Among the several alternatives examined as being available, it was considered that the six-cylinder in-line engine, with its very small frontal area per horse-power, offered the best solution.

The question of supercharging was thoroughly explored, but finally rejected. At first sight it is tempting to go for the greater speed which would be obtainable, but since one could hardly hope (for 2,900 miles' range) to use power at a greater rate than two-thirds of that available for lifting the petrol over the screen, the use of superchargers would have involved flying at a great height to obtain the above condition.

The governing considerations then, on which the design was commenced last March were something like this. The greatest weight of petrol per horse-power must be lifted over the screen, therefore, fixed pitch propellers were right out of the question. This petrol must be used as economically as possible, but at the highest speed obtainable, and in any case at not less than 200mph. The aeroplane must be large in order to lift the petrol and small in order to go fast economically. Without dealing with these things in any detail, it is evident that flaps are necessary when keeping the size down, and retractable undercarriage for keeping speed and economy up.

It will be seen that all the points which must be aimed at are merely those governing the design of fast commercial aeroplanes, and this is presumably what was aimed at by the promoters. No aeroplane specially designed for the Race could deviate very far from the main characteristics of a commercial aeroplane.

In the short time available for tests on the Comet, it was found that a consumption of less than 20 gallons/hour could be obtained at a speed of 225mph and an operational height of 10,000ft. The engines were tested for a greater duration than that of the Race under these conditions. The great importance of the mixture control will be evident. If the engines were run over-rich the range would be jeopardised; if over-weak, trouble would be expected from overheated pistons, etc.

There is, however, a tolerably wide zone of satisfactory operation, and greater range at a satisfactory mixture strength could be obtained by flying higher and slower. So far as can be ascertained, consumptions of 16 or 17 gallons/hour have been obtained in the Race, so it is likely that the 'higher-and-slower' policy was adopted.

Not a Racer

The Comet is generally referred to as a racing aeroplane, and to the extent that it was designed for the Australia Race, and has not a large and commodious cabin, this is true. It differs, however, very greatly in other respects from a racing machine. The operational conditions (which it is believed were adhered to) were that full throttle at 10,000ft or over should be employed. This means (at the rpm allowed by the propeller) that 75 per cent, or less of the type-test power for continual operation was used in the Race. The machine also complied with the International Airworthiness conditions on its full load. A further point is that in using only three-quarters of the 'continual operation' power a specific consumption could be used which was much lower than has ever been contemplated for racing purposes.

Perhaps the forced landings by night and full-load landings by night which took place during the Race might also be adduced to show that the rules of the Race do tend to exclude what is generally understood by the term 'racing machine'.

A recent article on Controllable Pitch Propellers in *Flight* indicated the gains which could be secured in designs which made full use of flaps, retractable undercarriages, c.p. propellers, etc. There is nothing new in these appliances, and the more recent American commercial aeroplanes have already made full use of them with results which have caused universal admiration.

Conditions in the British transport industry have not yet demanded this kind of synthesis, and perhaps the Comet may be looked upon as a full-scale demonstration (subject to certain qualifications, some of which have already been dealt with) that speeds may be increased by a very large amount without sacrifice of pay load by making the best use of known devices.

Without Sir MacPherson Robertson this demonstration might have had to wait."

The American press was quick to react to the success of the DC-2, pointing to the fact that although run under KLM colours, the airframe and engines were of American design and manufacture. Not a contrivance built especially for racing, the reports were keen to identify, but a commercial contribution which demonstrated that London was only four days from Australia.

The performance of the DC-2 which had

Mainstay of the Cairo to Baghdad sector of the Empire Air Route was the DH.66 Hercules, first flown in 1925. Sir Eric Geddes, Chairman of Imperial Airways, defended the continued use of the reliable but completely outdated DH.66 against what he obviously thought was a very cavalier attitude taken by KLM with their DC-2 operation during the Races.
(de Havilland Aircraft Co.)

flown what amounted to a basic airline schedule half way round the world, sparked criticism of the Imperial Airways' timetable published for the prospective Australian airmail service. On 5 November at Imperial Airways' Annual General Meeting held in London, Sir Eric Geddes, Chairman of the board, added an impromptu comment to his prepared Annual Report:

"No one must think that a concern of our size can change its policy suddenly because of a very gallant flight which has beaten all records between here and Melbourne, flown under conditions which could not possibly apply to commercial aviation. All machines in the Race flew free from such hampering regulations as Customs and Passport Control which would be applied in the ordinary way to passengers and freight. The Douglas DC-2 carried twice the amount of fuel she would normally carry; that saves stops. It flew by night over a route mainly unlighted, and no commercial company would care to do that on a regular passenger service, nor does the Douglas provide comfort for long flying such as that given by our Hercules."

The de Havilland DH.66 Hercules, a three-engined, fabric-covered biplane designed in 1925 in which the pilots sat together in an open cockpit, had inaugurated the Cairo-Baghdad section of the Empire Air Route in 1927. Geddes had chosen a strange example for comparison: actual stick and string against the new concept. When the airmail through-service was eventually agreed, the link in the chain from Singapore to Brisbane was operated by Qantas DH.86 aircraft, newly designed, four engined wooden biplanes with doped fabric covering and no protection for flying in bad weather. The first aircraft had been delivered to Brisbane only a week before the start of the MacRobertson Races in England. Qantas could have used the opportunity to deliver the aeroplane as a competitor with all the attendant publicity, but chose not to do so.

Those who continued to applaud the performance of the DC-2 were rewarded with much detailed analysis of the results of which the following was typical:

"Let us not forget that the *Uiver* made 13 stops more than the winning Comet with an average ground time of 30 minutes each. It is not unreasonable to assume that with the descents and climbs taken into account, the DC-2 lost at least one hour in total on each stop. If KLM had opted for extra fuel capacity instead of carrying passengers and mail, the aircraft might have been just six hours behind the leading aircraft in which case it would have missed the heavy weather short of Melbourne and the unplanned eight and a half hour delay at Albury, and probably would have been the winner!"

Captain W.E. Johns, Editor of his own paper, *Popular Flying*, wrote in the December 1934 issue:

"The Douglas, the so-called flying hotel, a beautifully finished job by the Douglas Aircraft Company of Santa Monica, California, has demonstrated very clearly to us the point reached by high-speed air transportation in America; conversely, it has shown just how far we are behind, and a great many people will be wondering why.

The answer is because Imperial Airways, who, until recently, have been the only people in this country interested in air transport in a big way, hold the view that it is more economical to run a big machine slowly than a smaller machine quickly. Holland and America take the opposite view. They say, and I believe they can prove, that in the end it is cheaper to use machines of the Douglas class. An aeroplane exists as a means of transport only by virtue of its speed. Given speed, more people will use the airway. It costs money, but time is money, and they are prepared to pay for speed because by saving time they save money. In effect, the fast machine results in more passengers, more frequent services and, therefore, a more economical working of the whole organisation."

Imperial Airways continued staunchly to defend its position and responded to such opinion with a letter generally circulated to the press in which comparisons in time taken to reach destinations in the Far East were made between themselves, KLM and Air France:

"So much has been written in the newspapers recently about the question of speed on a long-distance air route that a large number of people have gained the firm impression that the Imperial Mail Services take at least twice as long as foreign national services to reach a common destination. This belief is so widespread that I find even some of my own staff are becoming imbued with the same ideas.

It should be borne in mind that the magnificent performances in the recent speed contest to Australia were under racing conditions and cannot be regularly achieved by any air transport concern on this route under service conditions."

During the week-long preparation before the start of the Race at Mildenhall almost every British aircraft manufacturer was represented in the crowd. How ironic, following all the post-Race analysis and comparison of type performance, that this group-shot of de Havilland Aircraft Company Chairman Alan Butler and his wife Lois, Miss Betty Knox-Niven and Captain Geoffrey de Havilland, should have been taken with the Douglas Aircraft Company's DC-2 as a backdrop.
(Alan Butler.)

Immediately after the Races, Alan Butler recorded some personal thoughts covering the possible specification of a future airliner suitable for the England-Australia route. In May 1937 the prototype DH.91 Albatross flew from Hatfield, a project which had demanded enormous Company effort and resource. When presented for inspection by Imperial Airways the company's engineers were flabbergasted to discover that this elegant aeroplane was made almost entirely from wood.
(de Havilland Aircraft Company.)

The regular Airmail Service from England to Australia was inaugurated on 8 December 1934, and only two weeks later an announcement was made concerning the Empire Air Mail Scheme. The surcharge on air carried letters was to be abolished and a standard rate of a penny-halfpenny would ensure a letter was carried by air to any part of the Empire. The massive press speculation that Comets were to be adapted as express mail-carriers was never confirmed, and two and a half years elapsed before the Empire Scheme actually began.

The de Havilland Company had mostly shied away from government and military contracts, preferring to tread their own path at a pace they themselves could control. There were elements within the Company which were disappointed that the winning aeroplane had not been a new multi-seat airliner, but there was barely time to have designed a two seat racer, let alone a commercial transport. Company Chairman Alan Butler was minded to jot down some personal thoughts about a future project:

"My suggestions, October 1934, for a 20-30 passenger machine, after the Australia Race:

Four 450-500hp supercharged, horizontally opposed or Vee engines.

Monoplane, wing in one piece across the fuselage.

Fuselage and wing centre section metal.

Construction of wing of heavy aluminium stressed-skin plate.

But through his Editorials in *Popular Flying*, Johns persisted:

"America has led the way in high-speed air travel, as she was bound to, because America is a country which, by virtue of its size and absence of frontiers, is eminently suitable for air travel. Hitherto, American civil flying has had no relationship with air travel in Europe, which, bristling as it is with international obstacles and military considerations, is an entirely different proposition. But the Dutch, in the person of that old trouble-maker, Antony Fokker, have stepped in, and their cards are on the table for all the world to see. They are going to introduce American methods and machines into Europe and we must thank Mr MacRobertson for turning the spotlight on that significant fact. If, in the face of what they know now, Imperial Airways persist in their old methods, they will soon be flying empty machines."

Wing extensions from motors to tips wood, also stressed skin.

Motors one behind the other, two on each wing.

Cooling by fan if necessary.

Split flaps,

VP props.

Nose of machine perfect shape, pilot in front."

On 13 April 1935, Imperial Airways' first England-Australia through-passenger schedule left England. The service used different aircraft types on specific sections of the route and ran in competition with the DC-2s of KLM, which operated from Amsterdam through to Batavia. Imperial Airways lost traffic to KLM and a major shift in policy concluded that land based aircraft requiring fixed airfield installations should make way for flying boats.

In 1935 the DC-2 was the first design by Douglas Aircraft to receive the Collier Trophy for outstanding achievements in flight. When production ended in 1937 the company had built 156 DC-2s at Santa Monica, the type giving way to the natural development, the DC-3.

Towards the end of the decade with the threat of a European war becoming more of a probability, the de Havilland Aircraft Company unveiled a scheme for a fast, unarmed bomber made of wood. The type would again employ Comet technology, refined in the light of subsequent experience with the DH.91 Albatross, an aircraft conforming in some respects to Alan Butler's visionary specification penned on the back of an envelope in 1934.

Through its development of a series of Moth light aeroplanes, the DH.88 Comet and the DH.91 Albatross, the de Havilland Aircraft Company had evolved into a world-leader in wooden aircraft structures and technology. With the clouds of war gathering in the 'thirties, the Company's brainpower was directed towards the design of a new concept in bombers: compact, economical on crew and resources, so fast she required no defensive armament, and manufactured from wood. The result was the DH.98 Mosquito, perhaps the most versatile aircraft of the Second World War. The new aircraft was first flown in November 1940 and the night-fighter prototype, W4052, only six months later. Many of the de Havilland staff who worked on the type in great secrecy had been previously involved with the design and construction of the all-wood DH.88 Comet. (BAE Systems.)

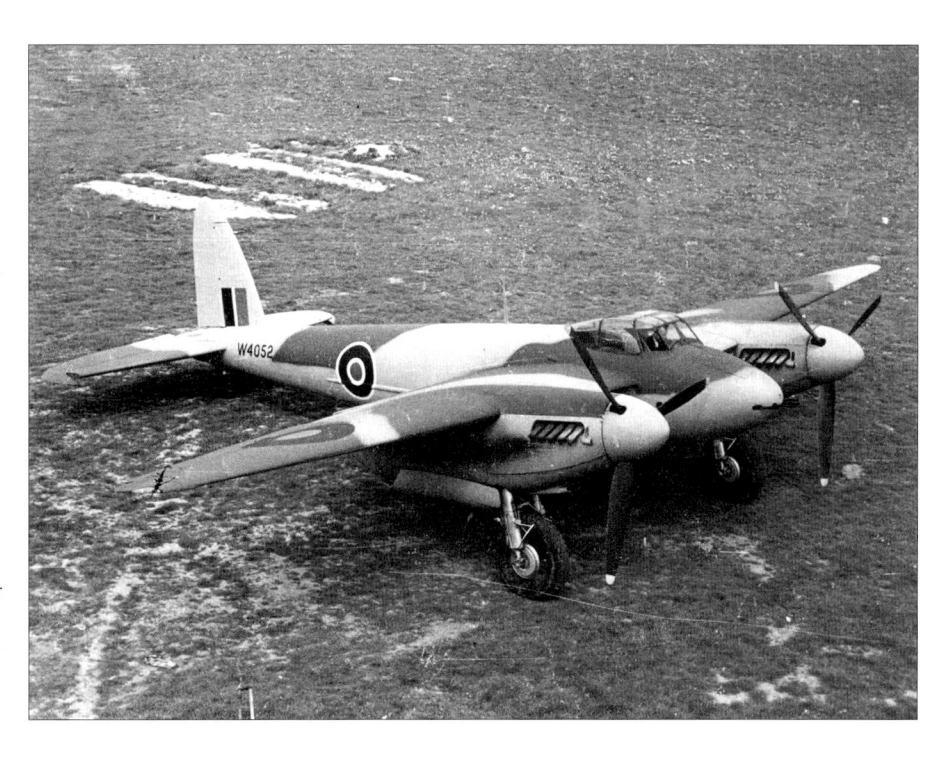

17

Final Straight

Tom Campbell Black proposed marriage to popular comedy actress Florence Desmond before the start of the MacRobertson Races but was told he would not receive an answer until he returned from Australia. He had to wait until a fortnight before Christmas to know he was accepted. The happy couple were married in London in March 1935.

After the Races the aircraft which had reached Melbourne gradually dispersed. Following maintenance and a degree of local activity some flew back to Europe and others continued on to New Zealand. The only American entry to reach Australia, the Boeing 247, was shipped home to the United States where quite separately she and her crew were to continue long flying careers. The winning Comet, the scarlet painted Grosvenor House, remained in Australia until she was shipped home to England at the end of the year to star in a round of static exhibitions. One of the great team successes of the Races was the Miles Hawk, her Gipsy Major engine and her unflappable New Zealand crew. The aircraft was dismantled in Sydney and shipped home to Wellington.
(National Library of Australia, Canberra.)

ON 14 NOVEMBER 1934, Royal Air Force Mildenhall was to have received its first operational aircraft as part of a re-organisation of day and night bomber 'Areas' within the British military defence plan but, due to thick fog it was not until the following afternoon that eleven Handley Page Heyfords of No. 99 Squadron flew in from Upper Heyford near Oxford to occupy their new base. To help cope with the increase in social activity occasioned by the influx of mostly young men at the new aerodrome, a cinema was opened in the village of Mildenhall in 1935. This rural picture palace was called the 'Comet'.

His Majesty King George V made a second visit to the site on 6 July 1935 for his Review of the Royal Air Force in honour of the Silver Jubilee, and his own 42nd wedding anniversary. On this occasion 356 aircraft were drawn up for inspection, and once again the surrounding roads became choked with spectators.

Throughout the Second World War, RAF Mildenhall was a home for heavy bomber squadrons and in peacetime was occupied by elements of the United States Air Force, who developed the station into a major centre for US military air transport operations in Europe.

For the MacRobertson Race aircraft and their crews that flew off from the quiet country aerodrome in October 1934 in search of gold and immortality, and for those that did not even manage to start, the future was to hold a bagful of mixed fortunes.

1. Messerschmitt Bf-108V1 D-ELIT.

Although no reason was ever offered for the non-appearance of the aircraft, an achieved level of performance below that expected was thought to have been a deciding factor. D-ILIT was reported to have been written off in an accident at Augsburg in 1935.

2. Miles Hawk Major ZK-ADJ.

The Miles Hawk Major was thoroughly cleaned in Melbourne and her cowling properly repaired before she was flown to Sydney. From here both the aeroplane and her crew were carried 'courtesy of the line' by the Union Steam Ship Company to Wellington, New Zealand, where the SS *Monowai* docked to a great welcome. The Hawk was shortly flying around the country as part of a publicity tour, and on one occasion McGregor could not resist executing a slow roll with Reg Tappenden of the Auckland Aero Club as passenger. As the Hawk passed through the inverted position, Tappenden was showered with dust and mud from the darker recesses of the cockpit, and growled at his pilot through the speaking tube. McGregor just chortled, explaining that his passenger had been doused in some of the world's most expensive debris accumulated in 16 countries!

The New Zealand Government voted £800 to be shared between all the New Zealand crew members who had competed in the Races, and the donation was welcomed by those who were now being called upon to settle accounts. The total cost of the

Airspeed Envoy G-ACVI was damaged in England before the Races started but in 1936 she was sold to Australia and as VH-UXM completed 10,000 hours of safe passenger carrying operations.
(Eddie Coates Collection.)

Roscoe Turner astride one of the engines of the Boeing 247D adjusting the propeller covers. The aircraft was carried as deck cargo on board the SS Mariposa *sailing from Victoria Docks, Port Melbourne, at the end of the year and was delivered to the Boeing factory at Seattle.*
(Reproduced by courtesy of Museum Victoria. Ref MM001528.)

Hawk's entry in the Handicap Race was finalised at £2,187.10s.1d, and guarantors were asked to contribute six shillings and nine pence in the pound.

Plans to sell the Hawk to the government lapsed when a year after the Races an election changed the political climate. However, the Wellington Aero Club offered £750 for the aircraft and this was accepted. In October 1936, 'Johnnie' Walker hired his familiar old machine for a flight to New Plymouth. Over high ground near Waverley the engine cut due to an airlock in the fuel line. The forced landing in a small paddock was arrested by a wire fence which tore off the port wing and broke the wooden fuselage midway between the rear cockpit and fin. The damage was considered too great for local repair and the gallant machine was returned to her base where subsequently her wounded remains were scrapped.

Malcolm McGregor was appointed Service Manager to Union Airways (NZ) Ltd. On Tuesday 18 February 1936, the company's Miles Falcon was chartered for a flight from Wellington to Hamilton. On the return leg the pilot landed at Milson Aerodrome, Palmerston North, due to heavy rain with poor visibility but McGregor decided it would be better if he took the aeroplane on to Wellington due to his instrument flying experience. At Rongotai Aerodrome the weather was also poor and on arrival the Falcon circled the airfield at 200ft prior to a landing planned to be as close to the hangars as possible. In poor visibility on final approach the aircraft collided with a metal anemometer mast which cut off the starboard wing at the root causing the Falcon to spin inverted into the ground. Although the passenger survived uninjured, Malcolm McGregor died of a fractured skull later that day. He was buried at Kelvin Grove cemetery, Palmerston North on 21 February 1936 and accorded full military honours.

Following his return to New Zealand, Henry Walker qualified for a commercial licence in April 1935 and until 1938 occupied the post of Aviation Officer with the Vacuum Oil Company based in Wellington. With the expansion of Union Airways he joined the company in June 1938 flying the Lockheed Electra and de Havilland DH86. On the outbreak of the Second World War he became an instructor with the Royal New Zealand Air Force at Ohakea flying Hinds and Oxfords and teaching navigation. Following a posting to No. 4 Squadron operating Hudsons where, as a Squadron Leader he became Commanding Officer in November 1942, he transferred to No. 3 Squadron at Guadalcanal in 1943. Promoted Wing Commander, Henry Walker returned to New Zealand in March 1944 to command the Training Wing at Ohakea and then No. 41 Squadron at Whenupai until the end of the war.

Rejoining Union Airways in January 1946, Walker became Chief Pilot of New Zealand National Airways Corporation the following year, later Flight Superintendent and Operations Manager. He retired in March 1973 with the title Project Development Manager and died at the age of 83 on 7 November 1991.

3. Airspeed AS.6 Envoy G-ACVI.

Although the Envoy never reached the starting line at Mildenhall, the aircraft did eventually get to Australia. In 1936 she was sold to Ansett Airways, registered VH-UXM and remained in use until 1951, logging over 10,000 hrs of safe operations.

5. Boeing 247D NR257Y.

Together with their Boeing 247D, Roscoe Turner and Clyde Pangborn left Station Pier, Port Melbourne on board the SS *Mariposa* bound for the USA where the aeroplane was delivered to the Boeing factory at Seattle for restoration to airline configuration. Turner was immediately engaged for a lecture tour and found himself gracing the cover of *Time* magazine's issue of 29 October. He claimed that the irreverent references to his aeroplane, or rather the engines, as the *'Nip'* and *'Tuck'* were completely justified and that he had banked heavily on prospective financial support from the H J Heinz Company but their late failure to endorse him led to heavy debts which took him five years to pay off.

The aeroplane was returned to United Airlines in January 1935 and later that year Turner flew her to his home town of Corinth where he overshot the runway five times and eventually was forced to divert to Memphis. The 247D remained in regu-

In 1953 the Boeing 247 was retired from flying duties and, recognising her historic associations, was donated to the Smithsonian Institute. The aircraft is currently displayed at the National Air and Space Museum in Washington DC. (The Aeroplane.)

lar service with the United Airlines until 1937 then, following two years as an executive transport with Union Electric Company of St. Louis, the US Department of Commerce Air Safety Board purchased the aircraft and during the next 14 years flew her on experimental work, during which time she was dubbed *Adaptable Annie* under the registrations NC11 and N11.

The Department presented the Boeing to the Smithsonian Institute in 1953, and in 1974 United Air Lines donated funds which permitted total restoration to static condition by CNC Industries of Camp Springs, Maryland. The aircraft is displayed in Washington carrying on one side the NR257Y registration and markings as flown by Turner and Pangborn in the MacRobertson Races and, on the opposite side, her later registration NC13369 and the colours of United Airlines.

During the summer of 1934, before the MacRobertson, Turner had flown a Wedell-Williams racer to first place in the Thompson Trophy Air Race. Knowing the type to be outclassed in 1936 he designed his own racing machine which, after a series of major modifications, was finally entered for the 1937 season as the Turner RT-14 Meteor. With this aircraft Turner achieved third place in the Thompson Trophy, first place in 1938 and again in 1939 for an unprecedented third victory after which he retired. He established the Turner Aeronautical Corporation, which in time of war trained more than 3,300 pilots for the United States forces. In 1952, Turner was awarded the Distinguished Flying Cross by the US Congress. In 1936 the first Municipal Airport at Corinth had been dedicated in his honour and in 1961 the new airport was also.

After Turner and his wife Carline sold their Aeronautical Corporation, he decided to build a museum at Indianapolis Airport within which to house his trophies and awards, the 'Turner Special', his Packard Phaeton and the stuffed remains of his pet lion Gilmore. Building work started in July 1969 but Roscoe Turner died on 23 June 1970. The museum was dedicated on 29 September that year on what would have been Turner's 75th birthday. Due to airport expansion the local authority forced closure and demolition of the museum in 1972 and the majority of the contents were donated to the National Air and Space Museum in Washington DC.

Clyde Pangborn filled various positions in civil aviation as a design consultant and test pilot, notably with Bellanca in Delaware and the Burnelli and Cunliffe-Owen companies in England. When war broke out in 1939, Pangborn joined the Royal Air Force and assisted in organising Ferry Command, recruiting crews from the USA and Canada for trans-Atlantic delivery operations.

After the war he continued his interrupted career as an engineer and ferry pilot until he died in New York on 29 March 1958 at the early age of 63, his accumulated flight time amounting to 24,000 hours. He was buried with full military honours at Arlington National Cemetery, Washington. The airfield at Wenatchee, Washington State, where he landed at the end of his trans-Pacific flight in 1931 was named Pangborn Field in his honour.

6. Pander S-4 PH-OST.

Following her accident at Bamrauli Aerodrome, Allahabad, as the result of a take-off collision with airfield equipment on 26 October, the Pander crew travelled by ship to Genoa where they arrived on 28 November returning to Holland from there by train.

Pronk returned to his duties with Radio Holland and disappeared into obscurity. Dirk 'Dick' Asjes continued his career as a military pilot and was commanding officer of a bomber squadron in the Netherlands' East Indies in December 1941, at the time of the Japanese invasion. He escaped to Australia and flew more than 50 missions

Lieutenant Michael Hansen and Lieutenant Daniel Jensen of the Danish Army Air Corps following their investiture by King Christian X at the Royal Palace in Copenhagen.
(National Library of Australia, Canberra.)

to New Guinea. Post war he retired from the air force with the rank of Commodore and held several important positions in business before he retired to Mexico where he died in 1997.

The vastly experienced Gerritt Geysendorffer returned to airline duties with KLM. He was killed when the DC-3 he was flying crashed on take off from Kastrup Airport, Copenhagen on 26 January 1947.

7. Desoutter II OY-DOD.

Michael Hansen and Daniel Jensen, roundly applauded for their friendly attitude, calmness and thoroughness, left Melbourne on Armistice Day and flew their Desoutter back to Copenhagen. Due to weight restrictions from Darwin the Danes were forced to jettison nearly all their souvenirs and gifts and a batch of airmail covers being carried on behalf of the Queensland Air Mail Society. In Denmark, Hansen used the aeroplane to establish a local airline, Nordisk Lufttrafik, in June 1935. Two years later he flew OY-DOD solo from Copenhagen to Cape Town and back, utilising the long range tank which had been installed in the cabin for the MacRobertson Race.

The Desoutter was presented to the Finnish Red Cross in November 1941, and Hansen delivered her to Helsinki. Transferred to the air force she was damaged at the end of her wartime service there and written off in 1944 but was rescued from a scrap heap by two young men who redesigned the undercarriage and windscreen and installed a Cirrus engine.

In 1947 OH-TJA, as she had become, was fitted with skis and following a protracted take-off from deep snow for a leaflet dropping mission, the port wing hit a telegraph pole, swinging her into an adjacent garden. The two would-be pamphleteers were not badly hurt but the Desoutter was wrecked.

In July 1938, Michael Hansen was invited to join a Danish expedition to North West Greenland when, at little notice, funds became available to support an aircraft. Hansen persuaded the military to loan him an Air Service Tiger Moth which was fitted with floats borrowed from Sweden. Hansen had no experience of flying floatplanes or of Arctic conditions but quickly learned and proved beyond doubt that the aircraft was an absolutely invaluable tool for reconnaissance in such an environment.

Michael Hansen returned to his duties as a flying instructor with the Danish Air Corps but when the occupying German power began to imprison Danish military personnel in 1943, he escaped to Sweden where he joined the staff of the Danish Embassy. After the war he returned to Denmark to help re-establish the air force and retired with the rank of Colonel.

8. Fokker XXXVI PH-AVA.

The aircraft was registered to the manufacturer in July 1934 but not to KLM, the operator, until March the following year. In September 1939 she was flown to Scotland and registered G-AFZR to Scottish Aviation Ltd at Prestwick where she was used for air experience flights for embryo RAF navigators. In May 1940, following partial engine failure during take-off, the aircraft over-ran the threshold, broke through the boundary fence and was destroyed by fire.

9. Keith Rider R-3 NR14215.

Following her fatal maiden-flight crash on 2 October 1934 which killed pilot James Granger the R-3 was repaired and successfully flown by Vance Breese who later took her in record time from San Francisco to Los Angeles. In July 1935 in the hands of Earl Ortman the R-3 set a record of 5 hrs 27 mins 48 secs for the flight from Vancouver, Canada to Agua Caliente, Mexico, a time which stood for two years. In 1936 the R-3 was acquired by Hal Marcoux and Jack Bromberg, both employed by the Douglas Aircraft Company who extensively modified the aircraft, renaming her the Marcoux-Bromberg Special. Flown by Earl Ortman the yellow-painted 'Special' was placed second in the Thomson Trophy in 1936.

The R-3 was stored in the General Airmotive hangar at Cleveland Airport throughout the Second World War until she was purchased in 1948 by an owner who intended to re-engine the aircraft with an R-1830, replacing the R-1535 then installed, and race her again. As part of the programme the wooden wing was removed and stored at a site off the airfield but the structure was badly damaged as the result of a leaking roof and the project was halted. The dismantled aircraft was discovered by Rudy Profant loaded onto a flatbed trailer parked in a Cleveland driveway but having rescued her, the R-3 was destined to spend a further 20 years in store in a local barn. Purchased by the Connecticut Air Historical Association and rebuilt to static display condition the R-3 is currently an exhibit of the renamed New England Air Museum situated at Bradley International Airport, Windsor Locks, Connecticut.

10. Lockheed Orion 9D F-AKHC.

Having been acquired by the French Air Ministry following withdrawal of the en-

Although there were plans to rebuild the Keith Rider R-3 to flying condition, the job was never completed and the aircraft was acquired by the Connecticut Air Historical Association. She is currently on display in the museum at Bradley International Airport, Windsor Locks. (New England Air Museum.)

try by owner Michel Détroyat, the Orion was sold to the Spanish Republican forces who eventually owned a total of 14 of the type. However, like many others, this aircraft did not survive the Civil War.

14. Airspeed AS. 5A Courier G-ACJL.

Third in the Handicap Race, the Airspeed Courier G-ACJL remained in Australia and was flown to Essendon for major fabric repairs to both wings and the fuselage, damage caused by severe weather which had also broken a panel in the windscreen. In November, following the return of his cousin to RAF duties in England, David Stodart was placed first in the Essendon-Cootamundra Air Race, winning £100. His continued stay in Australia was not without controversy for it transpired that the airframe, engine and some hydraulic equipment had been leased to him and were due for return to England after the Race. Stodart wished to retain the aircraft for aerial work in Victoria, but remaining under British registration. Subsequently, the Air Ministry contacted the local authorities and insisted on modifications which Stodart claimed he could not afford, and in any case, it was not his aeroplane.

During a flight from Ballarat to Sydney in April 1935, positioning for modification, demonstration and possible sale, she was forced down near Culcairn, NSW, due to strong headwinds and fading light, damaging the undercarriage. The aircraft was carried by road to Sydney where repairs were completed and Stodart made application for the Courier to be registered in Australia, but this was rejected as the legal owners in England, Aircraft Exchange and Mart Ltd., refused permission for the registration to be in Stodart's name. The aeroplane was further damaged in August 1935 when she was in collision with Avro Avian VH-UGA during a landing at Leeton, NSW.

Local registration letters were allocated on 9 October 1935 when she became VH-UUF, registered in the name of Aircraft Exchange and Mart Ltd. Based at Mascot, she was known to have been operated in the Cootamundra area. In June 1936 the British owners contacted the Australian authorities asking where the aircraft was, as they had expected the Courier to be returned to England in January 1935. Vickers also wrote asking the same questions in respect of their hydraulic pumps. With the expiration of the Australian Certificate of Airworthiness in October 1936, Stodart returned to England where he died at the age of 56 on 26 February 1938.

Kenneth Stodart returned to duties with No. 43(F) Squadron at Tangmere until April 1935 when he was commissioned as a Pilot Officer and posted to No. 17(F) Squadron flying Bulldogs at Kenley. Later that year he moved to the Meteorological Flight at Duxford. Promoted to Flying Officer in November 1936 he was attached to Station Flight at RAF Mildenhall where, in May 1937, he was awarded the Air Force Cross in the Coronation Honours. Posted to RAF Wyton in July 1938 he was flying the Station Flight's Miles Magister L8269, engaged in solo forced landing practice on 13 September, when the aircraft stalled on a turn and spun into the ground. Kenneth Stodart was severely injured and died in the RAF Hospital at Halton two days later.

What happened to the Courier is unclear but she may have been shipped back to Portsmouth to appease her legal owners, and there reduced to component parts.

15. Fairey IIIF G-AABY.

Following successful repairs to her header tank, G-AABY left Singapore on 19 November, more than a week after the conclusion of the official ceremonies in Melbourne. The crew guided her via Koepang and across the Timor Sea at a height of six feet to Darwin where they arrived on 20 November. Four days later they flew into Essendon Aerodrome, Melbourne, having made forced landings twelve miles from Camooweal and within 100 miles of Charleville. Davies revealed that he had to nurse the aircraft for the last few miles as the radiator had started to boil leaving only a three minute period before the engine seized. "The seven year-old aircraft has served us well," he added, "we had no engine trouble until we arrived in Australia when the cooling failed. As far as flying was concerned, it has been a cakewalk." It was generally agreed that engine overheating was almost certainly caused by the use of dirty water in the radiator which created a blockage of the tubes.

In London Harold Perrin received a number of letters from Clifford Hill's mother asking if he would forward news of her son having heard nothing from him since he left Mildenhall over a month previously.

News of their progress had been telegraphed forward so that when the biplane eventually arrived in Melbourne, she travelled the last few miles of the great adventure under escort by a trio of club aircraft which had flown out to greet her. On St George's Day 1935, Cyril Gordon Davies married the 22 year old Mavis Beryl Bell of Macedon, Victoria, in St. John's Cathedral, Brisbane. The bride was the girl who had been waving at Davies from the aerial es-

Clifford Hill and Cyril Davies with their Fairey IIIF G-AABY at Donmuang Aerodrome, Siam. They arrived at Darwin on 20 November.
(National Library of Australia, Canberra.)

Looking rather bedraggled but otherwise in fine spirits, Davies and Hill were met on their arrival at Essendon Aerodrome, Melbourne, on 24 November by Sir MacPherson Robertson who heard that for the last few miles the aeroplane was carefully nursed due to a boiling radiator.
(National Library of Australia, Canberra.)

cort on his port beam, "the highlight reflection on the Race," he wrote more than 30 years after.

Clifford Hill had been 'loaned' by the Royal Navy for the duration of the Race, during which he continued to be paid by the Service. He returned to England, survived the Second World War, and retired from a position in Naval Intelligence.

Davies flew the IIIF to New Guinea where official records indicate she was sold to Ray Parer on 3 April 1935, re-registered VH-UTT, and used on daily business about the goldfields until August 1936 when she was scrapped near Wewak. Davies remained in Australia as a freelance pilot, believing he was one of the first Britons to do so. After the war he continued flying in England until 1951 when he finally retired, sickened at the manner in which the small, post-war charter operators had been treated by the Labour Government.

16. DH.80A Puss Moth VH-UQO.

Jimmy Melrose's Puss Moth, which had now flown from Australia to England and back again, was sold to Harry Broadbent in January 1935 and renamed *Dabs III*. In May 1935 Broadbent flew the aircraft around the coastline of Australia, a distance of 8,000 miles, in just short of three days and ten hours, beating by two clear days the record set by Melrose in August 1934 in the same aircraft. Broadbent set off from Darwin in VH-UQO on 11 October 1935 on a solo record attempt to England but on 14 October the aircraft was damaged during a forced landing at Basra in Iraq and the flight was abandoned. The un-airworthy aircraft was sold to Flying Officer Edward Wheelwright, an RAF officer serving with No. 84 Squadron, flying Vickers Vincents at Shaibah and eventually found her way to England where she was registered G-AEEB to Wheelwright, quoting his address as RAF Eastchurch.

Application was made for a British Certificate of Airworthiness by an aircraft engineering company based at Gravesend, and was issued in June 1936. In December G-AEEB was sold to Henry Mitchell at Castle Bromwich and in May 1938 to Brigadier Arthur Lewin in Worcester. What happened after that is unclear. The C of A was not renewed after it had expired in October 1939, but the aircraft was not impressed into military service along with all others in her class at that time. One theory is that the Puss Moth was broken up in England, and another that she was flown to Arthur Lewin's farm in Kenya before war was declared.

Jimmy Melrose returned to England the year following the Races and in November 1935 delivered a Percival Vega Gull, VH-UVH, to Australia in a record solo flight. In all probability he was the last person to sight Kingsford Smith and Tommy Pethybridge on board the *Lady Southern Cross* when the aircraft overtook him over the Andaman Sea off the southern coast of Burma and disappeared. In April 1936 Melrose delivered a Heston Phoenix from England to Australia and it was in the Phoenix, VH-AJM, during the morning of 5 July 1936, when operating a charter flight from Melbourne to Darwin, the aircraft encountered an area of extreme turbulence near South Melton, 16 miles from Melbourne, and broke up, killing all on board.

Melrose Park in NSW, Melrose Park in South Australia and James Melrose Road at Adelaide Airport are all named in memory of 'The Young Australian'.

17. Caudron-Renault C550 Rafale F-ANAM.

The C.550 aircraft specified for the MacRobertson Races was never completed and the unfinished airframe became the core of the C.680, the last of the Rafale series, a miniature fighter and very successful racing aeroplane that disappeared during the Second World War.

James Melrose was given a warm welcome after the celebrations at Laverton when he flew home to Parafield where his Puss Moth was protected during presentations.
(National Library of Australia, Canberra.)

Following their return flight to England in record time in Comet G-ACSR, Owen Cathcart-Jones and Ken Waller were met by a substantial press corps at Lympne Aerodrome on Friday, 2 November. The aircraft carried news film of the celebrations in Melbourne recorded only five days previously.
(The Aeroplane.)

Lympne Aerodrome in Kent was the scene of many departures and arrivals of record breaking flights. The un-named green Comet G-ACSR was welcomed on her return from Australia by owner Bernard Rubin and his fiancée Audrey Simpson. The achievement of the out-and-return flight was overshadowed by the celebrations continuing in Melbourne.
(The Aeroplane.)

18. Potez 39.

The prototype Potez 39 was discovered to have poor aileron and rudder co-ordination and was refused certification by the French authorities. The aircraft, F-ALRL, was destroyed in a crash in October 1934 but a second prototype F-ALRS, was built at the same time and used to complete flight trials. It is not known which of these two aircraft was intended to take part in the Race, if either, and the entry may have been one of extreme optimism. Eventually a production contract allowed a series to be built which were delivered to the French Air Force as army observation aircraft and with whom they suffered heavy losses in 1940.

19. DH.88 Comet G-ACSR.

The first Comet to return to England after the Race, G-ACSR, arrived in the early afternoon of Friday, 2 November following a record flight home under less arduous conditions. The green Comet had left Melbourne on 27 October, two days after arrival from England, and landed at Lympne 5 days 6hrs 43min later, completing a journey of 23,000 miles in less than a fortnight. The scale of the achievement was eclipsed at the time due to the sharp focus on the Race winners still in Australia, but as the frenzy of publicity died down the out-and-return flight of Ken Waller and Owen Cathcart-Jones was finally recognised for the remarkable adventure and achievement it was and Waller was awarded the prestigious Seagrave Trophy by the Royal Automobile Club.

DH.88 Comet G-ACSR was sold to the French Government early in 1935 and de Havilland demonstration pilot Hugh Buckingham delivered her from Croydon Airport to Le Bourget Airport, Paris, on 11 April in an observed time of 52 minutes.
(Hugh Buckingham.)

In December 1934 with the rear petrol tank removed and a fuel tank and 16 cu.ft. cargo hold built into the long and elegant nose at Hatfield, G-ACSR, named *'Reine Astrid'* was used to carry 300lb of Christmas mail between Brussels and Leopoldville in the Belgian Congo with Ken Waller and Maurice Franchomme acting as crew. The aircraft was equipped with basic radio telegraphy, a direction-finding loop aerial and a generator and windmill in place of the landing light in the nose. In this configuration, the modified aircraft still had a safe range in excess of 2,000 miles and a top speed of 226mph. Cathcart-Jones had been hoping for a record breaking attempt to the Cape but early in 1935 G-ACSR was sold to the French Government and, registered F-ANPY, Hugh Buckingham's delivery flight from Croydon to Le Bourget on 11 April was made in a new record time of 52 minutes.

Record flights between Paris and Casablanca in August and Algiers in September, flown by Jean Mermoz, accompanied by a radio operator, M. Gimié, were in preparation for an anticipated South Atlantic air mail service which did not materialise. The ultimate fate of G-ACSR is not known but it is believed she was destroyed in a hangar fire at Istres in June 1940.

Bernard Rubin's delayed marriage to Audrey Simpson took place on 29 March 1935 but following surgery he died of pulmonary tuberculosis at the couple's new home in Kent on 27 June 1936 at the early age of 40.

Ken Waller returned to his job as a flying instructor with the Cinque Ports Flying Club at Lympne moving to Brooklands in 1936 where he remained until the outbreak of the Second World War when he joined the RAF as a flying instructor with No. 6 EFTS at Sywell operating Tiger Moths. From early in 1940 he was employed by Miles Aircraft at Woodley as a production and later as Chief Test Pilot and in 1948 emigrated to South Africa where he flew a Beech 18 as corporate pilot for the Strathmore/General Mining Company until he retired in 1965. He died in 1993.

Owen Cathcart-Jones published his book *'Aviation Memoirs'* in 1934 and emigrated to California where he became closely associated with some of the Hollywood set, notably Errol Flynn, acting as technical advisor and occasionally appearing in films where a British content was required. He opened a riding stable at San Ysidro Road in Montecito where he bred polo ponies and was known as Major Owen Cathcart to his pupils, one of whom later said, "He really scared the heck out of me. I remember being terrified of his drill-sergeant style!" He became an accomplished polo player in his own right and continued riding until he was over 80. Until his death at the age of 85 in February 1986, he maintained a vivid memory of the Race and was quick to criticise any published inaccuracies or misinformed opinion, remaining immensely proud of all his achievements.

20. Lockheed 5C Vega NC105W.
In 1935 Wiley Post was hired by American newspaper columnist and humourist Will Rogers for an aerial tour of Alaska in search of material. Post purchased a hybrid aircraft, an Orion-Sirius, made from a Lockheed Orion fuselage mated to Lockheed Explorer wings and fitted with second choice floats which were too heavy and badly affected the C of G. On 15 August 1935 the aircraft crashed on take-off near Point Barrow, Alaska, and both Post and Rogers were killed. Post's record-breaking Lockheed Vega NC105W was presented to the Smithsonian Institute in 1936 by his widow and is on display in Washington.

22. Northrop Delta SE-ADI.
Following airline service, the aircraft was rumoured to have been sold to Beryl Markham for a transatlantic flight and was delivered to Croydon in May 1937 but almost immediately returned to Sweden. In September, probably as a smoke-screen, the Delta was reported to have been sold to Iraq but was actually delivered to the Republican forces in Spain where she was destroyed on the ground during a bombing raid.

23. Fokker XXII PH-AJP.
Following pre-war civil airline operations with KLM the aircraft was delivered to Prestwick and registered G-AFZP to Scottish Aviation Ltd in September 1939. Used for flight experience duties she was formally impressed into the Royal Air Force as HM160 from October 1941 until June 1944 when she was returned to civilian service. Withdrawn from use, she was scrapped in July 1952.

25. Wedell-Williams 303 Landplane NR67Y.
Following the death of designer and acclaimed racing pilot Jimmy Wedell in a take-off crash during an instructional flight in his company's DH.60GM Moth NC924M on 24 June 1934, Harry Williams cancelled the MacRobertson project. The incomplete fuselage and wing were stored by the Delgado Trade School who continued work on other Wedell-Williams projects, notably 'The Flash' and 'The Maid'. After the Second World War, all remains of the 303 Landplane, jigs, drawings and models were destroyed when the remote warehouse in which they were stored was engulfed by fire.

26. Airspeed AS.5A Courier G-ACVF.
Following her pre-Race withdrawal, the aircraft was eventually sold and in Febru-

Airspeed Courier G-ACVF which was withdrawn from the Race before the start, flew as a civilian aircraft throughout the war but was painted in camouflage and carried Royal Air Force insignia.
(Ted Hawes.)

ary 1937 was registered to North Eastern Airways Ltd. at Croydon. She survived impressment with the RAF during the Second World War and was returned to civil ownership in February 1946 but was withdrawn from service and scrapped in December the following year.

28. Lockheed 8B Altair VH-USB.

Having decided not to take part in the MacRobertson Races due to the obvious lack of turn-round and preparation time which would be available in England, Kingsford Smith and Taylor flew the Altair from Brisbane across the Pacific to San Francisco in the hope of a quick sale. When this did not materialise plans developed to use the aircraft commercially back in Australia but the old problems of certification persisted. It seemed prudent to ship the Altair to England, where it was thought it would be easier to make the necessary arrangements to satisfy bureaucratic demands in Australia, then fly home in a record attempt for which financial support was more likely to be forthcoming.

Accompanied now by Tommy Pethybridge as co-pilot, Kingsford Smith flew the Altair to New York from where she was shipped to London only to discover that the British authorities were no more sympathetic than their counterparts in Australia. After many weeks of negotiation finally it was agreed to issue a Certificate of Airworthiness, albeit one carrying very severe limitations on maximum fuel uplift.

Operating within the terms of the C of A and whether by design or not, the Altair left Lympne on 20 October 1935, exactly one year after the start of the MacRobertson, determined to take on a full fuel load at the first stop. However, immediate plans were thwarted when the wooden wings were badly damaged whilst flying in a storm over Italy and following inspection at Brindisi she was ferried slowly back to England in search of repair facilities.

The second attempt began from Hamble on 6 November and the flight appears to have been at high weight as the aircraft arrived at Allahabad in a time only four hours longer than that taken by the Comet *Grosvenor House* during the MacRobertson Races. The aircraft left at 5.58pm on 7 November bound for Bangkok and Singapore but never arrived.

One of the last people to see the Altair was 'the young Australian', Jimmy Melrose, who reported that the aircraft had overtaken him about 150 miles south east of Rangoon. Melrose was flying a Percival Gull from England to Australia, ironically whilst he was attacking Kingsford Smith's own solo record time. In spite of an intensive air search conducted over many days, no trace of the aircraft was found. Not until May 1937 was a wheel, complete with inflated tyre and part of an undercarriage leg, discovered on a beach at Aye Island just off the west coast of Burma. The parts later were identified as coming from the Altair but in spite of a renewed search in 1938 by RAF pilots from Singapore, no further trace of the aircraft or crew was ever found.

29. Bellanca 28-70 Monoplane EI-AAZ.

Following the controversy at Mildenhall which sadly resulted in elimination of *The Irish Swoop* even before the Race began, Eric Bonar flew the aircraft back to Croydon. He searched for a suitable airfield with a smooth surface from where he was hoping to conduct performance trials observed by officials from the Air Ministry, and the machine eventually flew on to Portsmouth where thanks to the efforts of Rex Martin of the US Bureau of Air Commerce and Captain John W Monahan, United States' Air Attaché in London, tests were organised from the new airport's 'Hunterised' surface by permission of the Airspeed Company. On 25 October, with 500 gallons of paraffin in the main tank and 112 gallons of usable petrol on board, the aircraft took off in heavy drizzle in under 200 yards and was landed back on the wet grass aerodrome to stop well within the prescribed limits. As a result the Air Ministry issued a Certificate sanctioning operation at the high weights not permitted by the MacRobertson Race sub-committee. After some delay due to an engine oil leak, the aircraft flew along the South Coast to Lympne.

On Monday 29 October, Fitzmaurice and Bonar left England with petrol on board sufficient for a 3,000 mile non-stop sector. It was their intention to fly to Baghdad and from there to attempt to break the Comet's record time to Australia set only the week before. Levelling out at 15,000ft over Liege and allowing the speed to increase to 225 mph, they experienced severe buffeting.

It was later discovered that during the heavyweight landing trials at Portsmouth, the standard tyres inflated to 65psi had become deformed not allowing the undercarriage fully to retract into the tailored housings in the wing panels. Not realising this the crew had applied extra windings to the hand operated retraction mechanism and a wheel had been forced against a wing rib causing local damage and wrinkling the skin. Eddies which developed in the airflow then caused a wing root fillet to pulsate, dangerously loosening some of the securing screws.

One answer was to reduce speed by 25mph and proceed with the undercarriage locked down. Bonar thought that with their sophisticated radio apparatus which was working perfectly they should go on but Fitzmaurice disagreed. Bonar then suggested they should bail out over the English Channel and capitalise on the huge publicity that would follow, but Fitzmaurice would have none of that and after dumping fuel, a procedure with which Fitzmaurice had tried to influence the Stewards at Mildenhall as a practical means of

The unhappy story of 'The Irish Swoop', Bellanca 28-70 EI-AAZ, continued after she was returned to the USA. During a test flight on 23 November 1934 she overturned on landing. Rebuilt, re-registered NR190M and re-engined, the aircraft was sold to Jim Mollison who flew her from Harbour Grace, Newfoundland to Croydon Airport on 29 October 1936 in, at the time, the fastest ever solo crossing of the Atlantic, just over 13 hours. (John W Underwood Collection.)

controlling landing weight, the aircraft turned back to England and landed at Lympne before flying on to Croydon in search of repair facilities.

In view of the many difficulties encountered it was decided the best course of action was to ship the aircraft back to the USA. Fitzmaurice then left the project and in mid-November Bonar flew her from Croydon to Southampton where she was prepared for the voyage. Landed in New York, Eric Bonar ferried her from Floyd Bennett Field to the Bellanca works at New Castle, Delaware, arriving during the afternoon of Friday 23 November. Following overhaul the Monoplane was ready for flight test in February 1935 and rolling out of the first landing with Bonar as pilot, the heavy aircraft was hit by a gust of wind and flipped over onto her back.

Dumped in a hangar for a year, the Monoplane was rebuilt with a 900hp Pratt and Whitney Twin Wasp installed and many strengthening modifications. Re-designated a 28-90 she was declared airworthy carrying a new US registration, NR190M, late in 1936. The catalyst had been her sale to Jim Mollison who took her off from Harbour Grace, Newfoundland, on 29 October 1936 to make the fastest ever solo Atlantic crossing to Croydon in a little over 13 hours, an immense achievement that attracted little publicity. In England the Bellanca was registered G-AEPC and named *The Dorothy,* following Mollison's estrangement from his wife Amy and friendship with noted West End actress and star of musical comedy, Dorothy Ward.

It was perhaps ironic that Mollison, who did not complete the MacRobertson, together with Edouard Corniglion-Molinier whose Wibault 366 was withdrawn and whose alternative entry, the Blériot III/6, was damaged before the start, should together attempt an out-and-return record from England to The Cape in the Bellanca 28-90 Monoplane, an aeroplane which officials had banned from the MacRobertson! The attempt failed in November 1936 due to a series of problems with a leaking fuel tank. In 1937 Mollison flew her to Spain where she became LB-1 with the Spanish Republican Air Force. Her subsequent fate during the Civil War is unknown, although Bellanca received word that the aircraft had been lost in combat during a reconnaissance mission.

James Fitzmaurice spent much of the 'thirties living near New York but during the Second World War moved to London where he opened a club for servicemen. During the late 'forties he returned to Ireland where he died in hospital in Dublin on 26 September 1965.

Eric Bonar joined the staff of Personal Airways at Croydon and took charge in setting up the company's training operation at Luton. During the Second World War he became a test pilot with Rolls-Royce involved in the development programme of the Merlin engine and also spent time with Napier flying Sabre engined Tempests and Typhoons. After the war he operated his own air charter business at Croydon but retired from flying in 1951. Eric Bonar spent his latter years at the Royal Star and Garter Home at Richmond in Surrey where he died on 26 February 1991 in his 92nd year.

30. Northrop Gamma 2G NX13761.

Following her crash-landing in New Mexico whilst en-route for New York and a prospective voyage to England, the Gamma was rebuilt on the instructions of the insurance company, fitted with a Pratt and Whitney Twin Wasp engine and in 1935 was leased by Howard Hughes who, after extensive modifications, made several outstanding flights with her. She was raced, crashed and rebuilt many times for various owners in her long career.

Ruled out of the MacRobertson, the Curtiss-Wright Company reimbersed all Jackie Cochran's associated expenses totalling $20,000, including those for the trip to England, the Race entrance fee, the cost of manufacture of special refuelling docks at airfields en-route and for technical and personal effects which had been sent forward including all the fashions and accessories which were to be worn at the celebrations in Australia. The irony of the situation was a recognition of the fact that without the new supercharger, the Gamma would have been perfectly capable of winning the Race unaided.

31. Miles Falcon G-ACTM.

On the morning of 25 October, Harold Brook, his patient passenger Ella Lay and Miles Falcon G-ACTM left Rome for Tatoi Aerodrome, Athens, apparently oblivious to damage caused when a lorry had backed into the aircraft. During the transit the crew heard a loud report although the performance of the aircraft appeared to be unaffected. At Athens a Greek Air Force engineer discovered that the tip of one propeller blade was damaged together with part of the leading edge of the other, and a crack along one blade had developed as far as the hub. As a consequence the Falcon's crew spent ten days sightseeing, waiting for a replacement wooden propeller to arrive from England, together with another metal propeller ordered as an insurance. The wooden propeller was fitted to the engine and the other threaded with extreme difficulty into the cabin. They left for Nicosia and Aleppo after paying swingeing Customs charges levied on the imported spares!

On 2 November out of Aleppo, the Falcon landed with a blocked oil filter at Dayr az Zawr, a settlement in the Syrian Desert and Brook and his passenger were confronted by a patrol of the French Foreign Legion. They took-off heading for Baghdad, ignoring the French officer's instructions to fly in the opposite direction. Seven months later the Foreign Office in London received

Flown by Harold Brook with Ella Lay as passenger, Miles Falcon G-ACTM made landfall in Australia at Wyndham, some 300 miles south west of their destination. At Darwin next day, Brook studied maps laid out on the port wing of his aeroplane. The team arrived at Essendon Aerodrome, Melbourne, on 20 November to be met by nobody from the Race organisation although Sir MacPherson Robertson was advised and arranged accommodation.
(Northern Territory Archive.)

On arrival in Melbourne, Ella Lay was met by the cousins she had travelled to be with, members of the Lay and Sayle families. Miss Lay stayed in Melbourne and trained as a nurse. She was commissioned in the Australian Army during the Second World War and served overseas before returning to England in 1946.
(Nick Spencer.)

via the French High Commission and the British Consul General in Beirut two letters which had been left in the town for posting together with the Falcon's Race logbook. The items were accompanied by a formal complaint that the aircraft had taken-off without completing 'the usual formalities'. They arrived in Jask on Saturday, 3 November at 12.20pm and left for Karachi an hour later where they were entertained by RAF officers. On departure for Jodhpur, Brook mis-set the compass and in the evening was forced to land in the Sind Desert. He discovered they were close to Jodhpur Aerodrome but remained in the desert overnight waiting for sufficient light to illuminate their take-off path. Next morning they flew the short distance to Jodhpur where the local Air Race Committee, overjoyed at seeing their second arrival, entertained them royally for three days.

But their troubles were not yet over. Before they reached Calcutta on 13 November, they were forced down again at Gaya due to torrential rain. The monsoon had broken and the rains stayed with them as they transited Alor Star, the last of the six aircraft to land there, Victoria Point and Singapore, water leaking into the cockpit and soaking everything inside, not easing until they reached Koepang.

The Falcon crossed the Timor Sea at a steady 100mph, well throttled back, but following another error of navigation landed at Wyndham in Western Australia, 300 miles south west of Darwin, where after touch down, the aeroplane sank up to her axles in mud. They left next morning for Darwin and then proceeded along the chain of Checking Points, with additional stops at Bourke and Cootamundra, to the finishing line at Melbourne. When they arrived in the rain at Essendon Aerodrome on 20 November nobody from the Race Committee was there to welcome them. Harold Brook was both astonished and extremely depressed, attitudes not improved by references in the popular press which described his involvement in the MacRobertson as more like a grand tour rather than an air race to Australia.

Sir MacPherson was advised that the Falcon had arrived and arranged accommodation for Brook at his expense. After the Falcon was overhauled Harold Brook remained in Australia for some weeks during which time he earned his keep by operating the aeroplane on a joy-riding concession.

At 28 years of age Ella Marion Lay had answered Brooks' advertisement for two cost-sharing passengers. She had already decided to travel to Australia to visit her cousins and members of the Lay and Sayle families were on hand at Essendon to greet her, the only woman entrant to complete

Miles Falcon G-ACTM was overhauled in Australia, fitted with her metal propeller and re-painted red and white, retaining her racing number on the rudder. Before returning to England the product name 'Ovaltine' was painted on the port engine cowling and details of the aircraft's involvement in the MacRobertson Races on the starboard side whilst the fuselage attracted a number of badges of sponsoring organisations including Castrol. (National Library of Australia, Canberra.)

the course. She was born at Peasenhall in Suffolk, about 30 miles from Mildenhall and as a teenager had attended the all-girls' Bedgebury School near Goudhurst in Kent and finishing school in Switzerland. Whilst living in Bognor Regis she learned to fly on DH.60 Moths at Brooklands where she qualified for her 'A' licence on 17 March 1934.

Ella Lay stayed in Australia, qualifying as a nurse in 1938 and maintaining her flying interests as a member of the Royal Victorian Aero Club until the outbreak of the Second World War. in September 1941 she was commissioned as a Lieutenant in the Australian Army Nursing Service quoting her permanent address as Dunchurch in Warwickshire, England. Her wartime career included postings to North Africa, the Middle East and Far East until she was discharged from the Army in February 1946 when she returned to England. She maintained her nursing career and qualified as a Queen's District Nurse in Berkshire. Ella married Stephen Neate in 1954 and died in Somerset on 22 July 2005 at the age of 99.

Harold Brook advertised for a fare-paying passenger for the return journey to England but there were no applicants. The aeroplane was prepared in Arthur Butler's hangar at Cootamundra where the Falcon was sprayed red and the name *'Ovaltine'* prominently displayed on the engine cowlings beneath a declaration that the aircraft had been an entrant in the Centenary Air Race. Extra tankage fitted in place of a cost-sharing passenger, increased the range from 850 to 1,400 miles, Apart from a farewell lunch with Ella Lay, the pilot and his passenger who had shared so many adventures never met again. Early on the morning of 24 March 1935 Brook finally left Darwin after poor weather over the Timor had frustrated an earlier attempt. He had Jimmy Melrose's unofficial solo record to England in his sights and safely reached Lympne Aerodrome at 3.55pm on 31 March having taken 7 days 19 hrs and 50 mins for the journey, sleeping well each night and managing to cut more than a day off the previous official time set by Jim Mollison.

Brook was given a Civic Reception at his home town of Harrogate (where his father was Mayor and his sister Mayoress) and on 2 April received exceptional permission to land the Falcon on the famous Stray, an area of common land near the centre of the town. In an address to the town's Rotary Club he was able to explain that a number of letters he had carried from Australia, including one from the Lord Mayor of Sydney to the Mayor of Harrogate, had not been franked in Australia and were confiscated by British Customs on behalf of the General Post Office and its monopoly. After some delay the letters were released and finally delivered to their addressees.

G-ACTM was returned to Woodley for attention by Miles Aircraft and in July 1935, Harold Brook set off for Cape Town and a solo record attempt. He abandoned the flight at Mersa Matruh in Egypt after damaging the undercarriage during a night landing there on 18 July. The Falcon was returned to England, repaired and raced in August 1936 but the British registration was cancelled when the certificate of airworthiness expired in November citing that the aircraft had been withdrawn from use.

33. Lambert Monocoupe NR50IW.

Having withdrawn from the Race at Calcutta, John Polando returned to the USA while Wright made arrangements to ship the Monocoupe back to England where the pilot and aircraft arrived in November. Wright paid a hasty visit to the Royal Aero Club offices in Piccadilly to collect mail but left no forwarding address or details of his plans. On 17 November it was reported that the Monocoupe had made a forced landing at Grove Park near Bromley in Kent where she had been examined by the police. Wright arranged for the aircraft to be dismantled and hired a lorry from a local builders' merchant, Darby Brothers, who later confirmed they had delivered her to London's East India Dock where she had been loaded on board the SS *Ary Lensen* which on 24 November sailed for Rotterdam and the USA. Wright was believed to have accompanied her. Harold Perrin's particular interest was that the aircraft had been imported for the Race at Southampton without any Customs charge and he had to be certain that she had left again.

Wright loaned the aircraft to Helen MacCloskey of Pittsburgh soon after arrival back in the USA and she established a new record of 166.67 mph over the Miami 100 km closed circuit course in January 1935. The aircraft was sold to 25 year-old Ruth Barron in New York who entered her in a race at Denver, Colorado, but was killed when the Monocoupe crashed on landing at Omaha, Nebraska on 3 July 1935. The aircraft had caught fire in the air, a sad fact that was linked to the knowledge that Ruth Barron was a chain smoker. The metal airframe was salvaged and stored for 25 years until it was acquired by Jim Heim in California. Rebuilt by 1968 and flown for 11 hours until 1971, NR501W was sold to Al Allin, dismantled and carried by truck to Texas where again she was placed into store. In 1996 NR501W was sold to James L. White, an airline pilot in Chandler, Arizona who started the long task of restoring the aircraft to her MacRobertson Race configuration.

In May 1937 John Polando began a passenger and airmail service and a pilot training school at Plum Island Airport, Massachusetts, where he was appointed first airfield manager. In 1942 he left to fly aircraft for radar trials in association with the Massachusetts Institute of Technology before joining the United States Army Air Force as a bomber pilot flying B-17s and B-29s.

In recognition of his world record breaking flight from New York to Constantinople with Russell Boardman in 1931, on the 50th anniversary of the occasion Barnstable Municipal Airport, Hyannis, Massachusetts, was named Boardman-Polando Field in their honour.

In 1985 at the age of 83, John Polando died as the result of injuries sustained in the take-off crash of a light aircraft he was piloting.

34. DH.88 Comet G-ACSS.

Following her arrival at Laverton on 23 October G-ACSS was removed through early morning rain and mist, lashed athwartships onto a jury rigged transporter, for display in Melbourne's Alexander Gardens, where she could be viewed at close quarters for a small fee, all of which was donated to the crew. Subsequently she appeared at other locations in Australia before she was returned to England, carried fully assembled as deck cargo by a ship which docked at Hull. She was dismantled on the quayside by two de Havilland engineers and moved in sections to Yeadon Aerodrome, Leeds, where special cradles were fitted to allow wing and fuselage to be more readily transportable by road. The aircraft was exhibited around the country before returning to Hatfield in January 1935.

In the same month the FAI granted the first 'capital to capital' record to Scott and Campbell Black designating 'London to Melbourne in 71 hr 18 sec' although the Press release of 26 January which anticipated the confirmation quoted 72 hours. It was Harold Perrin who advised Scott of the ratification and who passed on the FAI invoice amounting to £2.0s.0d.

Tom Campbell Black offered to buy the aeroplane from Arthur Edwards who refused to sell. She was later purchased by the Air Ministry for £7,000 and despatched to the Aircraft and Armament Experimental Establishment at Martlesham Heath for trials, during which, in August 1935, the undercarriage retracted on touch down, substantially damaging both engine cowlings, but little else.

Under military ownership, the brilliant red of her civilian life had been oversprayed silver and serial number K5084 and RAF roundels applied. Wearing this scheme and sporting re-designed and enlarged cooling air intakes, the machine took part in the RAF Pageant at Hendon on 27 June 1936 but, during gross weight landing trials at Martlesham only nine weeks later, the undercarriage collapsed again and the aircraft was struck off charge and sold for the scrap value of her engines and metal fittings.

Frederick Tasker of Essex Aero Ltd. at Gravesend Airport tendered for the wounded Comet, realising her potential, and she was registered to him on 7 June 1937. Against all odds, Jack Cross completed the task of restoring her to airworthy condition during which he fitted Gipsy Six Series II engines and new de Havilland variable pitch propellers. Bicycle pump engineering was now obsolete.

The Comet was airworthy again by August 1937. Painted pale blue and re-named *The Orphan*, G-ACSS was entered for the Marseilles-Damascus-Le Bourget Race, and was placed fourth in the hands of Arthur Clouston and George Nelson. In September, Ken Waller flew *The Orphan* as No. 5 in the King's Cup Air Race from Hatfield, taking 12th position overall.

In November, *The Orphan* became *The Burberry*, sponsored by the tailoring company of that name, and Arthur Clouston with Betty Kirby-Green set a new record of 5 days 17 hrs from Croydon to the Cape and back, lowering the previous record by more than three and a half days.

Another name change in February 1938 accompanied her quest for a record return flight to New Zealand. As *Australian Anniversary* G-ACSS was flown by Arthur Clouston and Victor Ricketts but the attempt was terminated due to a flooded airfield in Turkey. The Comet was flown back to England, damaging its undercarriage in Cyprus on the way.

By March all was set for a second attempt. The same crew flew G-ACSS from Gravesend to Blenheim and back to Croydon, a distance of 26,450 miles, in under eleven days to create a record which has never been broken.

The Comet returned to Gravesend where her owners immediately removed an engine for installation in a racing Percival Mew Gull, and dumped the remainder of the largely gutted airframe outside the hangar, under the modest protection afforded by a large tarpaulin. It appeared that her flying life was at an end.

After Tom Campbell Black married actress Florence Desmond domestic bliss did not curtail his flying activities. On 8 August 1935 he and co-pilot John MacArthur left Hatfield in a new Comet, G-ADEF, the fifth and final aircraft, for an attempt on the Cape record, but an engine oil problem caused them to abandon the outbound leg

From Port Melbourne, Grosvenor House *was carried to Sydney on board the coaster* The Admiral *and offloaded onto the dockside at Wooloomooloo Bay from where she was towed through the streets for exhibition in the city.* (National Library of Australia, Canberra.)

After unloading at Wolloomooloo Bay, Sydney, Grosvenor House *was manhandled across a building site and up a makeshift ramp to position her onto the adjacent roadway. It might have been easier to fly her from Melbourne to Sydney if her Race crew had agreed but similar obstacles would have been faced when manoeuvring her into the city centre.*
(State Library of New South Wales.)

Even with a heavy police presence, a suitable towing vehicle and the necessary permissions to dismantle bollards and stop traffic, moving a racing aeroplane through a city centre was fraught with practical difficulties.
(National Library of Australia, Canberra.)

at Almaza in Egypt. A prospective flight to Canada and entry in the 1935 King's Cup did not materialise, but on 21 September, Campbell Black and MacArthur left Hatfield on another attempt to reach the Cape. This time they were beyond Almaza en route to Kisumu in Kenya when first one engine then the other lost power and the aircraft could not be persuaded to maintain height. Flying over unfriendly terrain unfit for a forced landing there was only one alternative, and the airmen bailed out. The Comet crashed and was destroyed by fire. The crew landed safely about 100 miles north of Khartoum and arrived in the city on camels.

In March 1936 the Gaumont British cinema business announced that Mr T Campbell Black had been engaged as Aviation Manager, promising that utilising a large fleet of aeroplanes their plan was to deliver newsreels to practically every town throughout the country within four hours of supplements being issued at the London film processing laboratories.

During the 1936 Easter holiday Campbell Black promoted his British Empire Air Display at two aerodromes near Sheffield and during the summer appeared in a press advertising campaign for Superlax armchairs.

Campbell Black is listed as a 'foreign volunteer' who ferried aircraft for the Nationalists during the Spanish Civil War soon after its outbreak in July 1936. He was an entrant in the Schlesinger Air Race from Portsmouth to Johannesburg starting on 29 September, an event with none of the charisma of the MacRobertson. On 19 September he flew his race aircraft, Percival Mew Gull G-AEKL, sponsored by the Littlewood Pools magnate John Moores, to Speke Aerodrome, Liverpool where she was named *'Miss Liverpool'* by the leader of the City Council. While taxying there for take-off, the tiny Mew Gull collided on the ground with RAF Hawker Hart K3044 of No. 611 Squadron being flown by the CFI, Flying Officer Peter Salter, which was at the end of its landing run. The Hart's rotating propeller sliced into the Mew Gull's wooden airframe which offered no protection to her pilot and Campbell Black was grievously injured. Although he was rushed to hospital he was pronounced dead on arrival. The relative positions of the two aircraft showed that a few inches either way, and the big wooden propeller blades would have missed the racer's cockpit altogether.

Charles Scott became a celebrity much loved by the press who rewarded him with many adulatory column inches and even his own regular newspaper features. Encouraged by his business manager John Leggitt he wrote the story of his long distance flights in a tome entitled merely *Scott's Book* which was published in November 1934, the thin final chapter covering details of the MacRobertson being cabled to Leggitt from Australia together with an Epilogue written by Campbell Black. He delivered his first lecture on the Race during the afternoon of Thursday, 7 February

Arthur Edwards sold his Comet G-ACSS to the British Air Ministry for £7,000 and she was sparingly flown at Martlesham Heath in a colour scheme of silver overall embellished with RAF insignia. It was not a happy association and following an undercarriage collapse during landing trials conducted at maximum gross weight the aircraft was sold for the scrap value of her engines and metal fittings.
(The Aeroplane.)

Rebuilt to flying condition by Jack Cross at Gravesend in 1937 and fitted with Gipsy Six Series II engines driving variable pitch propellers, G-ACSS, painted pale blue, raced under a series of different names. As 'Australian Anniversary' she was damaged during a force landing in Cyprus early in 1938 when returning to England from Turkey after an abortive attempt on the return record to New Zealand.
(BAE Systems.)

1935 in London's Queen's Hall for which Leggitt ensured everybody knew well in advance that tickets were for sale. Lord Wakefield presided and amongst the audience was HRH The Duke of York.

Flying a Vega Gull, Scott and Giles Guthrie won the 1936 Schlesinger Race in which Tom Campbell Black was to have participated. Speaking at a Foyles Literary Luncheon in 1938, Charles Scott said of the MacRobertson victory: "It was a flash in the pan that raised one temporarily to a sense of well-being. You had done something that nobody else had done."

As the world's political situation deteriorated, Scott's star rapidly waned, and deprived of the spotlight he took heavily to drink. Following declaration of war he joined the ARP as an ambulance driver, then the Royal Navy Volunteer Reserve before becoming an Atlantic ferry pilot. On 15 April 1946 whilst working in Germany with UNRRA, a United Nations' agency, lonely and depressed following a second divorce, Charles William Anderson Scott committed suicide by shooting himself.

In 1936, Arthur Edwards extended his business and property interests to South Africa and commissioned a motor racing circuit that was built in less than six months ten miles south of Cape Town city centre. There, on 16 January 1937, was run the first Grosvenor Grand Prix. The event was a financial disaster as fewer than half the anticipated 120,000 spectators put in an appearance. Arthur Edwards left England in 1939 to live in America where he developed significant property interests and where he died on 24 August 1960 at the age of 83.

In 1944, Martin Sharp, the eloquent and much respected Public Relations Manager of the de Havilland Aircraft Company, rediscovered G-ACSS mouldering under a tarpaulin at Gravesend and for £150 acquired sole rights to her mortal remains on behalf of the Company. Somehow, wartime transport was authorised, and the hulk was delivered to hangarage in daily use for apprentice training at the dispersed Mosquito design centre at Salisbury Hall, near London Colney in Hertfordshire, where the dismantled parts were stored. After the war, the aircraft was rebuilt to cosmetic *Grosvenor House* standard by students of the de Havilland Aeronautical Technical School for exhibition at the 1951 Festival of Britain held on London's South Bank. This entailed the removal of all metal parts to lighten the load, but more seriously the wing structure was drilled through to provide cable pick-up points in order that the airframe could be suspended from a light roof beam. Afterwards, she was returned to the de Havilland Engine Company's works at Leavesden Aerodrome where the airframe became part of a permanent exhibition, displayed in a nose-high vertical situation as though, like an insect in a collection, she was pinned to the hangar wall.

Hawker-Siddeley Aviation, having absorbed the de Havilland Enterprise by government decree, donated G-ACSS to the Shuttleworth Collection at a small ceremony held at Hatfield on 30 October 1965. She was not formally registered to Air Commodore Allen H Wheeler, Aviation Trustee of the Shuttleworth Collection, until 8 July 1975, two years after Shuttleworth's Board of Trustees had given their approval for the formidable task of restoring G-ACSS to flying condition. Work progressed slowly on the empty shell of the wooden airframe at the Trust's workshops at Old Warden Aerodrome, aided by generous financial and practical support from industry and individuals in Great Britain and Australia.

At Hatfield, Hawker Siddeley Aviation's Deputy Chief Engineer set up a group to handle the design and manufacture of the components required to make the aircraft airworthy and the Design and Training Departments who were both heavily involved were charged with ensuring that 'nothing went wrong with the Comet!'

Shuttleworth's programme consumed prodigious quantities of resource and cash and work was stopped in November 1982 following the discovery of serious faults in some of the work which had been completed on the fuselage, and also the wing, and pending an appeal for a further £100,000. The decision to discontinue restoration must have been particularly hard for the Chairman of the Trustees, Marshal of the Royal Air Force Sir John Grandy who, as a young pilot, had been a helper during the morning of the start of the Races at Mildenhall almost 50 years before.

In April 1983 G-ACSS was removed to the Apprentice School at RAE Farnborough where an intensification of effort was aimed at putting the Comet back into the air by the 50th anniversary of the most celebrated

Comet G-ACSS was rescued from almost certain destruction in 1944 and stored in a dismantled condition at Salisbury Hall. For the Festival of Britain in 1951 the aircraft was cosmetically restored but to reduce weight to enable her to be hung from a roof beam nearly all the metal fittings and fixtures were removed. Never thinking they might be needed in future, they were lost.
(BAE Systems.)

Prior to becoming part of a permanent showroom display of de Havilland engines at Leavesden Aerodrome, the stripped-out shell of G-ACSS was erected at Hatfield Aerodrome and photographed with the prototype of the new Comet, the DH.106, the world's first commercial jet airliner.
(BAE Systems.)

win in the history of air racing. Allen Wheeler even reviewed the prospects of flying the aircraft by gentle stages to Australia, a scheme which attracted a deal of popular support, but the idea was fairly quickly recognised as being impractical.

At Farnborough the effort was described as 'massive' working with design data and drawings supplied in a constant stream from Hatfield and work was sufficiently well advanced to permit the aircraft to be displayed in static condition in September 1984 at the SBAC Farnborough Air Show.

In the New Year the aircraft was gradually dismantled and, packed into a number of wooden crates, was delivered by road to Frankfurt. From here the cargo was flown to Melbourne, arriving on board a Lufthansa Boeing 747 Freighter chartered by Qantas, there to celebrate the 150th anniversary of the State of Victoria, starting with the Melbourne Airport Open Weekend on 23/24 February.

Positioned in a rig specially constructed by volunteers using Ansett Airlines facilities at Tullamarine, and mounted on a flat bed articulated lorry supplied by Saab-Scania, G-ACSS was the star exhibit in the

The New Guinea entry, Fairey Fox G-ACXO, suffered a major delay in France from the first day of the Race but, without any chance of any reward, the crew decided to continue to Melbourne just for the adventure. The aircraft was photographed at Darwin in February 1935, long after the infrastructure established for the Races had been dismantled. The reason for the presence of yellow sludge found in the fuel in Paris was never explained.
(Northern Territory Archive Service.)

annual Moomba parade through Melbourne city centre in mid-March, and later at the Open Day at RAAF Base Laverton from where, it was noted, G-ACSS had last been seen leaving the aerodrome on a lorry in 1934.

After the aircraft was returned to Farnborough where more work was completed, she was delivered to Hatfield where all the final details were added and from where on 17 May 1988 she took off in the hands of British Aerospace test pilot George Ellis for her first flight since 5 November 1938. Her first public appearance was, most appropriately, at the Air Fete organised by the United States Air Force at Mildenhall Airfield that same month.

The aircraft was damaged in landing accidents at Hatfield in 1988 and Old Warden in 2002 but was repaired on each occasion and remains potentially airworthy with the Shuttleworth Collection.

35. Fairey Fox G-ACXO.

The astonishing succession of incidents and misfortunes which dogged Ray Parer and 'Geoff' Hemsworth from the time they arrived in England to collect their old aircraft, purchased through the advertisement which claimed it to be 'The only British Plane with a chance,' continued unabated.

Following their enforced stay in Naples, the Fox was released after three days and flew to Brindisi where Parer landed at the military airfield by mistake. Instructions received from Rome were that the aircraft should be sent on its way and they took-off for Greece with an engine already showing signs of overheating. They were grounded at Tatoi Aerodrome, Athens, for 55 days while armies of mechanics checked the engine and fuel system and the local Fairey agent arranged for a brand new radiator from a Fairey III to be sent out from England.

Not until 10 January 1935 were they able to leave for Cyprus where for safety they landed in a field close to the waterlogged aerodrome at Nicosia. The journey through the Middle East was largely without incident but when they eventually arrived at Calcutta on 20 January they were forced to search for Dum Dum Aerodrome in the dark having endured an initial denial of Persian overflight rights. Cruising south along the Malay Peninsula, the engine faded. As Hemsworth took control, Parer pumped fuel and throttle in desperation and the engine responded sufficiently to allow them to return the 75 miles to Mergui landing ground. The plugs were fouled and the filters were found to be full of water which necessitated purging the complete system. Then a magneto failed.

Pressing on they passed through Batavia to Sourabaya only to discover the petrol was contaminated with water again and the compass had gone off-line. Between Batavia and Koepang they were obliged to divert to Rambang on Lombok Island where the Dutch authorities assisted in stripping down the recalcitrant carburettor before the Fox finally crossed the Timor safely to reach Darwin and, eventually, on 13 February 1935, Melbourne itself. Their arrival at Essendon Airport caused no excitement; all the celebrations had been concluded the previous year. Sir MacPherson immediately awarded both pilots a consolation gold medallion which was as big a surprise as any to the crew who had never given up hope.

Examination of the medallions revealed yet another surprise. The obverse carried a relief of a hatless Sir McPherson Robertson with the inscription 'International Air Race 1934'. The reverse side featured a map in relief covering the route between England and Australia and the legend 'London to Melbourne 1934-1935'. Under the Rules, generously interpreted, medallions were supposed to have been presented only to those crews completing the course within 16 days of the start. Some competitors received nothing apart from a signed scroll and the hard earned satisfaction that they had played their part in the creation of history.

Following their arrival in Melbourne, Ray Parer flew G-ACXO to Mascot Airport, Sydney, where she was completely overhauled and in January 1936 took up the registration letters VH-UTR, listed to R J P Parer, Salamaua, New Guinea, for communications duties with Parer's company, Pacific Aerial Transport, operating between the goldmines. The aircraft worked hard and consequently suffered from a combination of a lack of spare parts and a hostile environment. The Certificate of Airworthiness was suspended in November 1936 due to the alleged poor condition of the airframe. Abandoned at Maprik the registration was cancelled in February 1937 and the aircraft broken up for scrap.

After Pacific Aerial Transport was taken over by their main customer, W R Carpenter and Co. who introduced their own aircraft, Ray Parer left the company although he remained in the gold prospecting and aviation businesses in New Guinea until the Japanese invaded from the north when he returned to Australia to join the RAAF. However, he was classed medically unfit for flying duties and assessed himself temperamentally unsuited to office work, with the result that he was transferred to the Reserve and re-mustered as an engineer on

FINAL STRAIGHT

'Battling' Ray Parer and Godfrey 'Geoff' Hemsworth on arrival at Essendon Aerodrome, Melbourne, on 13 February 1935, 100 days after the final qualification date. Although their arrival caused no excitement, Sir MacPherson Robertson awarded them both a gold medallion on the grounds that, in spite of all adversity, they had never given up hope.
(National Library of Australia, Canberra.)

Lockheed Vega G-ABGK was shipped to Fremantle in January 1935 and transferred to Maylands Aerodrome, Perth, where extensive repairs were necessary following her inverted arrival at Aleppo. The aircraft was registered VH-UVK to Horace C Miller and was operated as a fast charter machine by the MacRobertson Miller Aviation Company.
(Eddie Coates Collection.)

a small coastal craft used to transport troops and supplies around the northern coast of New Guinea. After the war he was involved in a number of mostly unprofitable business ventures in Australia before he resumed his nautical interests again working with small boats around the North Queensland coast and New Guinea. In the 'sixties Ray Parer settled onto a smallholding at Mt. Nebo north of Brisbane, where he maintained agricultural machinery until he died at the age of 73 on 4 July 1967.

In 1975 on the eve of New Guinea's independence, the propeller from the Fairey Fox which had been rescued by a local businessman in Maprik was presented to Ray Parer's nephew, Robert.

After leaving Pacific Aerial Transport, 'Geoff' Hemsworth joined Qantas as a pilot on the flying boat service operating between Sydney and Singapore, and just before the outbreak of war was commissioned into the RAAF where he flew requisitioned Empire Class flying boats and later Catalinas. In May 1942 with the rank of Squadron Leader he was captain of a Catalina on patrol off Misimi Island in the Louisiade Archipelago, when his aircraft was attacked and shot down by Japanese fighters whilst shadowing a large enemy fleet. It is believed Hemsworth and his crew were rescued from the sea and taken as prisoners of war to occupied territory, but they were never heard of again.

36. Lockheed Vega G-ABGK.

Following the landing accident at Aleppo, Donald Bennett returned to England and in spite of three crushed vertebrae resumed his RAF duties within a fortnight. Jimmy Woods remained with the Vega until funds had been received to enable her to be shipped onwards to Australia. The aircraft was delivered to Fremantle in January 1935 and following major repairs she was airworthy again at Maylands Aerodrome in August when she was registered to Horace C Miller, Perth, as VH-UVK although he had made application for VH-BGK. As one of the fastest aircraft in Australia, for several years she was occasionally raced but mostly used for charter work by the MacRobertson Miller Aviation Co., flown frequently by Jimmy Woods, Operations Manager of the Western Australia service.

In November 1941 VH-UVK was impressed into the RAAF and allocated serial A42-1, acting as a communications aircraft. She was badly damaged in June 1942 when she ground-looped on landing at Cairns and was not repaired until November the following year. Post-war she was offered for disposal but based on misinformed reports that the aircraft was longitudinally unstable the Department of Civil Aviation insisted that the Vega was not suitable for civil operations and, in spite of protests by her former owner and from pilot Jimmy Woods who argued that he knew the aeroplane better than anyone, she stood in the weather at Mascot for many months until she was transported to RAAF Base Richmond and broken up for scrap.

Jimmy Woods returned to the life of a pilot with MacRobertson Miller Aviation and became a popular figure at towns along the West Australian coast. Following Japanese attacks on Broome he repeatedly flew refugee evacuation flights between Broome and Port Hedland, frequently cramming 22 passengers, mostly Dutch, into his ten-seater Lockheed Electra. For these services he was appointed to the Order of Orange Nassau by Queen Wilhelmina and censured by the Australian Department of Civil Aviation. In 1947 he set up his own business, Woods Airways, using two Avro Ansons for a scheduled service between Perth and Rottnest Island.

Jimmy Woods died in 1975 at the age of 82 having retired from flying when he sold his helicopter business only four years previously. In 1987 the new Rottnest Island passenger facility was named the 'Jimmy Woods Air Terminal' by the State Minister for Tourism.

Donald Bennett transferred to the RAF Reserve in 1935 and joined Imperial Airways where he was deeply involved in the development of flying boat passenger and mail services. During the war he was a major figure in the organisation of the Atlantic Ferry Service and later in Bomber Command he created and led the elite Pathfinder Force. Retiring as an Air Vice-Marshal, Bennett returned to civil aviation when he became General Manager of British South American Airways, later turning to politics in which field he was elected as an MP, becoming an independently minded activist and campaigner. Business interests included building sports cars and light aeroplanes and he owned and operated Blackbushe Airport. In 1984 he was guest of honour at the 50th anniversary celebrations of the MacRobertson Races held at RAF Mildenhall and died on 15 September 1986.

41. DH.60G Moth Major G-ACUR.

The red and silver Moth Major G-ACUR which was specially commissioned for the Race was said to have cost almost £1,000 ex works. Fuel tankage was increased from the standard 19 gallons to 58 gallons and she was fitted with navigation lights and luminous instrument dials which, with a fast cruising speed, made her ideal for long distance touring.

After the entry by W J Cearns was not progressed and the nominated pilot, Sydney Jackson, left England for Australia on a private trip the day before the scheduled start at Mildenhall, it was almost inevitable that the Moth Major would be advertised for sale.

With only flight test hours logged the aircraft was offered by de Havilland agent Brian Lewis and Company, firstly from Heston and later Aldenham, where she was noticed in the hangar by Mary du Caurroy, Duchess of Bedford, whose intention had been to buy a different aircraft altogether. Registered to the Duchess in August 1936, the aircraft disappeared over the North Sea on 22 March 1937 during a flight in which the Duchess would have logged her 200th hour of solo flight.

44. Douglas DC-2 PH-AJU.

The delay suffered by the DC-2 at Charleville had been well covered in the press and local correspondents added their own voices of complaint when they reported that no facilities had been arranged for them at the aerodrome and that other organisations, including the airfield catering service, had received no co-operation from members of the local Race Committee. The town's mayor, Mr O J Allen, convened a special meeting at the Town Hall two days after the DC-2 had departed at which he reported he had been completely unaware of the situation until he had read *The Telegraph*. He said he was left with a feeling of 'absolute disgust' and that the pressmen themselves were to answer charges that they caused to be published reports which were a slur on Charleville, a town famed for its hospitality.

After the Race, PH-AJU took a full load of passengers to Canberra where crews and officials were entertained at Parliament House. Later she flew a number of goodwill trips from Melbourne to a number of Australian towns and cities carrying invited guests. It was opportunistic promotion for the airline and its bid to extend its Eastern services on to Australia. At the start of her return flight to Europe, as she passed over Albury Racecourse, the crew dropped a silver cigarette case which was found by the Secretary of Albury Racing Club, Mr B Peacock. Inside, besides a small Dutch flag, was a hand-written note:

"To all our good friends at Albury. We salute you and say farewell. On board the *Uiver*, 1 November 1934, in the air."

The message was signed by all crew and passengers who had been on board during the drama of 24 October, an incident which, had it not been for the initiative of the people of Albury, would surely have ended in disaster.

On the return flight the aircraft again carried passengers and mail and was greeted with enthusiasm wherever she landed especially during transit through the Dutch East Indies. The two Dutch bank managers who had booked return fares were manifested but Thea Rasche had left Australia on board ship heading for a journalistic assignment in the USA. The *Uiver* arrived at Schipol Airport, Amsterdam just after 2 o'clock in the afternoon on 20 November 1934 to a huge reception headed by Dr Plesman, members of the government and the Lord Mayor of Amsterdam.

PH-AJU was soon pressed into standard commercial service. On 20 December 1934, while operating an express Christmas airmail service from Amsterdam to Batavia, she disappeared over the Syrian Desert. In dangerous weather conditions a thorough search was conducted by 24 Royal Air Force aircraft, one of which discovered wreckage near Rutbah Wells. All seven people on board were killed and were buried in the British cemetary at Baghdad. The cause of the accident was attributed to a sandstorm or lightning although there have been suggestions that bullet holes were found in the structure.

Most of the results of the Dutch investigation into the accident remained secret. The known facts were that the aircraft had hit the ground at high speed with both engines developing power and the wreckage distribution lay in a direction opposite from the aircraft's intended track. There was speculation about the flying qualities of the DC-2 in certain conditions, with particular reference to rudder control. It was also suggested that the last flight of *Uiver* was intended to be a KLM record attempt on the route but her commander, Captain Beekman, had protested that the crew was tired. He was allegedly threatened with dismissal by Dr Plesman if the flight was not continued. In the event, no conclusions were ever published.

The disaster had occurred one day after the Royal Aero Club's celebration dinner at Grosvenor House in London where her Race crew was feted. News of the crash was greeted with dismay in Albury, and as a result of public subscription there a memorial was erected in Holland.

Early in 1935, Ray Parer and Geoff Hemsworth, still on their way to Flemington, suffered one of their many delays in Baghdad. While waiting for the delivery of spares they took a day trip to Rutbah Wells, and there were shown the burned-out wreckage of their one-time fellow competitor.

Captain Koene Parmentier, commander of the KLM entry, the DC-2 PH-AJU, 'Uiver', photographed during acceptance trials of the aircraft in California. The reflection of the port engine in the brightly polished fuselage projects the illusion that the panels are badly dented. (The Aeroplane.)

Jan Moll continued with KLM and escaped to England during the Second World War to fly with the Netherlands Naval Air Service. Later, he was posted to Australia for duties with Dutch forces operating in the South West Pacific. After the war he returned to KLM and flew until he retired with the rank of Commodore, living well into his eighties at his home in Almere near Amsterdam.

Koene Parmentier similarly escaped to England when war was declared and from where he operated the dangerous civil airline service to neutral Portugal. He was promoted to be the company's Chief Pilot after the Second World War and was killed when the KLM Constellation PH-TEN hit high-tension wires during an instrument approach to Prestwick Airport in Scotland in October 1948 whilst en-route to New York.

Flight Engineer Bouwe Prins served with the Royal Netherlands Air Force in Australia, returned to duties with KLM after the war and retired to live in Amsterdam where he died in the 'sixties.

Wireless Operator Cornelis van Brugge was on board a British civil registered DC-3 flying from neutral Portugal to England on 1 June 1943 when the aircraft was attacked and shot down over the Bay of Biscay by German fighters, in the mistaken belief that Winston Churchill was a passenger.

46. Granville-Miller-de Leckner R-6H Monoplane NR14307.

Following her untidy arrival in Bucharest, Jacqueline Cochran decided she had seen enough of the aeroplane and returned to the USA. The Granville was repaired and by 27 October the machine was declared airworthy again and reportedly loaned to the Rumanian Air Minister, Radu Irimescu, who later became an American citizen. Flown by Wesley Smith with Prince Cantacuzene on board, the aeroplane established a number of new Rumanian Air Speed Records.

The aircraft was eventually flown back to England where the Customs at Lympne discovered on board Alan Goodfellow's two philatelic covers addressed to Lady Spencer in Australia. The covers were seized but following correspondence with the General Post Office they were returned to the sender who signed them in his capacity as Clerk of the Course. The aeroplane was delivered to Southampton and loaded on board the Cunard liner RMS *Berengaria* which sailed for New York on 28 November.

Following her return to the USA, the Gee Bee was entered in the 1935 and 1938 Bendix Races and the 1936 Thompson Trophy Race but, due to engine problems, failed to start in any of the events. The Monoplane was demonstrated to the Chilean Air Force, but crashed on landing at Newark Airport, New Jersey, whilst being flown by Wesley Smith and a representative of the Chilean Government, ensuring there would be no production order and no royalties for Miss Cochran.

Following repairs the aircraft was acquired by dealer Charles Babb and re-painted from her 'Lucky Strike' green to white. Sold to the Mexican Government, in May 1939 the Granville was flown from Mexico City to Washington in record time by Captain Francisco Sarabia. Just after take-off for the return flight the aircraft dived into the Potomac River. It was discovered that a rag which had been left under the cowling had blocked the carburettor air intake causing the engine to stop. Sarabia was not wearing a shoulder harness and thrown forward in the impact he was trapped by the canopy frame and drowned before he could be freed. On hearing the news, Jackie Cochran's reaction was, "I think I was one of the few owners of a Gee Bee who did not lose my life in one. They were squatty, fast and most unstable."

The relatively undamaged aircraft was recovered and transported to Mexico where it was rebuilt to flying condition and presented to a one-aircraft museum at Ciudad Lerdo, Captain Sarabia's home town.

With 25 other American women pilots, Jackie Cochran joined the Air Transport Auxiliary in England in 1942 but returned to the USA in 1943 to lead a new organisation, the Women's Air Force Service Pilots (WASPS). In 1946 she returned to air racing, set a new speed record for propeller-driven aircraft in 1950 and flew a Canadian built F-86 Sabre at Mach 1 in 1953, the first woman pilot to exceed the speed of sound. She was decorated by the American Government and recognised by many other international organisations. By 1970, with over 15,000 flying hours logged, Jacqueline Cochran retired to live on her ranch at Indio, California, but suffered from deteriorating health and died on 9 August 1980.

At the time of her death Jackie Cochran had been awarded over 200 trophies for air racing and she held more speed, altitude and distance records than any other pilot in aviation history, male or female.

In 2004, the wartime airfield near Thermal, 20 miles south east of Palm Springs, California, adopted as Thermal Airport in 1948, was renamed *Jacqueline Cochran Regional Airport*.

47. Klemm Eagle G-ACVU.

The Eagle suffered irreparable undercarriage damage while landing at Bushire on Tuesday, 23 October, and was withdrawn from the Race by her solo pilot, Flight Lieutenant Donald Shaw, who arranged for her to be shipped back to England. In April 1936, during a flying holiday around the Mediterranean, the aircraft fell into the sea off Corsica and was lost.

54. Blériot III/6 F-ANJS.

Having suffered undercarriage problems at Villacoublay on the very point of departure for Mildenhall, the Blériot III/6 was repaired and completed a number of show appearances before she joined the growing stream of aircraft being delivered through France to the Republican forces in Spain where she disappeared during the Civil War.

55. Waugh and Everson Monoplane.

The Races were long since competed and won by the time the Evo III Monoplane was finished, but with Ernie Everson in command she did fly at Churchill on the Waikato River and was then flown to Hobsonville for inspection by the AID. Certification was denied as, amongst other reasons, the clearance of the propeller tips from the sides of the cockpit was only a few inches and considered to be insufficient. The group lost heart and Ivan Waugh, a garage proprietor who had provided finance, stored the dismantled aircraft in his premises at Waikato where the fuselage at least is known to have survived well into the 'sixties.

58. Airspeed Viceroy G-ACMU.

The Viceroy was retired at Athens and flown back to England. A year after the Race, Stack and Turner filed a law-suit against Airspeed, alleging that the company had been 'negligent in failing to ensure that the Viceroy was fully fit for a flight to Australia, and seeking 'the rescission of a Hire Purchase agreement and repayment of £2,448 paid by them for the Viceroy'. It was the contention of the manufacturer, however, that Stack had begun the Race mentally ill-prepared and that under normal circumstances he would easily have coped with the weather and what were minor technical difficulties. To protect their reputation Airspeed took the matter to law but the case eventually was settled out of court. The terms required the crew to return the machine to the Airspeed Company together with a cash payment of £1,850.

After some time languishing in the company's hangar at Portsmouth, Airspeed was approached with an offer to purchase, allegedly by a film distribution organisation, Yellow Flame Distributors Ltd., who asked for under-wing bomb racks to be fitted in order to carry their highly inflammable nitrate stock in relative safety outside the cabin. Obviously suspicious, the company discovered that the aircraft was actually destined for the Abyssinian Air Force (together with three Gloster Gladiator fighters) to assist in that country's war against Italy by initiating a surprise bombing attack on the oil storage tanks at Massawa during a seasonal lull in the fighting. £5,000 was paid for the Viceroy but before any delivery could be effected the war ended. The aircraft remained in storage at Portsmouth, owned by the Emperor of Abyssinia who had been rescued by the Royal Navy and was living in exile in Cheltenham. G-ACMU was then sold and resold, eventually becoming the property of Ken Waller who was intending to fly her, with Max Findlay as co-pilot, in the London to Johannesburg Race of October 1936.

Meanwhile, the Spanish Civil War had broken out, and Waller was besieged by a Spanish emissary who wanted to buy the aeroplane. He offered what Waller had paid Airspeed plus a sum equal to first place prize money in the Johannesburg Race, plus the anticipated cost of petrol and oil, a total of £9,500. A deal was done and the aeroplane released into Spanish custody whereupon she was flown immediately to a destination in France and never heard of again.

Neville Stack was familiar with the Indian sub-Continent having made the first flight from England to Karachi in a DH.60 Cirrus Moth in 1927. After the Second World War Stack was Managing Director of Orient Airways, a Pakistan Government subsidised company based in Karachi but he had been interned by police whilst the 'improper sale' of a company DC-3 Dakota was being investigated. On Tuesday, 22 February 1949, on one of his approved daily visits to a priest at Mauripore Airport, Stack threw himself into the path of an air force lorry and was run over and killed.

60. DH.89 Dragon Six ZK-ACO.

Following the celebrations in Melbourne the aircraft was flown to Mascot Aerodrome, Sydney, where she was overhauled, wireless equipment installed and additional fuel tanks to extend her range to 1,600 miles. She was positioned to RAAF Base Richmond on 14 November where she took on 276 gallons of petrol and at 5.55am the following day took-off for New Zealand, maintaining wireless contact with Richmond and Auckland. In spite of her additional long range tanks it was necessary during the flight to top up the feeder tanks in the fuselage from a supply of two gallon tins carried on board, a most hazardous operation.

Hewett flew the aeroplane and acted as navigator but they struck bad weather three hours out and were forced to fly above cloud for two hours during which time they were unable to assess their drift. After a flight of ten hours the aircraft arrived over the New Zealand coast south of Westport where they got a bearing from a lighthouse and set course for Auckland where an official reception and a large crowd awaited them but oncoming darkness compelled them to land at Palmerston North at 6.04pm, just over 12 hours after take-off. Following the immense care taken by the Race crew with their brand new aeroplane, on landing the aircraft was damaged when she collided with a fence. However, the Dragon Six created history as the first aircraft to fly from London to New Zealand where ZK-ACO became the most modern aeroplane in the country. That same day the crew posted the 100 postal covers they had uplifted at Mildenhall, re-addressed to themselves c/o the Post Office at Palmerston North, the first, but unofficial, airmail flown from Great Britain to New Zealand.

The anticipated sale of the aeroplane came to nothing and in May 1935 with 167 flying hours logged, *Tainui* was dismantled, put on board the SS *Wanganella* and shipped back to Sydney where her tail was damaged again as it was swung against the wall of the wharf during unloading on 27 May. The following day, ZK-ACO was towed through the streets to the de Havilland factory at Mascot Aerodrome, and from there despatched by lorry to Melbourne.

As VH-UUO she served with West Australia Airways, then several other lines until the war, when as A33-1 she became the first Dragon Six (now re-named Dragon Rapide) taken on charge by the RAAF. Upon demobilisation, and civilian registered once more, the Rapide continued with her interrupted duties in Queensland and New South Wales, until the afternoon of 23 May 1952. In the service of Butler Air Transport when attempting a forced landing in poor weather, the aircraft crashed on Warrumbungle Mountain, NSW, and although the five occupants escaped without serious injury, the aeroplane caught fire and was destroyed.

Due to the lingering depression in New Zealand James Hewett emigrated to Australia and became a pilot with Qantas. Returning to New Zealand in 1938 he joined the Territorial Air Force and later the Reserve before taking up a commission in the Royal New Zealand Air Force in 1940 when he commanded No. 42 Squadron until August 1945. He was demobilised with the rank of Wing Commander and died at his home at Taranui Bay, Anakiwa, South Island, on 27 October 1955.

Cyril Kay was about to return to England in 1935 when he was offered and accepted a short service commission in the New Zealand Permanent Air Force. Four years later he was one of a number of Royal New Zealand Air Force crews awaiting delivery of new Wellington bombers which were to be ferried home. In the event, war was declared and the aircraft and crews remained in Europe, evolving into No. 75 Squadron RNZAF. Continuation of hostilities in the Pacific took Kay back to New Zealand where in June 1956 he was appointed Chief of the Air Staff wih the rank of Air Vice-Marshal. He retired in July 1958 and died in London on 20 June 1993 at the age of 91.

62. Fairey Fox G-ACXX.

The aircraft crashed during an attempted forced landing near Foggia, southern Italy, and both crew members were killed. The

The tangerine, green, black and silver DH.89 Dragon Six ZK-ACO was unable to find a buyer in New Zealand where she arrived after an epic flight across the Tasman Sea on 15 November. As a consequence the aircraft was dismantled and shipped back to Australia.
(James Dyson. Eddie Coates Collection.)

ZK-ACO was re-registered VH-UUO with West Australian Airways and others until she became the first Dragon Six (Rapide) to be enlisted with the RAAF as the result of impressment at the beginning of the Second World War.
(Geoff Goodall. Eddie Coates Collection.)

funeral for New Zealanders James Baines and Harold Gilman was conducted at the English Church in San Pasquale on 26 October 1934, after which the airmen were buried with full military honours at the British cemetery in Naples.

63. DH.88 Comet G-ACSP.

The morning after retirement from the Race at Allahabad, Jim Mollison is alleged to have offered *Black Magic* for sale to a young Indian pilot, probably Mr J R D Tata. Jehangir Tata ('JRD' to his friends) did not take up the offer but was soon to make a name for himself as the founder of Tata Airlines which evolved into Air India. A Maharaja (probably the air-minded Maharaja of Jodhpur) is also reported to have been offered the aeroplane and to have asked a locally based Australian pilot to seek terms on his behalf. In the circumstances it might or might not have been the most appropriate time to attempt to negotiate any business arrangement with Jim Mollison who was already anticipating having to dismantle the aircraft, have her carried to Calcutta and shipped back to England from there, but it was quickly agreed with the de Havilland Company that for many reasons it was best that she should be repaired on site and then flown home to Hatfield.

Having no financial backers Jim Mollison was desperate to sell *Black Magic* as quickly as possible in order to raise his half of the capital. Both he and his wife Amy had put all their financial reserves into the project and effectively were flat broke. Spare cylinder heads and low compression pistons were flown to India on board an Imperial Airways' scheduled service and fitted to the damaged engines by Rex Pearson, an engineer positioned at Allahabad by Bernard Rubin with the sole intention of assisting Ken Waller and Owen Cathcart-Jones on their east and later westbound transits.

Black Magic was to remain at Allahabad for three weeks after her stuttered arrival before she was declared airworthy for a less arduous return flight to England. During her enforced stay she was robbed to provide *pukka* Gipsy Six R spares for the port engine of G-ACSR, work entrusted to Pearson and other engineers drafted in from the de Havilland works at Karachi. The precious parts were transferred when Cathcart-Jones and Waller passed through heading west during their attempt to establish a record for an out-and-return flight.

With none of the pressures of racing to affect her, Amy Mollison had cancelled her earlier plans to fly home by commercial airline, leaving Jim to cope with the Comet, and she crewed the westbound flight as far as Cairo where they again met Rex Pearson, homeward bound via KLM.

Amy complained about the cramped cockpit of the Comet and eventually Pearson agreed to fly home with Jim Mollison while Amy returned to Croydon arriving on the scheduled KLM service via Rome. The captain was Ivan Smirnoff, nominated by his company as a Race entrant until the airline scratched all entries in favour of maximum support for the DC-2.

It was not without further adventure that *Black Magic* landed safely at Lympne on 11 November where she remained until 'collected' by de Havilland in January 1935, and was flown to Hatfield for a complete overhaul and re-sale at an undisclosed price.

The Comets had been given considerable publicity as potential high-speed mail carriers, and after the Race G-ACSP was acquired by the Portuguese Government.

Having languished at Allahabad for three weeks after her forced exit from the Races, Black Magic *was ferried home to England in easy stages, with Amy Mollison crewing for her husband as far as Cairo. During a stop-over at Baghdad there was a second opportunity for ministers from the Iraqi Government to make the acquaintance of the famous couple and in less frenetic circumstances than previously.*
(Alan Hartley via David Luff.)

Still recovering from his gashed head and awaiting funds for the repatriation of Lockheed Vega G-ABGK from Aleppo for repair in Australia, Jimmy Woods was on hand at Baghdad to offer a word of encouragement as the Mollisons passed through on their way back to England.
(National Library of Australia.)

Flown by de Havilland's Portuguese agent Carlos Bleck and Lieutenant Carlos Costa Macedo on the first direct flight between England and Portugal, she left Hatfield for Lisbon on 26 February 1935 in preparation for an experimental postal service across the South Atlantic to Rio de Janeiro. For this purpose, the aircraft was re-registered CS-AAJ and named *Salazar*.

Little changed from her Race configuration apart from her markings she was damaged on take-off from Sintra on 14 March 1935 when crewed by Costa Macedo and Bleck and was returned to Hatfield where, after repair, she was test flown by Geoffrey de Havilland Jnr on 26 June. The aircraft returned to Hatfield again during the late summer of 1935 and after a test flight by Geoffrey Jnr in the evening, his father requested the opportunity of logging some dual instruction before flying the aircraft solo. On his first solo landing an observer reported that Captain de Havilland's approach was too high and too slow and he allowed the aircraft to stall heavily onto the ground causing substantial damage. The date was recorded as Friday 13 September.

There is some evidence to suggest that by 1936 Amy's personal finances had recovered for her to offer to buy CS-AAJ from the Portuguese Government for the projected race from London to Johannesburg, but the asking price was too high and her interest waned.

CS-AAJ eventually was re-delivered in another direct flight from Hatfield to Lisbon but no transatlantic mail trips were ever undertaken and the aircraft was given to the Portuguese Army Air Corps with whom she is believed to have crashed into the sea. Small portions of the airframe were salvaged from a Portuguese smallholding and appeared in England in 1978. An attempt to rebuild a Comet around the few remains was frustrated when a hangar fire at Chirk destroyed most of the newly constructed parts. A further serious attempt to 'restore' G-ACSP was made in 1987 following the successful flight of G-ACSS, but the Civil Aviation Authority was not convinced there was a sufficiency of original parts for G-ACSP to become a 'restoration' rather than a 'replica'. Her 'identity' was traded through several owners, each of whom added something to the project, before acquisition by the current syndicate of enthusiasts, 'The Comet Racer Project Group' at the Derby Aero Club and Flying School, Egginton Airfield, Derby, took on the task in August 2004.

For Amy and Jim Mollison the MacRobertson Race was to be the last great record breaking enterprise undertaken as a team. Shortly after returning to England their marriage finally broke down and Mrs Mollison reverted to using her maiden name: Amy Johnson. The couple were divorced in 1938.

Amy joined the Air Transport Auxiliary in 1940. In January 1941 she was ferrying an Airspeed Oxford from Blackpool to Kidlington in bad weather, but appears to have run out of fuel miles off course over the Thames Estuary and parachuted into the freezing water where she drowned. The exact circumstances of her death will forever remain a mystery.

Jim Mollison was a pilot with the Atlantic Ferry Service and later the ATA. He survived the war but suffered a series of broken marriages and became an alcoholic. To provide an interest in life his estranged Dutch wife bought him the Carisbrooke Hotel, a temperance establishment in Surbiton, but his alcoholism and failing health restricted his lifestyle and he died in Priory Hospital, Roehampton, on 30 October 1959. His ashes were scattered over the Atlantic Ocean from a light aircraft flying a few miles off the West Coast of Ireland, near to where his heroes Alcock and Brown had made landfall in 1919.

Every person, whether passenger or operating crew, who flew into Melbourne on board a Race aircraft which completed the course within the stipulated 16 day schedule was presented with a solid gold medallion.

A bronze medallion was presented to the City of Launceston, Tasmania, as consolation after a much trailed suggestion that Sir Macpherson Robertson had agreed to pro-

Collected from Lympne by a de Havilland pilot and ferried to Hatfield for overhaul, Black Magic was sold to the Portuguese Government for an undisclosed sum. Re-registered CS-AAJ the aircraft was to be used for experimental trans-Atlantic mail services although none was ever attempted.
(BAE Systems.)

vide 'a sum of money' for aircraft post-MacRobertson to embark on a series of races between Australian cities and in particular from Melbourne to Hobart, was cancelled. Launceston's medallion which was displayed at the city's Victoria Museum and Art Gallery was stolen in 1941 and never recovered. Another bronze medallion was presented to Mr Rex Allison, an able-seaman whose professional duties were performed in the torpedo rooms of the Royal Australian Navy battleship HMAS *Canberra*.

Rex Allison developed a passionate interest in the MacRobertson Races and compiled an outstanding collection of memorabilia from along the route which was inspected on board ship by several senior civic officials and by Sir MacPherson himself. Rex Allison's powers of persuasion were such that in 1936 Sir MacPherson had agreed to instruct Hardy Bros. to make a scale replica of the gold trophy for his collection but 'owing to trade difficulties' the prospect was abandoned.

In January 1937 Sir MacPherson Robertson confirmed that the casts for the medallions had been destroyed.

The gold trophy awarded to the winning Comet simply disappeared. Unsubstantiated rumour has it that the trophy, presented to Charles Scott and Tom Campbell Black at Laverton, was handed on to Arthur Edwards and was melted down as part of an exercise to recover costs and provide funds but precisely when and where, if such was the case, is unknown. Exactly what did happen to the trophy remains a mystery. But even without an icon the events triggered by Sir MacPherson Robertson and all the organising teams in 1934 have become an indelible part of world aviation history and well deserving of the accolade: 'The World's Greatest Air Race'.

18
Postscript

In 1936 an Eric Kennington sculpture was commissioned for erection in the car park of The Comet Inn at Hatfield. (Michael Ramsden.)

THE de HAVILLAND aerodrome at Hatfield was opened in 1930 and the whole of the Company's aeroplane manufacturing business was transferred across from Stag Lane by the end of 1934 leaving that site to the expanding Engine Division.

At the confluence of the St. Albans Road West and the Barnet bypass, just a short distance from Harpsfield House and Highlands, two of the private houses occupied as offices by directors of the de Havilland Aircraft Company and situated with views across the aerodrome, a 'roadhouse' was built in 1934. The establishment was constructed in the shape of a racing aeroplane and called The Comet Inn.

In 1936 the acclaimed sculptor, Eric Kennington, was commissioned to design and craft a 13ft high pillar from Portland stone which was to be erected in the car park. Kennington executed 18 rhomboid shaped panels depicting various means of flight from mythological to supernatural, fish, frogs and fairies, and mounted on the top of the pillar was a metal scale model of the DH.88 Comet *Grosvenor House*.

In spite of many changes of ownership and proposals to alter the displayed name of the hostelry to reflect the 'brand image', driven by public opinion the old roadhouse, now a listed Art Deco building serving as a popular restaurant and hotel, retains and prominently displays the name 'The Comet', the classic sculpture retained in the car park with the iconic model mounted at its summit.

In 1934 the Australian writer and director Beaumont Smith made a film entitled *'Splendid Fellows'*. The plot sees Monty Ralston, an amateur aviator and dissolute son of an English shipping magnate, sent to Australia in an effort to reform him. Invited to stay on a farm the man discovers the farmer has a sprightly daughter, Eileen. A friend of the family is the local vicar, Rev. Arthur Stanhope, who was a fighter pilot during the First World War and who tours his parish by light aircraft. Ralston and Stanhope decide to build an aircraft to enter the MacRoberston Races and Ralston promises to win after Eileen says she will marry him if he does. The film includes newsreel clips from the actual Races and a guest appearance by Sir Charles Kingsford Smith. One review of the film says: "The story of an English 'new chum' finding his future in the colonies was already hackneyed in 1934 and audiences stayed away!"

In 1933 the Impex Print and Carton Company of Hawke Street, Melbourne, a business with a division based at Utrecht in Holland, marketed a board game on which the route of the 'Impex Centenary Air Race' between London and Melbourne was picked out in a regular series of small holes across a map of the world. Three electric light bulbs were connected to a '3 volt Impex Junior Flat Battery'. The object of the exercise was for one of the two players to be first to arrive at Melbourne by moving a peg along the route, the number of holes traversed dictated by the roll of a dice. At certain points, and without warning, one of the lights would be illuminated, indicating the player had to forfeit a turn or return to a previously negotiated 'Air Port', some of which, because the game was marketed as soon as the Races were announced, were not actually on the final route. Impex did have the grace to acknowledge that a £10,000 prize had been offered by a 'Melbourne citizen', even though they managed to mis-spell his name.

Three quarters of a century after the conclusion of the MacRobertson International Air Races, substantial prices continue to be paid at auction for items from the great selection of postal covers flown by Race competitors and which are regularly offered for sale. An equal popularity is enjoyed by what appears to be almost endless supplies of press photographs, programmes, copies of Rules and Regulations, menu cards and other paraphernalia, much of which was generously autographed on an appropriate occasion by all the principal players.

RAAF Base Laverton, the assembly point for the first finishers in the Races, and where the prize-giving ceremony took place on 10 November 1934, was closed to flying operations in the 'eighties and subsequently much of the aerodrome was sold and given over to housing and industrial development for the new district known as Williams Landing. From 1999, administration of what remained of the old base and its current ground training programmes, was merged with those of the nearby RAAF Base at Point Cook, now known collectively as RAAF Base Williams. The main street in Laverton is called Aviation Road.

Flemington Racecourse, about four miles north west of Melbourne city centre, and reputed to be one of the most beautiful courses in the world, is host to the famous Melbourne Cup held annually on the first Tuesday in November. Apart from its close proximity to the City, the racecourse is equipped with a 'straight' of well over a mile which made it an almost perfect choice for a finishing line for the Races, even though no Race aircraft were expected to land there. In 1934, the first Tuesday in November was the 6th, a date which might have been considered when the 16 day time limit was imposed for qualifying arrivals.

On 27 October 1984, as part of the celebrations of the 150th anniversary of Victoria and the 50th of the Races, the Antique Aeroplane Association of Australia organised a dinner at Flemington Racecourse attended by 250 guests. For the occasion a Tiger Moth and a Sopwith Pup replica were permitted to land and were floodlit overnight. The Association arranged for a brass plaque to be fixed to the wall of the Crown Room which was unveiled by Mr Norman Robertson, son of Sir MacPherson, commemorating the site of the finishing line.

At Mawson, ACT, a thoroughfare is named MacRobertson Street. In Melbourne, the crossing of the Yarra River at Grange Road is via the MacRobertson Bridge.

To commemorate the flight of the KLM DC-2 in the Races, the Australian composer Peter Flanagan wrote the 'Flying Dutchmen' suite.

Henry Walker who, with Malcolm McGregor, crewed Race entry No. 2, Miles Hawk ZK-ADJ, consented to become Patron of the Aviation Historical Society of New Zealand and on the 50th anniversary of the Races in October 1984 contributed to an exhibition of memorabilia which was mounted in Wellington.

To mark the 75th anniversary of the Australian Broadcasting Commission's 2CO/ABC Goulburn Murray station the ABC Regional Production Fund commissioned local artist John Walker to write a radio play covering the events of the night of 23/24 October 1934 at Albury. The Albury-Wodonga Theatre Company presented his script 'Flight of the Uiver' before a full-house at the Albury Performing Arts and Convention Centre and which was broadcast live. Stuart Baker, then Mayor of Albury, played a cameo part taking on the role of 1934 Mayor, Alfred Waugh, and relatives of other characters in the play who helped in real life were in the invited audience.

Sir MacPherson Robertson donated £100,000 towards Victoria's Centenary Celebrations, a sum which included £40,000 for the establishment of a Girls' High School in Melbourne. He also funded the MacRobertson Fountain in front of the Shrine of Remembrance on St. Kilda Street and the Herbarium at the Botanic Gardens.

In 1925, Sir MacPherson Robertson presented a mounted silver shield for international competition in the game of croquet, one of his passions, between England and Australia. New Zealand was invited to join the competition in 1930. In 1979 'England' became 'Great Britain' and the USA was invited to make a quartet in 1993. Known as the MacRobertson International Croquet Shield the competition represents the premier croquet team event in the world.

With the growth of the town of Mildenhall, a purpose-built cinema was erected in North Terrace Road in 1935, taking over nightly programmes lately provided by temporary facilities in the Town

The Comet cinema was built in North Terrace Road, Mildenhall, in 1935 and survived until 1996. The site is acknowledged by a plaque.
Nick Spencer. Mildenhall Museum.

Hall. In recognition of the events of the previous year, the new cinema was called the 'Comet' and was to remain a central part of the community for 50 years until the building was converted into a bingo and snooker hall and finally closed and demolished in 1996. The site was acquired by the Jehovah's Witnesses for their Kingdom Hall and an exterior wall of the new building carries a stone plaque which acknowledges the site as the previous location of the Comet cinema.

Near RAF Mildenhall a sign on a brick plinth identifies the development as the 'Douglas Park Estate' and credits the layout as having 'streets and roads named after aircraft, places and flyers connected with the MacRobertson Air Race from Mildenhall to Melbourne in 1934'. The names listed are: Amy Johnson Court; Boeing Way; Charles Melrose Close; Comet Way; Courier Close; de Havilland Court; Fairey Fox Drive; Granville Gardens; Grosvenor House Court; Jim Mollison Court; MacPherson Robertson Way; Miles Hawk Way. Older established byways in the same area are named: Campbell Close; Darwin Close and Melbourne Drive.

In 1997 a new shopping experience, the Galleria Outlet Centre, was opened at Hatfield in a futuristic building opposite what had been the main entrance to the now-closed de Havilland factory site. In order to show some appreciation of the aviation history of the immediate area the management commissioned a film-set construction company, Acorn Scenery of Feltham, to build a replica of DH.88 Comet G-ACSS which would be hung from the high-domed ceiling. The result, in true film tradition, was extraordinary. The full-size model, built from wood around a steel girder core, was completed in just six weeks. After ten years of suspension a new owner of the business decided to scrap the attraction. Word reached the de Havilland Aircraft Heritage Centre at Salisbury Hall who were presented with the 'aircraft' hoping it could be pressed into service as a gate-guardian but the structure was found to be unsuitable for outside display in England.

KLM, Royal Dutch Airlines, bought 18 DC-2 aircraft all of which carried the name of a bird, 'Uiver' (stork), being the first. The Dutch Aviodrome Museum currently owns and operates an American registered DC-2, N39165, based at Lelystad, faithfully restored to represent the original PH-AJU. In celebration of the 50th anniversary of the MacRobertson Races, funded partly by the Uiver Memorial Foundation and the Netherlands Broadcasting Corporation, the aircraft was flown from Schipol Airport, Amsterdam, to Melbourne between 18 December 1983 and 5 February 1984. Captain for the flight was Boeing 747 skipper Jan Plesman, grandson of the founder of KLM, and he and his three crew members all were dressed in 1934-style uniforms. The aircraft flew low over Albury Racecourse before landing at the town's airport where the team was enthusiastically welcomed by a crowd of thousands. A three-part documentary film was later shown on Dutch television.

In 1993 a 'flight simulation' computer game was published which enabled players to 're-create the 1934 MacRobertson Race'. Amongst the images displayed as part of the advertising blurb was a DH.85 Leopard Moth which was a mis-identification or just wishful thinking. The following year, 1994, Dutch author Floris Wouterlood published a program which allowed PC users to simulate the memorial flight made by the replica *'Uiver'* as part of the 50th anniversary celebrations.

On 14th February 1984, the Australian Broadcasting Commission's *'This Australia'* programme included a number of interviews with Albury residents involved with the DC-2 emergency. One said that had the racecourse surface not been so muddy the aircraft would not have stopped on landing before crashing through a fence. Another recalled how Cornelis Van Brugge had told him that a short religious service had been held on board as they circled.

In Holland commemorative postage stamps featuring the *'Uiver'* were issued and newborn babies were Christened *'Uiver'* or *'Uiverjay'*, meaning 'little stork'. The Dutch people subscribed to a plaque which was presented to Albury City Council and when news came of the aircraft's crash in Syria the people of Albury contributed to a memorial erected in Holland.

In 1990 a two-part film was made for Australian television entitled *'The Great Race'* or in some areas, *'Half a World Away'*. The programme was described as 'A fictional story about the 1934 Air Race but built around actual events and people'.

Largely filmed at Maryborough, Victoria, it was necessary to enrol several different types of aircraft to play the part of competitors and the producers commissioned three full-size, non-flying, replica DH.88 Comets, one of which was capable of fast runs with the tail up and a second could taxi. Most of the flying shots were achieved with models. The finished effort was described by one critic as being 'full of stereotypes' and failed to sell in Britain.

The post-production clear-out of effects included the three replica Comets which were offered for sale at grossly optimistic prices and there was little interest. The one machine capable of speed-runs, 'G-ACSS', was shipped to England where there was thought to be a better market but it failed to achieve its high reserve price at auction and was eventually sold by private treaty to an enthusiastic collector of static aircraft.

Unlike the real aeroplane, the film replica had been built with the ease of dismantling, transport and re-erection much in mind. The premature death of her new British owner, who operated a transport company, occurred after the aircraft had been exhibited on only a handful of occasions and led to the machine's acquisition by the de Havilland Aircraft Heritage Centre.

In October 1984, to celebrate the 50th anniversary of the Races and the commissioning of RAF Mildenhall, the de Havilland Moth Club was granted permission by the United States Air Force Base Commander to re-create the start of the Races following a massive hangar-dance organised by the station on 20 October. Guest of honour and race starter the following morning was, in 1934, the co-pilot of the Lockheed Vega, Race No. 36. A Flying Officer in the RAF then, Donald Bennett was now a retired Air Vice-Marshal whose name will forever be associated with the wartime 'Pathfinder Force'. Twenty aircraft took off in order carrying appropriate racing numbers and turned over the village of Melbourn in Cambridgeshire before landing at the army base which was previously RAF Oakington. Only one aircraft had a different planned agenda. DH.80A Puss Moth G-AAZP, flown by Tim Williams and Henry Labouchere, flew past Melbourn in England and just like Jimmy Melrose 50 years earlier, went on in easy stages to land in Melbourne, Australia, where they arrived on 15 December.

Due to operational requirements at RAF Mildenhall in October 2009, it was only possible to mark the 75th anniversaries during a special day set aside for the purpose on 15 May. A large number of aircraft types had been organised to create a fleet as close to the original starting list as was practicable, but strong winds and storms around the country reduced the overall numbers if not the spirit, and kept the new *'Uiver'* firmly on the ground at Lelystad.

In the USA, the thoughts of Bill Turner, proprietor of Repeat Aircraft, a company specialising in the manufacture of one-off replicas of very significant aeroplanes, and Tom Wathen, an enthusiast for such things and with a degree of financial stability to support his dreams, coincided when the possibility of building a new DH.88 Comet was first mooted. In December 1991, with considerable help from those who had been involved with the airworthy restoration of G-ACSS, work started at Flabob Airport, Riverside, California, to build an exact replica, using the same drawings, materials and methods of construction initiated by the de Havilland Company in 1934. Only slight deviation from absolute authenticity was allowed when considered to be in the interests of safe and practical operation in the modern environment and the longevity of the project. The first flight of the new aeroplane, NX88D, was undertaken from Flabob by Bill Turner on 28 November 1993, less than two years after cutting the first wood. The plan to fly from London to Melbourne was never progressed and the anticipated appearance of the aircraft in Europe in 2009 was thwarted due to what were considered excessive insurance demands.

To celebrate the 75th anniversary of the founding of the Rotary movement in 1905 the West Albury Rotary Club purchased the derelict DC-2 No. 1286 languishing in Sydney. Following refurbishment to display standard in the colours of PH-AJU, the aircraft was mounted outdoors at Albury Airport where she was unveiled on 2 March 1980 by the Governor General of Australia, Sir Zelman Cowan. The aircraft was a gift to the people of Albury in recognition of the events of October 1934. Requests for the Council to provide covered accommodation for the aircraft were parried over several years and as its physical condition had deteriorated, the aircraft was dismantled in August 2002 for a detailed structural survey. In 2006 the Uiver Memorial Community Trust was established with a brief to raise funds and the DC-2, now owned by the Council, currently stands under cover where it will be 'preserved' together with associated memorabilia.

Commander Harold Ernest Perrin, on whose shoulders much of the organisation of the MacRobertson Races fell, was Secretary of the Royal Aero Club from 1903 until his retirement in 1945 at the age of 67. Not until two years after the Races was he afforded official recognition for his efforts when he was appointed CBE. He died in April 1948. Apart from awards made by and on behalf of aviation-related organisations, the crew of the victorious Comet, G-ACSS, received no public honours.

Sir MacPherson Robertson, whose patriotism and above all else, personal generosity, had made the Races a practical proposition, died on 20 August 1945. The confectionery business he started from nothing, and whose brand name bore his favoured form of address, was acquired by the Australian branch of Cadbury Bros. of Bourneville, England, in 1967.

Index

Abbeville 112
Adelaide 10, 39, 52, 156
Aircraft Exchange and Mart 25, 43, 61, 83, 88
Airspeed Company 43, 59, 60, 66, 180, 192
Airspeed Courier 25, 37, 43, 59, 111, 112, 113, 118, 124, 128, 145, 150, 152, 158, 161, 178, 179
Airspeed Envoy 25, 31, 60, 86, 173
Airspeed Viceroy 25, 60, 61, 85, 88, 97, 108, 111, 112, 118, 124, 127, 165, 166, 192
Akyab 125, 151
Albury 139, 140, 141, 142, 190, 197, 198
Aleppo 19, 20, 22, 114, 115, 119, 121, 124, 128, 129, 146, 181, 188
Allahabad 14, 19, 20, 21, 22, 46, 47, 120, 121, 122, 123, 125, 126, 127, 128, 129, 144, 145, 148, 150, 151, 180, 193, 194
Allison, Rex 195
Almaza 185
Alor Star 19, 20, 21, 125, 134, 150, 152, 155, 158, 182
Amalgamated Wireless 22, 86, 136, 146
Amsterdam 26, 46, 56, 75, 190
Anderson, Flt Lt David 31, 87
Anderson's Park 43, 45
Argyle, Sir Stanley 16, 162
Armstrong Siddeley Motors 38
Armstrong Whitworth 21, 98, 99
Asjes, Dirk 35, 89, 123, 152, 174, 175
Athens 19, 20, 29, 114, 117, 118, 119, 120, 124, 125, 127, 128, 181, 188, 192
Automobile Association 25, 80, 120
Baghdad 19, 20, 22, 46, 63, 76, 95, 109, 110, 112, 114, 115, 116, 119, 120, 121, 122, 124, 125, 128, 129, 145, 146, 151, 169, 180, 181, 190, 194
Baines, James 62, 63, 92, 97, 110, 112, 113, 124, 125, 127, 128, 193
Balchen, Bernt 45
Ballantyne, Flt Lt HBS 78
Bangkok 19, 20, 114, 144, 151, 180
Barnato, Woolf 53
Bassett, Walter 11, 12
Batavia 19, 20, 22, 26, 35, 46, 64, 127, 134, 146, 149, 150, 152, 155, 166, 188
Batten, Jean 29, 137, 141
Bay of Bengal 22, 144, 151
Beauvais 113
Beck Row 9, 76, 107
Bedford, Duchess of 190
Beech Staggerwing 60, 96
Bellanca 46, 50, 63, 80, 88, 105, 110, 111, 165, 180, 181
Bennett, Flg Off Donald 24, 52, 53, 87, 112, 118, 119, 188, 189, 198
Bergamaschi 80
Bergamo 62
Berlin 45, 56, 94
Bernard Monoplane 53
Bingham, Senator Hiram 27
Birdum 84, 130, 153
Blackpool 27, 194
Bladin, Sqn Ldr FR 127

Bleck, Carlos 194
Blériot Company 27, 37, 59, 94, 191
Boardman, Russell 50, 184
Boeing 247 32, 52, 57, 75, 85, 88, 90, 96, 99, 105, 114, 121, 123, 125, 127, 139, 146, 147, 148, 156, 159, 161, 166, 167, 173, 174
Bonar, Eric 45, 46, 47, 180, 181
Bonney, Mrs Lores 137
Bourke 134, 141, 147, 155
Bower, Sir Alfred 108, 109, 110, 135
Bradley, Sqn Ldr 25
Brancker, Sir Sefton 58
Bremerhaven 46
Brindisi 29, 188
Brisbane 44, 130, 169, 180
Bristol aircraft 44, 109
Bristow, Colonel 107
British Klemm Eagle 58, 110, 111, 114, 118, 124, 146, 191
Broad, Hubert 63, 65, 69, 71, 72, 73, 92, 95, 110, 119, 166
Brodie, John 67
Brook, Harold 48, 49, 86, 98, 111, 113, 124, 128, 181, 182
Brooklands 31, 48
Broome 126, 189
Brunette Downs 154
Buchanan, Maj John 15
Bucharest 22, 114, 116, 117, 191
Buckingham, Hugh 179
Bushire 19, 20, 128, 145, 146, 191
Butler, Alan 61, 72, 164, 170
Cadbury Bros. 198
Cairo 125, 169, 193, 194
Calcutta 14, 19, 20, 120, 122, 124, 125, 144, 145, 150, 151, 182, 188, 193
Camooweal 157
Campbell Black, Tom 25, 51, 69, 73, 96, 110, 122, 125, 127, 130, 131, 132, 133, 134, 136, 137, 138, 139, 159, 163, 164, 165, 172, 184, 185
Canberra 18, 160, 190
Caproni 62, 94
Cathcart-Jones, Owen 20, 40, 41, 52, 72, 73, 81, 82, 91, 94, 95, 97, 99, 105, 110, 112, 115, 116, 119, 121, 127, 148, 149, 158, 160, 161, 164, 178, 179, 193
Caudron 40, 68, 177
Cearns, WJ 54, 190
Cessna A.W. 53
Challe, Capt Leon 37, 59
Chamberlin, Clarence 59
Chance Vought Corsair 99
Chandi, VL 43
Chanteraine, D'Estailleur 40
Chapman, John 94
Charleville 19, 20, 21, 22, 126, 13, 131, 132, 133, 134, 135, 139, 140, 146, 147, 149, 154, 155, 156, 157, 161, 190
Clarkson, Flt Lt Christopher 16, 76, 80, 83, 101, 109
Clarkson, Richard 66, 69, 71, 72, 95, 132, 163
Clennell, Walter 79, 109, 184
Cloncurry 19, 20, 45, 133, 135, 139, 146, 149, 153, 154, 155, 156, 157, 160
Cobby, Wing Cmdr AH 11
Cobham, Sir Alan 17, 25, 42, 43, 60
Cochran, Jacqueline 32, 35, 47, 48, 50, 57, 58, 89, 90, 109, 110, 116, 117, 122, 123, 130, 139, 147, 181, 191
Cole, Wing Cmdr Adrian 10, 11, 14, 15, 18, 23, 34
Collett, Sir Charles 109
Comper aircraft 54, 87, 141
Conder, Maj WT 11
Constantinople 50, 112
Cootamundra 140
Copenhagen 81, 159

Cord Vultee 63, 64
Corniglion-Molinier, Capt Edouard 37, 59, 181
Courtenay, William 27, 42
Cowan, Sir Zelman 198
Cowper, Maj GA 28
Cross, Jack 184
Croydon 39, 44, 47, 54, 59, 60, 79, 85, 88, 180, 181
Curtiss Co. 25, 34, 48, 181
Daly Waters 84, 157
Dancy, Capt Wilfred 80, 93
Darby, Lt Col Maurice 15
Darwin 18, 19, 20, 21, 22, 29, 45, 59, 110, 126, 127, 130, 131, 133, 135, 146, 148, 149, 152, 153, 155, 156, 188
Davies, Flg Off Cyril 38, 39, 82, 97, 110, 113, 124, 128, 151, 176, 177
de Havilland Aircraft Company from page 24
de Havilland, Capt Geoffrey 65, 68, 71, 159, 163, 166, 168, 170, 194
de Havilland, Geoffrey Jnr 194
de Havilland, Peter 119
de Sibour, Vicomte Jacques 37
Desmond, Florence 96, 158, 165, 172, 184
Desoutter 35, 62, 80, 81, 97, 111, 112, 113, 118, 124, 128, 145, 150, 155, 160, 165, 175
Détroyat, Michel 36, 37, 176
DH.106 Comet 187
DH.50J 130
DH.60 Moth 28, 29, 40, 42, 137, 138, 141, 155, 179, 189, 190
DH.61 Giant Moth 9, 10
DH.66 Hercules 169
DH.71 Tiger Moth 66
DH.77 66
DH.80A Puss Moth 39, 40, 49, 52, 53, 54, 82, 83, 85, 86, 87, 88, 94, 98, 101, 111, 112, 118, 128, 150, 154, 158, 165, 177, 198
DH.82 Tiger Moth 77
DH.83 Fox Moth 43
DH.84 Dragon 61, 82, 92, 97, 100
DH.85 Leopard Moth 40, 68, 69, 80, 83, 107
DH.86 65, 166, 169
DH.88 Comet from page 40
DH.89 Dragon Six 61, 62, 65, 67, 87, 97, 111, 113, 117, 118, 124, 125, 128, 144, 157, 158, 160, 162, 165, 192, 193
DH.91 Albatross 170, 171
DH.98 Mosquito 171, 186
Dizful 115, 116
Domenie, Roelof 57, 99, 143
Doolittle, Jimmy 24, 36, 90
Doret, Marcel 37
Douglas Aircraft 52, 73, 143, 169, 171
Douglas DC-1 31, 32, 74
Douglas DC-2 from page 25
Douglas DC-3 143, 171, 192
Dublin 45, 181
Dundas, RK 55
Eastleigh 46, 47, 76
Edwards, Arthur 51, 63, 65, 68, 69, 73, 80, 133, 137, 158, 164, 184, 186, 193
Essendon 26, 182, 188
Everard, Lindsay 12, 15, 79, 82, 95, 100, 161
Everson Bros 59, 60
Fairey Aviation Company 95, 97
Fairey Fox 52, 62, 92, 98, 109, 110, 112, 113, 118, 124, 125, 128, 151, 188, 189, 192
Fairey IIIF 39, 82, 97, 110, 113, 118, 124, 128, 151, 176, 177

Farnborough 97, 187, 188
Ferguson, Miss Nell 59
Ferris, Lyle 140
Fielden, Flt Lt Edward 41
Fitzmaurice, Col James 25, 45, 46, 47, 105, 106, 180, 181
Flabob Airport 198
Flemington Racecourse 14, 17, 25, 130, 131, 134, 135, 137, 138, 139, 142, 148, 149, 155, 156, 157, 160, 162, 196
Foggia 127, 192
Fokker Aircraft 24, 25, 36, 42, 54, 67, 75, 144, 175, 179
Frye, Jack 31, 73
Furness, Lord Marmaduke 51, 165
Gatty, Harold 31, 32, 35, 41
Geddes, Sir Eric 169
Gelly, GB 21, 122
Gengoult-Smith, Sir Harold 9, 10, 11, 138, 162
Geysendorffer, Capt Gerritt 35, 89, 123, 124, 152, 175
Gilissen, JJ 56, 143
Gilman, Harold 62, 92, 97, 110, 112, 113, 124, 125, 127, 128, 193
Gloster Aircraft 46, 50, 159
Goetz, Carlos 54
Goodfellow, Maj Alan 80, 81, 83, 99, 102, 161, 191
Gorrell, Lord 80
Gouge, Arthur 95
Goulburn 139
Grandy, Flg Off John 109, 187
Granger, James 36, 175
Granville Monoplane 24, 35, 48, 57, 89, 90, 114, 116, 117, 123, 165, 191
Gravesend 76, 184, 186
Grey, CG 14, 16, 25, 39, 81, 94, 107, 108, 159
Guiet, André 40
Hagg, Arthur 65, 66, 67, 69, 71, 73, 132, 163, 168
Halford, Maj Frank 66, 67, 70, 71, 73, 82, 95, 96, 131, 134, 163, 168
Hamble 34, 50, 53, 58, 61, 81, 180
Handley Page, Sir Frederick 159
Hansen, Michael 35, 80, 81, 86, 97, 111, 112, 113, 124, 155, 156, 159, 175
Hanworth 52, 53, 58, 61, 63, 88
Harkness Monoplane 54, 55
Harmondsworth 76, 95
Hatfield 40, 61, 62, 63, 65, 66, 68, 69, 71, 72, 73, 76, 77, 83, 86, 88, 91, 94, 96, 119, 136, 159, 163, 184, 185, 194, 196, 197
Hawker Aircraft 99, 159, 185
Hawks, Capt Frank 60
Hearle, Frank 67, 68, 71, 77, 163, 168
Helmore, Sqn Ldr W 105, 111
Hemsworth, Godfrey ('Geoffrey') 51, 52, 63, 88, 100, 110, 113, 124, 127, 151, 188, 189, 190
Hendon 99, 100, 184
Henry, Flt Lt A 156, 157
Henshaw, Alex 107
Heston 25, 34, 50, 53, 54, 61, 80, 87, 89, 90, 99, 107, 119, 155
Hewett, James 61, 62, 87, 111, 113, 144, 157, 192
HH Maharajah of Jodhpur 122, 127, 193
Hill, Lt Cdr Clifford 38, 39, 81, 82, 97, 110, 113, 124, 128, 151, 176, 177
HM King George V 12, 100, 158, 172
HM Queen Mary 12, 100, 101
HM Queen Wilhelmina of The Netherlands 142, 159, 189
HMAS *Canberra* 138, 195
HMAS *Moresby* 20, 22, 131, 152, 153
HMNZS *Diomede* 138

HMS *Glorious* 39
HMS *Silvio* 22
HMS *Sussex* 101, 138, 161
Hosler B 58
Howard, HB 80, 85
HRH Duke of York 186
HRH The Duke of Gloucester 101, 161
HRH The Prince of Wales, 51, 100, 101
Hughes, Howard 181
Hutchinson, Lt Col GR 59
ICAN 15, 16, 17, 27, 72, 105, 106
Imperial Airways 21, 46, 85, 105, 124, 152, 159, 164, 169, 193
Ingalls, Laura 26, 59
Ilrimescu, Radu 116, 191
Istres 113, 124
Jackson, Sydney 54
Jask 19, 20, 21, 115, 128, 144, 145, 151, 182
Jeffs, James 79, 80
Jensen, Lt Daniel 35, 80, 81, 86, 97, 111, 112, 124, 155, 159, 175
Jillard, R 140
Jodhpur 19, 20, 21, 122, 127, 128, 182
Johns, Capt WE 169
Johnson, Amy 50, 63, 194
Johnson, Flt Lt Hugh 60
Johnston, Capt EC 11, 15, 16, 23
Karachi 14, 19, 20, 21, 114, 119, 120, 121, 122, 126, 127, 128, 144, 145, 155, 182, 192, 193
Kay, Cyril 61, 62, 87, 99, 111, 113, 152, 157, 192
Kennington, Eric 196
Kidston, Glen 52
Kidston, Lt HRA 30
Kihikihi 62
King Ghazi of Iraq 114
Kingsford Smith, Sir Charles 22, 24, 26, 43, 44, 45, 49, 63, 167, 180
Knight, Sqn Ldr FF 11
Koepang 14, 19, 20, 22, 94, 126, 130, 146, 149, 152, 155, 188
Labouchere, Henry 198
Laird Racer 36, 59
Lambert Monocoupe 49, 81, 97, 101, 110, 112, 118, 119, 124, 145, 146, 183
Lamplugh, Capt AG 95
Laverton 10, 135, 136, 137, 138, 141, 144, 148, 149, 156, 157, 160, 161, 162, 188, 196
Lay, Miss Ella Marion 48, 49, 86, 128, 181, 182
Leggitt, John 164, 185, 186
Lelystad 197
Leonard, Royal 48, 122
Lewin, Arthur 177
Lindholm, Marshall 41
Lindow, Georg 41
Lindsay Lloyd, Col F 80
Linnell, Wing Cmdr FJ 78
Llewellyn, Harold 97, 98, 164
Lockheed Altair 26, 43, 44, 45, 49, 59, 180
Lockheed Orion 27, 36, 37, 49, 175
Lockheed Sirius 44
Lockheed Vega 31, 41, 52, 87, 98, 99, 112, 118, 119, 120, 166, 179, 188, 189, 198
Lombardi, Francis 59, 62
Londonderry, Lord 95, 99, 100
Los Angeles 37, 44, 47
Loughborough, Col AH 80
Lympne Aerodrome 29, 40, 48, 82, 162, 180, 181, 191, 193, 195
Lyon 112, 151
Lyon, Capt Harry 49
MacArthur, John 184, 185

Maier, Joe 139, 147
Mann, HH 153
Marseilles 19, 20, 29, 109, 112, 113, 114, 117, 118, 124
Martlesham Heath 48, 56, 61, 72, 85, 88, 89, 184, 186
Maryborough 200
Mascot 45, 54, 132, 188, 189, 192
Mayo, Maj Robert 12, 14, 16, 30
McClure, Ivor 80, 85
McGregor, Sqn Ldr Malcolm 28, 30, 31, 82, 112, 117, 119, 121, 124, 128, 144, 150, 154, 155, 160, 172, 173
McIsaacs, 'Mac' 85
McNicoll, Capt RR 130
Melbourne from page 9
Melrose, James 39, 82, 83, 85, 86, 87, 97, 99, 101, 111, 112, 124, 128, 145, 150, 154, 155, 156, 158, 160, 161, 177, 178
Mersa Matruh 183
Messerschmitt Bf-108 30, 172
Mildenhall from page 9
Miles Falcon 48, 86, 98, 111, 113, 118, 124, 128, 181, 182, 183
Miles Hawk 24, 28, 30, 31, 48, 82, 111, 112, 117, 119, 120, 121, 128, 144, 145, 150, 152, 153, 166, 172
Miller, Horace 58
Miller, Mrs Jessie (Chubbie) 59
Mitchell, Eric 95, 109
Mitchell, Henry 177
Mitchell, Reginald 46
Mohammerah 145
Moll, Capt Jan 55, 56, 136, 141, 142, 191, 193
Mollison, James 20, 24, 26, 34, 42, 44, 45, 49, 63, 65, 68, 69, 72, 73, 82, 83, 88, 91, 96, 97, 98, 100, 101, 102, 105, 108, 109, 110, 112, 114, 120, 121, 126, 164, 181, 193, 194
Mollison, Mrs Amy 25, 26, 63, 73, 91, 96, 97, 100, 101, 105, 109, 114, 117, 120, 121, 193, 194
Moore, Guy 11
Moore-Brabazon, Col 106, 163, 164
Morane Saulnier 36
Morris, William 31, 86
Mosul 115
Mott, Clifford 140
Moult, Eric 67, 95
Naish, Alan 25, 37, 43
Narromine 19, 20, 140, 154, 157
Neill, EVE 10, 11
Nelson, George 184
New Guinea 51, 52
New York 33, 34, 44, 45, 46, 47, 50, 53, 56, 58, 74, 97, 124, 180, 191
Newcastle Waters 19, 20, 130, 152, 153, 154, 155, 156, 157
Newmarket 78, 92, 107
Newnham, Arthur 140, 141
Nichols, Reeder 32, 120, 123, 146
Nichols, Ruth 59
Nicosia 119, 121, 124, 129, 181, 188
Northrop, Jack 48
Northrop aircraft 41, 44, 48, 57, 58, 179, 181
O'Driscoll, Inspector 132, 133
Orton Bradshaw, Stanley 40
Palazzo di San Gevasis 127, 128
Pander Postjager 26, 35, 85, 88, 89, 110, 112, 123, 124, 152, 165, 174
Pangborn, Clyde 32, 35, 48, 57, 90, 99, 114, 126, 127, 141, 147, 173, 174
Parer, Ray 51, 52, 63, 88, 97, 100, 110, 113, 124, 127, 151, 188, 189, 190, 191
Paris 37, 44, 112, 113, 117, 124, 127
Parmentier, Capt Koene 55, 56, 115, 122, 125, 134, 139, 140, 141, 142

Pearson, Rex 193
Pelambang 146
Penang 150
Percival Gull 41, 44, 177, 180, 185
Percival, Edgar 24, 41, 129
Perrin, Harold 12, 16, 20, 21, 25, 27, 28, 30, 31, 35, 45, 50, 58, 76, 78, 80, 86, 87, 100, 101, 107, 110, 164, 165, 176, 183, 184, 195, 198
Phuket Island 22
Pisa 124, 151
Plesman, Capt Jan 197
Plesman, Dr Albert 25, 99, 140, 143, 158, 159, 164, 190
Point Cook 135, 196
Polando, John 49, 50, 81, 112, 124, 129, 145, 183, 184
Pond, Lt Cdr George 63
Ponte San Pietro 62
Poole, Flt Lt 139, 140
Portsmouth 31, 61, 76, 180, 185, 192
Post, Wiley 24, 31, 35, 41, 179
Potez 39, 40, 178
Prins, Bouwe 55, 56, 98, 141, 143, 191
Pronk, Mr P 35, 89, 174
Quatremaître, Charles 37, 59
Rambang 19, 20, 134, 135, 151, 152, 155, 188
Rangoon 19, 20, 125, 126, 127, 144, 150, 151
Rasche, Thea 56, 143
René Couzinet 37
Reynolds, George 80, 108, 110
Rider, Keith 36, 175, 176
RMS *Berengaria* 191
Robertson, Norman 196
Robertson, Sir Macpherson 9, 10, 11, 12, 13, 22, 25, 30, 44, 52, 137, 138, 142, 160, 161, 165, 168, 188, 195, 197 198
Rollason, Bill 17, 46, 47
Rome 19, 20, 29, 109, 112, 113, 114, 117, 118, 119, 124, 128, 181
Rose, Franklin 37
Rossi, Capt M 37
Rotterdam 53, 56, 75
Rowarth, Fred 80, 93
Royal Aero Club 14, 45, 50, 63, 76, 80, 83, 94, 102, 163, 164, 183, 190
Rubin, Bernard 22, 26, 40, 41, 65, 68, 69, 70, 106, 130, 160, 162, 178, 179, 193
Rutbah Wells 190
San Feliu de Guixols 114
San Francisco 37, 44, 45, 146, 180
Santa Monica 36, 52, 56, 63, 74, 169, 171
Saul, Capt JP 24, 45
Saville, Flg Off D 54
Savoia Marchetti 59, 94
Schwortz, Henry 46
Scott, CWA (Charles) 20, 25, 50, 51, 69, 73, 82, 96, 100, 105, 110, 119, 122, 125, 126, 127, 130, 131, 132, 133, 134, 136, 137, 138, 139, 156, 158, 159, 163, 164, 185, 186
Sharp, Martin 186
Shaw, Flt Lt Geoffrey 58, 81, 110, 124, 146, 191
Shelmerdine, Lt Col 80, 100
Short aircraft 53, 75, 105
Shute Norway, Neville 38
Siddeley, Sir John 46
Simpson, Miss Audrey 178, 179
Singapore 15, 19, 20, 21, 22, 35, 46, 120, 122, 125, 126, 127, 130, 134, 146, 148, 150, 151, 152, 155, 169, 180, 182
Smith, Wesley 47, 48, 57, 58, 89, 90, 102, 110, 116, 117
Southampton 33, 34, 46, 50, 57, 80
SS *Ary Lensen* 183

SS *Majestic* 64
SS *Mangola* 130
SS *Mariposa* 45
SS *Olympic* 49, 50, 57, 58
SS *President Roosevelt* 34
SS *Statendam* 56, 74
SS *Volsella* 130
SS *Wanganella* 192
SS *Washington* 33
SS *Westernland* 33
St Barbe, Francis 25, 51, 106
Stack, Neville 24, 25, 61, 84, 88, 112, 124, 127, 192
Stag Lane Aerodrome 25, 37, 44, 51, 65, 66, 67, 69, 163, 196
Stewart, Frank 62, 87, 111, 151, 152
Stodard, Sgt Kenneth 38, 43
Stodard, Sqn Ldr David 37, 43, 88, 111, 112, 113, 124, 129, 145, 150, 152, 153, 156, 158, 176
Surabaya 19, 56, 152, 188
Swinbourne, Flt Lt TA 12, 16, 80, 110
Sydney 41, 43, 51, 62, 63, 64, 130, 141, 192
Tasker, Frederick 184
Tata, Jehangir 193
Taylor, Gordon 43, 44, 45, 180
Thaden, Herbert and Louise 60, 96
Thompson, Freda 28, 29
Tiltman, Hessell 66
Tuckett, Roy 94
Tuckfield, Charles 131, 132, 133
Turner, Bill 198
Turner, Roscoe 24, 27, 32, 34, 42, 57, 90, 96, 97, 99, 101, 105, 109, 114, 120, 123, 126, 127, 141, 144, 146, 159, 161, 173, 174
Turner, Sydney 25, 60, 127, 192
Tyson, Geoffrey 60
USS *Augusta* 138
Van Brugge, Cornelis 55, 56, 191, 198
Vance Viking 57
Vance, Donald 59
Varney, Walter T 37
Vickers Company 15, 38
Victoria Point 150, 182
Villacoublay 37, 59, 94, 191
Vultee VI 59
Waalhaven 35, 53, 75
Walker, CC 24, 66, 71, 73, 159, 167, 168
Walker, Henry ('Johnnie') 28, 30, 31, 82, 84, 117, 119, 124, 144, 150, 155, 160, 173, 197
Waller, Ken 26, 40, 72, 73, 81, 97, 99, 110, 112, 115, 119, 121, 122, 127, 148, 149, 158, 160, 161, 164, 178, 179, 184, 192, 193
Wathen, Tom 198
Waugh and Everson 59, 60, 191
Waugh, Alfred 140, 197
Wedell, Jimmy 24, 42, 43, 179
Wellington 28, 172, 197
Westbrook, Trevor 34
Wibault 37, 59
Willebrandt, Mabel 57, 58
Williams, Air Cmdr 11, 12
Williams, Tim 198
Woodley Aerodrome 28, 48, 82, 84, 183
Woods-Humphery, George 159
Woods, James 52, 53, 87, 98, 112, 119, 120, 121, 188, 189, 194
Wooloomooloo Bay 184, 185
Wright, Dudley 132, 133, 134
Wright, Jack 49, 50, 58, 81, 82, 97, 101, 110, 112, 124, 129, 145
Wyndham 126, 130, 182
Young, Don 109
Young, Harold 96